Educational Research and Innovation

AI and the Future of Skills, Volume 1

CAPABILITIES AND ASSESSMENTS

OECD
BETTER POLICIES FOR BETTER LIVES

Please cite this publication as:
OECD (2021), *AI and the Future of Skills, Volume 1: Capabilities and Assessments*, Educational Research and Innovation, OECD Publishing, Paris, https://doi.org/10.1787/5ee71f34-en.

ISBN 978-92-64-48530-3 (print)
ISBN 978-92-64-84306-6 (pdf)
ISBN 978-92-64-58032-9 (HTML)
ISBN 978-92-64-50743-2 (epub)

Educational Research and Innovation
ISSN 2076-9660 (print)
ISSN 2076-9679 (online)

Revised version, February 2022
Details of revisions available at: https://www.oecd.org/about/publishing/Corrigendum_AI-and-the-Future-of-Skills-Volume1.pdf

Photo credits: Cover © Shutterstock/LightField Studios; © Shutterstock/metamorworks; © Shutterstock/Rido; © Shutterstock/KlingSup.

Corrigenda to publications may be found on line at: www.oecd.org/about/publishing/corrigenda.htm.

Foreword

In a world in which the kinds of things that are easy to teach and test have also become easy to digitise and automate, we need to think harder how education and training can complement, rather than substitute, the artificial intelligence (AI) we have created in our computers.

While there is much debate around the potential impact of AI on our economic and social lives, surprisingly little is known about the actual capabilities of AI, and how we can anticipate their further evolution.

This is where the novel methodology of *OECD's Artificial Intelligence and the Future of Skills (AIFS) project* seeks to make a difference. It enables an understanding of the potential as well as the limits of AI capabilities at a detailed task level so we can describe more precisely how humans and AI are complementary. The project builds on a pilot which the OECD conducted in 2017, that compared the performance of adults on the literacy tests of OECD's Survey of Adult Skills with expert judgement on the capabilities of AI on the same test. This pilot showed that, already in 2017, AI outperformed the majority of humans on basic information processing tasks. However, information processing skills are just a small slice of human capabilities, and the AIFS project now extends this methodology to a wide range of human skills that capture essential dimensions for the success of humans in modern economic and social life. This will deliver valuable insights for the future of education, in terms of directions for the design of instructional systems, the delivery of educational content and the preparation of educators, and the ways in which we recognise and certify knowledge and skills.

Optimists will say that throughout history, education has always won the race with technology. However, there is no assurance it will do so in the future, and humans have always been better at inventing new tools than to use them wisely. In the past, when technology evolved slowly, education had long periods of time to adjust. When fast gets really fast, and when AI competes directly with the cognitive ability that have long been the focus of education, it is important to build rapid, robust and ongoing intelligence that can help us track and anticipate the capabilities of AI.

The AIFS project is therefore designed to provide a baseline against which the OECD can then systematically monitor the evolution of AI capabilities in the longer term. With the technical approaches described in this initial report, the AIFS project is taking the first steps towards building a "PISA for AI" that will help policy makers understand how AI connects to work and education - and how it will transform both of these foundational institutions of human society in the years ahead.

I would like to thank the CERI Governing Board for its leadership in developing and guiding the programme and the German Ministry of Labour for a generous grant to support the work.

Andreas Schleicher

Andreas Schleicher
Director for Education and Skills
Special Advisor on Education Policy to the Secretary-General

Acknowledgements

This publication was planned and developed by the OECD's Artificial Intelligence and Future of Skills project team – Stuart Elliott (Project lead), Nóra Révai, Margarita Kalamova, Mila Staneva, Abel Baret and Matthew Gill.

This publication would not have been possible without the invaluable contributions of the renowned computer science and psychology experts who are supporting the project.

Firstly, we would like to express our gratitude to the authors who contributed to this publication (in order of their chapters): Kenneth D. Forbus, Patrick Kyllonen, Sylvie Chokron, Filip De Fruyt, Anita Woolley, Samuel Greiff and Jan Dörendahl, Phillip Ackerman, Britta Rüschoff, David Dorsey and Scott Oppler, José Hernández-Orallo, Ernest Davis, Richard Granger, Anthony Cohn, Guillaume Avrin, Yvette Graham, Lucy Cheke, Marta Halina and Matthew Crosby, Art Graesser, and Eva Baker and Harry O'Neil.

Kenneth D. Forbus would like to thank Stuart Elliott, John Laird, Toby Walsh, Peter Norvig and Bryan Pardo for helpful feedback.

Britta Rüschoff would like to express her special thanks to Prüfungsaufgaben- und Lehrmittelentwicklungsstelle (PAL) and GACC Midwest (German American Chamber of Commerce of the Midwest, Inc.) for kindly providing the materials used within her chapter.

José Hernández-Orallo thanks Fernando Martínez-Plumed for very insightful comments and discussion on an early version of this report.

Anthony Cohn is partially supported by a Fellowship from the Alan Turing Institute, and by the EU Horizon 2020 under grant agreement 825619 (AI4EU). He would like to thank the participants of an online expert meeting, held in October 2020 and organised by the OECD AI and the Future of Skills project, who gave useful feedback on a draft of his chapter.

Guillaume Avrin wishes to thank the Metrological Evaluation and Testing of Robots in International Competitions (METRICS) partners for their valuable input and collaboration.

Secondly, we would like to thank the AI and the Future of Skills project's "Critical Friends" – Lucy Cheke, Kenneth D. Forbus, José Hernández-Orallo, Patrick Kyllonen and Michael Schoenstein. They have been a crucial sounding board to the project throughout the development of this workshop and report.

Additionally, we wish to thank our colleagues in the Centre for Educational Research and Innovation (CERI). Dirk Van Damme, former Head, provided oversight, direction and valuable advice during the process. Matthew Gill, Project assistant, who made an enormous contribution to both formatting and the preparation of the publication, along with colleagues within the Directorate for Education and Skills communications team and the Public Affairs and Communications Directorate.

Our thanks are extended to Mark Foss, who made substantive and structural editing to the publication, ensuring for coherent, comprehensible reading.

We are grateful for the encouragement and support of the CERI Governing Board in the development of the project and this volume.

This publication contributes to the OECD's Artificial Intelligence in Work, Innovation, Productivity and Skills (AI-WIPS) programme, which provides policy makers with new evidence and analysis to keep abreast of the fast-evolving changes in AI capabilities and diffusion, and their implications for the world of work. The programme aims to help ensure that adoption of AI in the world of work is effective, beneficial to all, people-centred and accepted by the population at large. AI-WIPS is supported by the German Federal Ministry of Labour and Social Affairs (BMAS) and will complement the work of the German AI Observatory in the Ministry's Policy Lab Digital, Work & Society. For more information, visit https://oecd.ai/work-innovation-productivity-skills and https://denkfabrik-bmas.de/.

Table of contents

FIGURES

TABLES

Abbreviations and acronyms

ADL	Activities of daily living
AGI	Artificial general intelligence
AI	Artificial intelligence
ANN	Artificial neural networks
ASI	Artificial social intelligence
ATENE	Automatic transcription evaluation for named entities
ATP	Automated Theorem provers
BART	Balloon Analogue Risk Task
CB	Cognition battery
CBR	Case-based reasoning
CCC	Computing Community Consortium
CERI	Centre for Educational Research and Innovation
CHC	Cattell-Horn-Carroll
CI	Collective intelligence
CIG	Computational Intelligence in Games
CMS	Children's Memory Scale
CPT	Continuous performance test
DCF	Detection cost function
DER	Diarisation Error Rate
DET	Detection error trade-off
DTVP	Developmental Test of Visual Perception
EER	Equal error rate
EGER	Estimated global error rate
EPIC	Executive Process Interactive Control
EQF	European Qualifications Framework
ESCO	European Skills, Competences, Qualifications and Occupations
ETER	Entity Tree Error Rate

ETS	Educational Testing Service
EVA	Evaluation of visuo-spatial abilities
GLUE	General Language Understanding Evaluation
GRE	Graduate Record Examination
ICAR	International Cognitive Ability Resource
MAE	Mean Absolute Error
MD	Maximum discrepancy
ML	Machine learning
MECE	Mutually exclusive and collectively exhaustive
NAEP	National Assessment of Educational Progress
NIH	National Institutes of Health
PDDL	Planning Domain Definition Language
PIAAC	Programme for the International Assessment of Adult Competencies
PISA	Programme for International Student Assessment
POGS	Platform for Online Group Studies
RL	Reinforcement learning
TCI	Test of Collective Intelligence
TEA	Test of Everyday Attention
TIMSS	Trends in International Mathematics and Science Study
TPTP	Thousands of Problems for Theorem Provers
TRL	Technology readiness levels
SAMR	Substitution Augmentation Modification Redefinition
SER	Slot error rate
SJT	Situational judgement tests
SME	Subject matter experts
STEM	Situational Test of Emotion Management
STEU	Situational Test of Emotional Understanding
VET	Vocational education and training
VGDL	Video Game Description Language
VPR	Verbal, perceptual and image rotation
WIL	Word Information Loss
WMS	Wechsler Memory Scale
WSC	Winograd Schema Challenge

Executive summary

Artificial intelligence (AI) and robotics are major breakthrough technologies that are transforming the economy and society. To understand and anticipate this transformation, policy makers must first understand what these technologies can and cannot do. The OECD launched the *Artificial Intelligence and the Future of Skills* project to develop a programme that could assess the capabilities of AI and robotics and their impact on education and work. This report represents the first step in developing the methodological approach of the project. It reviews existing taxonomies and tests in psychology and computer science, and discusses their strengths, weaknesses and applicability for assessing machine capabilities.

Assessing AI and robotics capabilities is a necessary foundation for understanding their implications for education, work and the larger society.

An ongoing programme of assessment for AI and robotics will add a crucial component to the OECD's set of international comparative measures that help policy makers understand human skills. The Programme for International Student Assessment (PISA) describes the link between the education system and the development of human skills, while the Programme for the International Assessment of Adult Competencies (PIAAC) links those skills to work and other key adult roles. A programme for assessing AI and robotics capabilities will relate human skills to these pivotal technologies, thereby providing a bridge from AI and robotics to their implications for education and work, and the resulting social transformations in the decades to come.

There are a number of taxonomies and tests of human skills. These provide different perspectives and opportunities for understanding AI capabilities.

Taxonomies stemming from the cognitive psychology literature are hierarchical models of broad cognitive abilities, such as fluid intelligence, general memory/learning, visual and auditory perception, assessed by factor analysis of cognitive ability tests. These tests have been widely used and validated for assessing human skills.

Research interest in social and emotional skills is growing and their testing is advancing. These skills are focused on individuals' personality, temperament, attitudes, integrity and personal interaction. Recent research considers not just individual abilities but also collective ones. This emerging literature studies the factors of "collective intelligence" and is developing tests to measure them.

Education research has also contributed to defining and shaping the understanding of human skills. This domain focuses on subject-specific knowledge (e.g. in mathematics, biology and history), basic skills such as literacy and numeracy, and more complex transversal skills such as problem solving, collaboration, creativity, digital competence and global competence. A wide range of tests is available from international and national large-scale educational assessments.

Skills can be linked to work tasks and occupations, and measured through complex vocational tests.

Another major area – industrial-organisational psychology – links abilities to tasks specific to particular occupations. The resulting comprehensive occupation taxonomies classify occupations by work tasks, and the required skills, knowledge and competences. The most widely used classifications are the Occupational Network (O*NET) database of the US Department of Labor and the European classification of skills, competences, qualifications and occupations (ESCO). Assessments in this domain comprise a variety of vocational and occupational tests.

Healthy human adults share some basic skills that AI systems do not have.

Many taxonomies for assessing skills overlook ubiquitous low-level or basic cognitive skills. These are rarely assessed in human adults because there are few meaningful individual differences in the absence of severe disability. However, AI systems do not necessarily have these skills (e.g. navigating in a complex physical environment, understanding basic language or knowing basic rules of the world). Taxonomies and assessments for these skills are found in the fields of animal cognition, child development and neuropsychology. A recently emerging field assesses basic (low-level) skills of AI systems drawing on these fields of psychology.

Evaluating AI and robotics systems is challenging and applying human tests can be misleading.

AI assessment focuses on functional components of intelligent mechanisms, such as knowledge representation, reasoning, perception, navigation and natural language processing. These are strongly linked to the underlying technique used by the mechanism. Many components overlap with the ability categories developed in psychology for humans, but the match is not exact. In addition, many capabilities that AI is developing – such as language identification and the generation of realistic images – are not well covered by human skill taxonomies or tests.

Moreover, the design of human tests takes for granted that the test takers all share basic features of human intelligence, which might be radically different from AI. For example, integrating basic skills, such as natural language understanding and object recognition, is easy for humans. However, most AI systems are trained to perform a specific narrow task, but they are not (or are rarely) able to integrate and apply these to perform a different type of task. This makes it difficult to generalise from an AI system's performance on a specific human ability test to an underlying AI skill, let alone infer general intelligence.

Different types of empirical assessments can gauge AI capabilities, but these are scattered and not systematic.

A multitude of benchmarks and competitions assess and compare AI systems empirically. However, these have not yet been systematically classified. Increasingly, more institutions carry out rigorous evaluation campaigns to assess the capabilities of AI and robotics systems. These include the evaluation of individual functions, i.e. self-contained units of capability, such as self-localisation. They also include evaluation of complete tasks that constitute a meaningful activity, such as autonomous driving and text summarisation. Evaluation of AI systems is particularly well developed in certain areas, such as language understanding. Machine translation, in particular, is a field that holds many lessons for assessing AI.

A systematic assessment of AI demands a comprehensive framework that covers all human skills necessary for work and life.

Providing valid, reliable and meaningful measures of AI and robotic capabilities requires a comprehensive approach that brings together different research traditions and complementary methodologies. The goal should be to address the full range of relevant human capabilities; the extra capabilities needed to consider for AI (because they are difficult for AI and often neglected in lists of human skills); and the full range of valued tasks that appear in education, work and daily life.

A robust methodology involves understanding how AI and robotics systems are assessed and bringing together different assessment approaches.

A multidisciplinary approach needs theoretical underpinning that considers the challenges linked to assessing AI and robotic capabilities with regard to human skills. The different disciplinary approaches can be organised across two dimensions. One relates to whether skill taxonomies and tests measure *primarily human or primarily AI capabilities*. The second dimension is whether they measure *single (isolated) capabilities or complex tasks* that require multiple capabilities. Future systematic assessment of AI capabilities should bring together different assessments along these two dimensions and skilfully integrate their potential to draw valid implications for the future of work and education.

Part I. Setting the scene

1. New approaches to understanding the impact of computers on work and education

Nóra Révai, OECD

Mila Staneva, OECD

Abel Baret, OECD

This chapter describes the background and purpose of the OECD's *Artificial Intelligence and the Future of Skills* project, which is developing an approach to assessing the capabilities of artificial intelligence (AI) and robotics and their impact on education and work. This report represents the project's first step to identify the capabilities to assess and the tests to use for the assessment. The chapter provides an overview of the approaches applied to date to predict the impact of technology on the future of work. It sets out a new approach and presents the project's stages of developing a sound methodology for a systematic assessment of AI capabilities in the future. The chapter ends by presenting the structure of this report.

Introduction

Policy interest in the impact of artificial intelligence (AI) has sprung up in the past few decades as AI technologies are developing and being integrated into more and more aspects of life. A deeper and more precise understanding of this impact for the economy and society is fundamental for strategic planning in various policy areas. With regard to employment and education, this understanding can provide the basis for realistic scenarios about how jobs and skill demand will be redefined in the next decades. It can also demonstrate how the education system needs to be reshaped to prepare today's students for these possible futures.

However, this understanding of the impacts of AI and robotics fundamentally rests on an understanding of the technology's capabilities. What can AI and robotics do and what can they not do? How do the capabilities of AI and robotics compare to those of humans?

The OECD's Centre for Educational Research and Innovation (CERI) launched the *Artificial Intelligence and the Future of Skills* (AIFS) project in 2019 to address the questions above. The project builds on pilot work carried out in 2016 that explores AI capabilities with respect to literacy, numeracy and problem-solving skills using the OECD's Survey of Adult Skills. The project aims to develop a new set of measures to serve as a foundation for research and policy on how AI and robotics will transform skill demand and educational requirements in the decades ahead. It addresses the following concrete questions:

- What human capabilities will be too difficult for AI and robotics to reproduce over the next few decades?
- What education and training will be needed to allow most people to develop some work-related capabilities that are beyond the capabilities of AI and robotics?

Studies that have attempted to gauge the impact of computer capabilities on employment, skill demand and education demonstrate that predictions on work and society are by no means straightforward. Technological development affects the labour market in diverse ways that are sometimes hard to predict. It usually involves the transformation of jobs and tasks rather than their full replacement by machines.

Despite these complexities, an understanding of that transformation must begin with an understanding of the computer capabilities themselves. This can provide a basis for reflecting on potential transformations. This report does *not* discuss the implications of technological change, or even reflect on how to identify those implications. Rather, it aims to build a methodology that can provide valuable and robust data for policy makers and researchers who want to consider those transformations from a solid understanding of the technology itself.

Accordingly, the first stage of the project focuses on constructing *valid*, *reliable* and *meaningful* measures of AI capabilities. First, to be valid, measures must not mislead and indeed assess capabilities of AI. Second, to be reliable, measures must ensure consistency over time. To that end, they must rely on recognised experts, and a transparent and robust process that is reproducible. Reliability also involves addressing convergences and divergences of experts' judgements transparently and appropriately (inter-rater reliability), and using consistent items for measurement. Validity and reliability make the measures credible, ensuring that the measurement avoids basic methodological pitfalls. Third, for measures to be meaningful, particularly to the policy community, the constructs and comparisons should enable decision makers without AI expertise to understand likely implications. For this to happen, the set of measures should be comprehensive, covering the full spectrum of relevant capabilities. A straightforward comparison with human capabilities would help interpret constructs and help produce meaningful measures.

Such a set of measures requires scientific thoroughness and broad partnerships to be effective and comprehensive. Consequently, the project dedicates substantial effort to build a robust methodology and involve a wide range of experts from around the world. Developing the AIFS approach will take place over six years, which began with a planning process in 2019.

This first volume explores the methodology for the project. It is a technical report, which provides the outcomes of an expert workshop held in October 2020 on "Skills and Tests". This initial workshop provided the direction for pilot work in 2021 on different types of assessment tasks. The pilot will be followed by an initial systematic assessment in 2022 and 2023 across the full range of capabilities. An analysis of the potential implications for work and education will be produced in 2024. The project will conclude with a proposed approach for a regular programme to update the assessments.

This first chapter of the technical report situates the project in the broader literature on evaluating the progress of computer technology and its impact on the world of work. It starts with a snapshot of the broader effort of the OECD to gauge the impact of AI. Next, it discusses the various methodologies adopted to date and their challenges. It then presents the pilot work underpinning the AIFS project, showing how a new approach can address some of the methodological caveats and gaps. The report structure is presented at the end of this chapter.

OECD's work on the impact of artificial intelligence

Over the past decade, advances in big data, computational power, storage capacity and algorithmic techniques have dramatically accelerated the development and deployment of AI systems. The OECD has been increasingly engaged in supporting countries to understand this technological development. A major achievement was the adoption of the OECD Principles on Artificial Intelligence in May 2019, which sets international standards for the responsible stewardship of trustworthy AI (OECD, 2019[1]). In 2020, the OECD launched the AI Policy Observatory, which brings together information, analysis and supports dialogue to shape and share AI policies (OECD, 2020[2]).

A recent cross-directorate effort of the OECD is the AI programme on Work, Innovation, Productivity and Skills (AI-WIPS) supported by the German Federal Ministry of Labour and Social Affairs. AI-WIPS incorporates several streams of work to provide a comprehensive analysis of the different aspects of AI and their implications for society. The AIFS project represents the "Assessing AI and robotics capabilities" work stream. In parallel, work has started on building a framework for classifying AI systems and mapping their development in various fields (Baruffaldi et al., 2020[3]; OECD, 2019[4]). The OECD has been contributing to analysing the impact of AI on workforce skills in the past few years. As part of this work, analyses have been conducted on the extent to which machines can automate jobs and substitute for workers (Arntz, Gregory and Zierahn, 2016[5]; Nedelkoska and Quintini, 2018[6]). A more recent publication reviews literature on the impact of AI on employment, wages, the work environment and the ways in which AI transforms jobs and skill needs (Lane and Saint-Martin, 2021[7]). As the impact of technology on work and society also depends on the speed of its development and diffusion, the OECD is also working on assessing the speed of AI diffusion (Nakazato and Squicciarini, 2021[8]). Finally, the organisation has been facilitating policy and societal dialogue that brings together experts, researchers, policy makers, social partners and civil society to discuss and contribute to AI-related topics.

In the domain of education, CERI engages in several aspects of understanding the impact of technology on education (Vincent-Lancrin and van der Vlies, 2020[9]; van der Vlies, 2020[10]). Work on educational innovation explores the uses of digital devices and software for enhancing learning inside and outside the classroom [see also (Verhagen, forthcoming[11])]. This work extends to understanding how data – whether collected in formal educational settings or through other means – can be used to personalise learning, improve people's educational experience, and inform decision making and policies in education.

Box 1.1. Artificial intelligence: Definition, use cases, scope

There is no commonly agreed definition of AI systems (OECD, 2019[4]). While machine learning has become popular (see Chapter 12 for more information), computer scientists stress that the understanding of AI should extend beyond this technique. The OECD's AI Experts Group (AIGO) defines an AI system as:

> a machine-based system that can, for a given set of human-defined objectives, make predictions, recommendations or decisions influencing real or virtual environments. It uses machine and/or human-based inputs to perceive real and/or virtual environments; abstract such perceptions into models (in an automated manner e.g. with machine learning (ML) or manually); and use model inference to formulate options for information or action. AI systems are designed to operate with varying levels of autonomy *(OECD, 2019, p. 15[4])*.

The ability to "make predictions, recommendations or decisions" makes AI applicable in various tasks and domains. A recent OECD report presents a number of use cases across a range of areas (OECD, 2019[4]). In health care, for example, AI is used to make more accurate and faster diagnoses of diseases than humans. In e-commerce, AI is applied to recommend products that better fit the needs of potential buyers. In finance, AI is used to predict the credibility of loan applicants more accurately. AI can also automate numerous other tasks where the role of prediction is less obvious. One example is autonomous driving, where cars rely on AI to anticipate the right trajectories and manoeuvres. Each of these areas has important limits with respect to AI's capabilities, but the wide applicability in different economic sectors makes AI a "general-purpose technology" (OECD, 2019[4]). Like the steam engine and electricity, AI has the potential to raise productivity across vast parts of the economy (Bresnahan and Trajtenberg, 1992[12]).

Although AI is being applied to increasingly more areas, it still cannot perform the full range of tasks of humans and lacks some basic human skills. Therefore, one way to describe the types of AI that currently exist is that they represent artificial narrow intelligence, meaning that current AI systems are designed to perform specific, narrowly defined tasks (OECD, 2017[13]). This state-of-the-art is contrasted to a hypothetical artificial general intelligence (AGI) (OECD, 2017[13]). In AGI, machines are human-like in how they can abstract, generalise, perceive, judge, create and make decisions. Such skills are still out of reach for AI systems (OECD, 2019[4]).

Parallel to progress in AI, robot technology has continued to advance (OECD, 2021[14]). Indeed, both these technological developments are intertwined as many new AI applications involve the sensory and motor control capabilities that are fundamental to robotics. In addition, both AI and robotic technologies enable the automation of tasks typically executed by humans. As a result, they have a similar impact on the world of work. For all these reasons, advancements in AI and robotics and their implications for the future of work are commonly studied together in the literature. The current report is centred on AI but encompasses robotic technologies as they are also likely to affect work and the future demand of skills.

Attempts to measure artificial intelligence capabilities and impact

There is general agreement that AI is a major breakthrough technology that will transform the economy and society (see Box 1.1 for definitions, use cases and the scope of AI). However, to unravel this impact, studies must first understand what computers can and cannot do. Most of the work prominent in the policy discourse stems from economics and the social sciences. Its driving concern is that AI may lead to technological unemployment, while there are also views holding that AI is well placed to augment workers' capabilities and, thus, raise productivity and enable innovations. Accordingly, this literature focuses mainly on AI's potential to substitute for labour in the workplace and measures AI capabilities with regard to occupations and work tasks (the so-called task-based approach). Other strains of research from computer science and psychology analyse AI from the perspective of skills and abilities. They measure which computer capabilities are now available and how they will eventually relate to human skills.

Studies analysing AI capabilities may also differ with respect to the crudeness of their measures. Some measures are based on vague notions of the capabilities of computers. Others rely on ratings of whether AI can perform (more or less specifically defined) tasks. Yet others draw on actual AI performance.

The next few paragraphs provide a snapshot of the most prominent studies in each of these areas.

The task-based approach to measuring AI capabilities and its impact on jobs

Studies following the task-based approach analyse the extent to which AI can displace workers by focusing on occupations and their task content and by studying the susceptibility of tasks to automation. The goal is to determine the share of tasks within an occupation that can be performed by computers and, ultimately, the share of jobs in the economy that can be largely carried out by machines [see Frey and Osborne (2017[15]) or Brynjolfsson et al. (2018[16])]. In this literature, the focus is less on AI and more on its impact on work and employment. AI capabilities are rated with regard to broad descriptions of occupations or job tasks. These are usually derived from occupation taxonomies, such as the Occupational Information Network (O*NET) database of the US Department of Labor. The resulting measures of occupations' automatability are then matched to micro-level labour market data to further examine the characteristics of jobs at high risk of automation (e.g. industry, region and wage level), as well as the characteristics of workers who are at risk of displacement (e.g. age, gender and education level).

The origins of the task-based approach

The task-based approach has its origin in the seminal work of Autor, Levy and Murnane (2003[17]). Their study stipulates that machines can substitute for workers only in particular tasks. These are typically routine cognitive and manual tasks that follow exact repetitive, predictable procedures. As such, the tasks can be readily formalised and codified. By contrast, tasks that follow tacit, inexplicable rules, such as those involving flexibility, creativity, problem solving or complex interaction, are not apt for computerisation. Both types of tasks are operationalised loosely by using broad occupational descriptions. These include the involvement of direction, control and planning of activities at work or the extent to which finger dexterity is required.

Autor, Levy and Murnane (2003[17]) study how labour input in routine and non-routine tasks develops over time. The premise is that declining prices of technology should reduce labour demand for routine tasks and increase demand for non-routine tasks. That is, employers would increasingly replace workers in routine tasks with cheap machines. At the same time, they would employ workers for complementary non-routine tasks, such as developing, managing and monitoring machines.

The approach of Frey and Osborne

Since the Autor, Levy and Murnane study in 2003, computers have advanced significantly. They can now perform many of the tasks previously thought as "uncodifiable", such as driving or translation. This was possible mainly through progress in AI and in machine learning, in particular. To account for these technological advancements, Frey and Osborne (2017[15]) take a new perspective on measuring computer capabilities. Instead of asking what computers can do in the workplace, they concentrate on tasks that computers still cannot perform. As such tasks are declining in number, they are becoming easier to characterise and thus to assess. Specifically, the authors identify three engineering bottlenecks to AI-driven automation: perception and manipulation tasks, such as navigating in an unstructured environment; creative intelligence tasks, such as composing music; and social intelligence tasks, such as negotiating and persuading. According to Frey and Osborne (2017[15]), computers can perform any task not subject to one of the three bottlenecks.

To quantify computers' ability to automate work, Frey and Osborne (2017[15]) first asked AI researchers to rate the automatability of 70 (of 700) occupations in the O*NET occupation taxonomy. O*NET contains systematic information on occupations' task content and skills requirements. Based on the task descriptions, the experts labelled occupations as automatable or non-automatable. In a second step, Frey and Osborne (2017[15]) approximated the three engineering bottlenecks with nine O*NET variables, available for the full set of occupations in the database. For example, the variable "originality", which describes the degree to which an occupation requires unusual or clever solutions, serves as a proxy for creative tasks. Finally, the authors estimated the relationship between the automatability of the initial 70 occupations (derived from the subjective expert assessments) and the bottlenecks (measured with the nine O*NET variables). They used the obtained estimates to predict the probability of automation of all 700 occupations.

The work of Frey and Osborne (2017[15]) stimulated much research but has several limitations. Most importantly, the information from O*NET used by experts to rate occupations' automatability involves simple one-line task descriptions. These descriptions provide little indication of task difficulty or specific examples. Based on the different tasks of an occupation, experts provided judgements for the entire occupation. This raises questions as to how they dealt with essential and less essential tasks of the job. Similarly, it is unclear whether they included abilities needed for the occupation but not listed in these databases because all (healthy) human beings have them (e.g. vision and common sense reasoning). In addition, such occupational-level judgements do not recognise that jobs within the same occupation may differ in their task mix and, hence, in their amenability to automation. Furthermore, only occupations where experts were most confident were considered; only occupations in which all tasks were rated automatable were labelled as automatable. However, the study does not specify how inter-rater agreement was determined, and how high it was. Neither does the work address the issue of those occupations that are not fully automatable but have a high number of automatable tasks. The lack of such information raises questions about the validity of the exercise.

Two studies supported by the OECD – Arntz, Gregory and Zierahn (2016[5]) and Nedelkoska and Quintini (2018[6]) – address one of these points: they estimate the risk of automation at the level of jobs instead of occupations. More precisely, the studies map the expert judgements on the automatability of the 70 occupations from the study of Frey and Osborne (2017[15]) to micro-level data from the Survey of Adult Skills of OECD's Programme for the International Assessment of Adult Competencies (PIAAC). They then estimate the link between automatability of occupations and various work tasks at the level of individual jobs assessed in PIAAC. This contrasts with Frey and Osborne (2017[15]), who model occupations' automatability as a function of hard-to-automate, bottleneck tasks at the occupation level. Analysing how job-level characteristics are linked to automation makes it possible to reflect the variation of jobs within occupations in the overall estimate of automatability across the economy.

Further efforts to assess the automation of tasks

Following the task-based approach, some major consultancies have issued reports on the impact of technology on employment and the economy. McKinsey Global Institute studies automation by linking 2 000 work activities to 18 key performance capabilities needed to execute them (such as sensory perception and retrieving information) and by establishing the level of performance that AI has with regard to these capabilities (Manyika et al., 2017[18]). However, the study does not describe the methodological approach in further detail. It remains unclear how capabilities are defined and matched to abilities. Similarly, it is unknown how their susceptibility to automation is determined. In addition, PricewaterhouseCoopers (2018[19]) examines the global economic impact of AI by studying both its potential to enhance productivity through automation and to improve product quality. To assess how AI can automate jobs, the report adopts the approach of Arntz, Gregory and Zierahn (2016[5]) and PwC (PwC, 2017[20]).

Other studies in this literature focus more narrowly on AI. Brynjolfsson, Mitchell and Rock (2018[16]), for example, define key criteria for whether a work task is suitable for machine learning applications. These criteria include, for example, clearly definable rules and goals or the availability of large digital datasets for training algorithms (see also Brynjolfsson and Mitchell (2017[21])). The authors then rate 2 069 work activities linked to 964 occupations in O*NET against these criteria to measure occupations' suitability for machine learning. By contrast, Felten, Raj and Seamans (2019[22]) use objective AI metrics made available by the Electronic Frontier Foundation to track progress of AI across major application domains, such as image and speech recognition. They map these AI progress measures to information from O*NET on key abilities required in occupations. To that end, they ask gig workers on a freelancing platform to rate the relatedness between both. In this way, the authors assess the extent to which occupations are exposed to computerisation.

While all these studies rely in some way on the subjective judgement of computer experts, economists or "laypersons", the study of Webb (2020[23]) aims at developing an objective measure of the applicability of technology in the workplace. To derive such a measure for AI, the study first identifies patents of AI technologies by scanning patent descriptions for keywords such as "neural networks" and "deep learning". It compares verb-noun pairs in patents' texts and occupations' task descriptions available in O*NET. In this way, it quantifies the overlap between such AI patents and occupations.

Squicciarini and Staccioli (forthcoming[24]) adopt a similar approach to Webb (2020[23]) but with a focus on robotics. They identify patents associated to labour-saving robotic technologies by applying text-mining techniques. Subsequently, they connect these patents to occupation descriptions available in ISCO08 – a standardised classification of occupations. They use a text similarity algorithm to measure occupations' exposure to such innovations.

Again, these studies rely on broad occupation descriptions available in occupation taxonomies. They focus more on quantifying the extent to which AI can automate the economy than on developing a comprehensive measurement of what computers can do.

Remaining gaps in methodology

To sum up, research following the task-based approach has so far mostly focused on how evolving AI and robotics alter the workplace by automating tasks within occupations. Most studies following this approach share some (or all) of the following methodological gaps:

- judgements are based on vague descriptions of skills or tasks, which omit important details needed to evaluate if AI can do them
- judgements about entire jobs mix tasks that AI can and cannot do
- judgements about entire jobs require knowledge of AI capabilities and knowledge of job design at the same time, but no experts have both
- information about the experts' identity, selection and domains of expertise is lacking
- information about the exact methodology and the rating process is lacking.

In addition, the task-based approach offers a narrow view of humans: workers are seen as displaced from tasks or from entire jobs that have been overtaken by computers. This leaves a number of key questions largely unexplored. How do people's skills and abilities compare to AI performance? Which of these skills are reproducible? Which can be usefully complemented or augmented by machines? Which skills are hard to automate and, thus, worth investing in? Skills-based approaches in computer science and psychology offer a promising way forward to address these gaps. This new approach is discussed next.

Skills-based assessment: A new approach

Apart from the work in economics that assesses AI capabilities as an initial step to studying its potential impact on jobs, there are efforts carried out to assess those capabilities in the field of computer science (Martínez-Plumed et al., 2021[25]; Clark and Etzioni, 2016[26]; Crosby, Beyret and Halina, 2019[27]; Ohlsson et al., 2016[28]; Davis, 2016[29]). In addition, a wide set of assessment approaches has been developed over the past century to measure human capabilities that could potentially be applied to understanding AI. These tools offer a deeper way of measuring and understanding AI capabilities that goes beyond judgements about the automatability of entire occupations in the economics literature.

In 2016, an OECD/CERI pilot project assessed computers' capabilities using an assessment of human competences linked to the workforce. This was an initial exploration of how to connect a more skills-based assessment of AI capabilities to the kinds of economic questions that concern policy makers. The pilot assessed computers' literacy, numeracy and problem-solving skills using the PIAAC test. This test assesses skills in the adult population and is linked to information about work and other adult activities (Elliott, 2017[30]). It is part of the OECD's effort to evaluate educational outcomes through assessing skills such as literacy and numeracy. Some of these assessments, including both PIAAC and the Programme for International Student Assessment have been used for a long time at a large scale across different contexts. As such, they provide robust information on the distribution of these skills in the population. Assessing computer capabilities through such tests potentially allows for comparing AI and robotics skills with humans at various proficiency levels.

To gauge the changes that technology may bring, the pilot study asked a group of computer scientists to rate the ability of current computer technology to answer each test question. In addition, a subset of experts also provided projections for the evolution of technology in the next ten years. The aggregate rating across specific test questions was then used to place AI capabilities on the PIAAC test scale. This helped understand how AI capabilities compare to the human population in these three skill areas.

Strengths of the new approach

The pilot study revealed a number of strengths of this new approach.

First, *rating with regard to specific test items provides a more precise estimation* of computer capabilities. As discussed in the section above, prior estimations related to general descriptions of work tasks or activities, or even broader descriptions of occupations. Experts judging computer capabilities do not know exactly what granular tasks are required to carry these out. Their judgements thus inevitably involve assumptions that are likely to vary greatly across experts (Elliott, 2017[30]). By contrast, questions in standardised tests are precise and contextualised. This allows computer scientists to analyse the information processing required to answer a specific question based on the information provided (Elliott, 2017[30]). This level of specificity implies greater reliability across raters and greater reproducibility.

Second, *using human tests makes it possible to compare computer and human capabilities*. In particular, when large-scale data on human skills are available across different contexts, different age groups and occupations, these data can be used to conduct fine-grained analyses of skill supply and future demand. Simply put, employers could rearrange job tasks to use AI for its capabilities and human workers for the capabilities that AI still lacks. This means that skill demand for human workers should shift towards the (aspects of) capabilities that AI still lacks. A fine-grained comparison of human and computer capabilities across the full range of skills required for work and life will help avoid jumping to faulty conclusions on job automation. Instead, this connection provides information about AI's impacts that extends beyond the definition of current occupations. It will be useful for thinking about AI's implications for employment more generally, including occupations that do not yet exist, as well as education.

The pilot study illustrates this potential through a number of figures and analyses. PIAAC questions are grouped in five levels based on their difficulty for the adult human population. Expert ratings can indicate whether computers can perform at a certain level of difficulty as human adults. Figure 1.1, for example, shows the distribution of workers based on whether they use general cognitive skills daily in their work and, if they do, how they compare to computer capabilities. Work tasks of the part of the workforce that does not use any of these skills on a daily basis will not be substantially affected by the computer capabilities examined in the pilot study (as the first two bars in the figure show). Workers who use one or more of these skills regularly and have proficiency above the projected level of computer capabilities will likely continue to have regular tasks using these skills that are not substantially affected by computer capabilities in these areas (as the last two bars in the figure show). However, automation will likely affect those who use one or more of these skills on a daily basis but have proficiencies only at the level of projected computer capabilities (middle bars) (Elliott, 2017[30]).

Figure 1.1. Distribution of workers by use of general cognitive skills and proficiency compared to computers (Results from the pilot study)

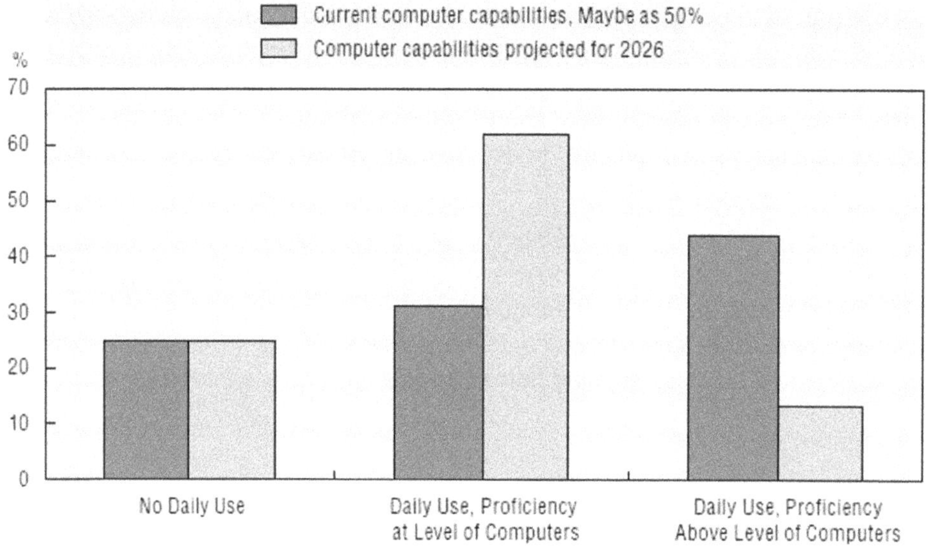

Source: (Elliott, 2017, p. 92[30]).

Challenges and further development of the new approach

The pilot study also identified a number of challenges of the approach. Overfitting is a commonly cited danger of assessing whether computer technology can answer a particular test item. It is often possible to train computers for a specific task. However, this does not mean the computer can perform a range of similar tasks (Elliott, 2017[30]). Overfitting relates to the question of generalisability, i.e. the possibility to infer that computers have an underlying capability from their performance on specific items. Another challenge is whether the same ability is tested through a test item for computers and for humans. For example, an item that requires counting objects in a picture tests a simple numeracy skill for humans. However, the challenge for AI is visual processing rather than counting (Elliott, 2017[30]).

To sum up, despite the challenges identified in the pilot study, the new approach for assessing AI capabilities is promising both in terms of its credibility (validity and reliability) and in producing measures that are meaningful for policy. Further developing and extending this approach could enlarge the conversation about AI capabilities. On the one hand, it could estimate automatability indices for current occupations (which has been done before). However, it would also provide information on how occupations are likely to transform, what new occupations may emerge and what this all means for developing people's skills within and outside formal education. The following section discusses what this development and extension involve.

Purpose and structure of the report

The pilot work explored assessing AI capabilities with one example test that involves just a few skills. As a result of its preliminary success, CERI decided to expand the work to a more comprehensive set of assessments. However, the methodology in the pilot study needs to be refined and the assessment extended to a comprehensive list of capabilities. This is necessary to establish valid, reliable and meaningful measures for an ongoing systematic assessment of machine capabilities. This work involves two steps:

- reviewing taxonomies of human skills and capabilities[1], and identifying an appropriate taxonomy to use for the project that spans the full range of skills used in the workplace
- reviewing available tests of human skills and establishing criteria for their suitability for assessing AI capabilities.

The project needs a comprehensive framework of skills that would fulfil three requirements. First, it would include all the skills that people need for their work and life. Second, it would be suitable for analysing AI capabilities, including both those similar to and different from human skills. Third, it would be suitable for comparing these capabilities to human skills and draw implications of AI progress on the world of work and education.

As demonstrated by the challenges described earlier, assessing humans and machines is a different exercise. Therefore, the project needs to bring together different disciplines and interpret their findings for one another. This technical report is a first step in this process.

The report builds on an online meeting of the AIFS project held on 5-6 October 2020 with experts from various domains of psychology and computer science. The meeting sought to explore different domains of psychology (cognitive, personality, industrial and occupational, developmental and neuropsychology) and to review existing taxonomies of human skills in these domains. The meeting also aimed to identify tests of these skills and discuss their strengths, weaknesses and applicability for assessing machine capabilities.

Psychologists were asked to present a taxonomy (or taxonomies) of the skill domain of their expertise and describe the types of tests available to assess these skills. Experts were requested to discuss the challenges and opportunities of assessing these skills based on the research literature. The experts then provided sufficient examples of actual test questions to illustrate the kinds of tasks typically included and the criteria used to evaluate performance. Computer scientists were invited to reflect on the progress of AI and robotics technologies. They presented types of empirical evaluations and benchmarks in the field, and outlined the main considerations for assessing AI and robotics capabilities against human capabilities. Experts then compared the different types of skill taxonomies and tests with the objective to work towards some broadly supported guidelines to govern the project's choice of a skill taxonomy and a set of tests.

This volume contains papers prepared by psychologists and computer scientists who attended the meeting. They each present research from their specific field of expertise and reflect on the project based on the exchange in the meeting and their personal-professional views. Content sometimes overlaps between the chapters. There is also some repetition both in the arguments and in how these are illustrated. Views across some chapters may also either complement or conflict with each other. Such recurrence, complementarity and conflict of arguments are an invaluable resource for the project and are necessary to elicit guidelines for the way forward.

Report structure

Part I. Setting the scene

The first part sets the scene for the report with two introductory chapters.

The current **Chapter 1** presents the background and rationale for this work.

In **Chapter 2**, *Kenneth Forbus* analyses the progress in AI over the past four decades. The author describes three ongoing revolutions – deep learning, knowledge graphs and reasoning – and foresees a fourth revolution: that of integrated intelligence. The chapter discusses the implications of these revolutions for efforts to derive relevant measures of AI's progress with respect to human capabilities.

Part II. Taxonomies and tests of human skills

The second part explores taxonomies of human skills in different branches of psychology and reviews existing measures of these skills. It distinguishes two major areas: cognitive psychology discussed in Chapters 3 to 7 and industrial-organisational psychology addressed in Chapters 8 to 10.

Cognitive abilities and their extensions

Chapter 3 by *Patrick Kyllonen* provides an overview of widely known taxonomies of human cognitive abilities, presents the history of their development, and discusses their strengths and weaknesses. The author describes measures of cognitive abilities and their quality characteristics such as reliability, validity, fairness and measurement invariance. Finally, the chapter discusses the prospects and feasibility of using these tests as the basis for evaluating machine intelligence.

In **Chapter 4**, *Sylvie Chokron* presents the neuropsychological perspective of capturing weaknesses and strengths of cognitive abilities in children. The author describes the most commonly used neuropsychological tests, as well as their limits and caveats in understanding the cognitive profile of children. The chapter also considers the opportunities and challenges in using such tests for assessing machine capabilities.

Chapter 5 focuses on social and emotional skills. *Filip De Fruyt* reviews the theoretical conceptualisations of social and emotional skills, and discusses how these relate to educational and labour-market outcomes. The paper presents taxonomies of these skills, including the widely known Big Five framework, and describes different types of items through which such skills can be assessed. Importantly, the author discusses recent developments in the field. These attempt to provide more objective measures of social and emotional skills, such as situational judgement tests and behavioural residue indicators.

Chapter 6 by *Anita Woolley* explores the recent concept of collective intelligence, i.e. the ability of a group to perform a wide range of tasks. The paper presents task batteries to measure collective intelligence and describes how these can be used to elicit the factors that predict team performance. Ongoing research also includes understanding how AI capabilities can enhance collective intelligence in a mixed team of machines and humans. Finally, the author illustrates how these measures can provide a vehicle for assessing artificial social intelligence.

In **Chapter 7**, *Samuel Greiff* and *Jan Dörendhal* describe two major skill domains typically measured in large-scale educational assessments: core domain skills such as mathematics, reading and science literacy, and transversal skills such as problem solving, collaboration and creativity. The chapter presents the theoretical underpinning and measurement of these skills and examines their role in occupational settings. The chapter concludes with recommendations regarding the use of education tests for assessing AI capabilities.

Occupational assessments

Chapter 8 proposes an assessment strategy that draws on tests developed for jobs subject to licensing examination. *Phillip Ackerman* presents the foundations of human intelligence tests. The author discusses a number of methodological challenges including those arising from tacit knowledge, humans' use of tools, differences in learning between humans and AI, and the inaccuracy of skills assessments at high performance levels. Based on these challenges, the chapter argues that the OECD project should focus on domain knowledge and skills – in the context of specific jobs – rather than higher-order cognitive abilities. .

Chapter 9 provides a vocational perspective to skills assessment. *Britta Rüschoff* reviews the methods of skills assessment in German vocational education and training (VET). The chapter defines vocational competences and presents instruments to assess them in VET examinations through concrete examples. The author also describes how these examinations are developed and administered, and discusses the validity and reliability of the instruments. Finally, the chapter indicates the advantages of using VET tests for assessing AI capabilities. It concludes with considerations for applying these instruments to machines.

In **Chapter 10**, *David Dorsey* and *Scott Oppler* propose an approach for comparing human and AI capabilities based on comprehensive occupational taxonomies. The authors start with clarifying the structure and underlying concepts of occupational databases. The chapter then outlines a number of methodological recommendations and describes four major steps of the proposed approach: identifying an occupational taxonomy, sampling occupations from the taxonomy, collecting expert judgement on AI capabilities and analysing data from expert interviews.

Part III. AI capabilities and their measures

The third part of the volume explores the perspective of computer scientists on the evaluation of AI and robotics capabilities. Chapters 11 to 13 discuss major challenges in assessing AI capabilities with human tests, while Chapters 14 to 17 focus on existing empirical evaluation efforts of machines.

Challenges of assessing AI capabilities with human tests

Chapter 11 connects the second and the third part of the report. After situating AI measurement in the context of the roles AI can play in the future, *José Hernández-Orallo* provides an overview of human skill taxonomies and links these to the world of AI. The chapter explores types of human tests used in recruitment and education, and contrasts these with the evaluation of machines. The author discusses the challenges of using human tests for assessing AI capabilities and identifies guidelines for devising tests that can compare the capabilities of humans and AI reliably.

Chapter 12 focuses on the specificities of machines and their striking differences from humans. *Ernest Davis* presents areas in which computers excel and illustrates their weaknesses compared to humans through examples of sometimes "grotesque" failures. The author proposes consequences of these various strengths and limitations for using human tests for assessing AI.

Chapter 13 also brings attention to the ways in which AI is similar to and utterly different from human intelligence. *Richard Granger* first discusses how the architectures of artificial neural networks relate to networks of neurons in the human brain. The author then compares the behaviour and computational abilities arising from artificial neural networks to human behaviour and abilities, illustrating both with entertaining examples. The chapter points to current efforts to overcome shortcomings and concludes with a number of implications for understanding machine capabilities.

Efforts to assess AI and robotics capabilities

In **Chapter 14**, *Anthony Cohn* presents approaches and methods of the AI community to measuring and evaluating AI systems. The author first presents tests proposed for measuring AI, then describes some competitions created to compare AI systems, as well as a few benchmark datasets. The chapter discusses some of the benefits and limitations of these approaches.

Chapter 15 focuses on empirical evaluations of AI systems as performed by the French Institute of Metrology (Laboratoire national de métrologie et d'essais: LNE). *Guillaume Avrin* first presents the characteristics and process of these evaluations. The author then proposes a high-level taxonomy of AI capabilities and generalises it to other AI tasks to draw a parallel with human capabilities. The chapter then discusses the relevance of existing evaluation methods for comparing AI and human capabilities. It concludes with recommendations for the project approach.

Chapter 16 provides an overview of evaluation techniques applied in the domain of natural language processing. *Yvette Graham* describes methods that offer fair and replicable evaluations of system performance in this domain. The author shows how longitudinal evaluation can capture progress in AI language processing capabilities and how these methods allow for comparison with human performance. The chapter also discusses human-machine hybridisation in tasks and its implication for understanding the potential for machines in society.

In **Chapter 17**, *Lucy Cheke*, *Marta Halina* and *Matthew Crosby* focus on basic or common sense skills that all healthy human adults have but in which machines still often fail. The authors propose a taxonomy of these skills identifying two major domains: spatial and social skills, and describe tests used in the fields of animal and developmental psychology. The chapter presents examples of workplaces and situations that might require the use of such skills, and explore limitations and opportunities for assessing common sense skills in AI.

Part IV. Reflections and a pragmatic way forward

The last part of the report attempts to synthesise the discussion and draw conclusions as to the way forward for the AIFS project.

In **Chapter 18,** *Art Graesser* discusses three questions relevant to the AIFS project: What is the value in identifying ideal models when comparing humans and AI and robotic systems? How might we conduct a systematic mapping between skill taxonomies, tasks, tests and functional AI components? How can we handle major differences in the skills we target, the different occupations and changes in the worlds we live in? The author offers suggestions on next steps in addressing these questions.

Chapter 19 provides guidance for setting up a general analytical framework for assessing AI capabilities with regard to human skills. *Eva Baker* and *Harry O'Neil* offer recommendations on operative aspects of the project, on the selection of tests for comparing AI and human skills, and on the selection and training of expert raters. The chapter concludes with a summary of considerations made for planning the study.

Finally, in **Chapter 20**, *Stuart Elliott* reflects on key considerations from the expert contributions in the field of psychology and computer science. The author proposes to bring together the different domains of psychology to benefit from the strength and relevance of each domain. The chapter also suggests a pragmatic way forward for the project to address the concerns formulated by the AI community and develop AI-specific assessment approaches where those are required.

References

Arntz, M., T. Gregory and U. Zierahn (2016), "The Risk of Automation for Jobs in OECD Countries: A Comparative Analysis", *OECD Social, Employment and Migration Working Papers*, No. 189, OECD Publishing, Paris, https://dx.doi.org/10.1787/5jlz9h56dvq7-en. [5]

Autor, D., F. Levy and R. Murnane (2003), "The skill content of recent technological change: An empirical exploration", *The Quarterly Journal of Economics*, Vol. 118/4, pp. 1279-1333, http://dx.doi.org/10.1162/003355303322552801. [17]

Baruffaldi, S. et al. (2020), "Identifying and measuring developments in artificial intelligence: Making the impossible possible", *OECD Science, Technology and Industry Working Papers*, No. 2020/05, OECD Publishing, Paris, https://dx.doi.org/10.1787/5f65ff7e-en. [3]

Bresnahan, T. and M. Trajtenberg (1992), *General Purpose Technologies "Engines of Growth?"*, National Bureau of Economic Research, Cambridge, MA, http://dx.doi.org/10.3386/w4148. [12]

Brynjolfsson, E. and T. Mitchell (2017), "What can machine learning do? Workforce implications", *Science*, Vol. 358/6370, pp. 1530-1534, http://dx.doi.org/10.1126/science.aap8062. [21]

Brynjolfsson, E., T. Mitchell and D. Rock (2018), "What can machines learn and what does It mean for occupations and the economy?", *AEA Papers and Proceedings*, Vol. 108, pp. 43-47, http://dx.doi.org/10.1257/pandp.20181019. [16]

Clark, P. and O. Etzioni (2016), "My Computer Is an Honor Student — but How Intelligent Is It? Standardized Tests as a Measure of AI", *AI Magazine*, Vol. 37/1, pp. 5-12, http://dx.doi.org/10.1609/aimag.v37i1.2636. [26]

Crosby, M., B. Beyret and M. Halina (2019), "The Animal-AI Olympics", *Nature Machine Intelligence*, Vol. 1/5, pp. 257-257, http://dx.doi.org/10.1038/s42256-019-0050-3. [27]

Davis, E. (2016), "How to Write Science Questions that Are Easy for People and Hard for Computers", *AI Magazine*, Vol. 37/1, pp. 13-22, http://dx.doi.org/10.1609/aimag.v37i1.2637. [29]

Elliott, S. (2017), *Computers and the Future of Skill Demand*, Educational Research and Innovation, OECD Publishing, Paris, https://dx.doi.org/10.1787/9789264284395-en. [30]

Felten, E., M. Raj and R. Seamans (2019), "The variable impact of artificial intelligence on labor: The role of complementary skills and technologies", *SSRN Electronic Journal*, http://dx.doi.org/10.2139/ssrn.3368605. [22]

Frey, C. and M. Osborne (2017), "The future of employment: How susceptible are jobs to computerisation?", *Technological Forecasting and Social Change*, Vol. 114, pp. 254-280, http://dx.doi.org/10.1016/j.techfore.2016.08.019. [15]

Lane, M. and A. Saint-Martin (2021), "The impact of Artificial Intelligence on the labour market: What do we know so far?", *OECD Social, Employment and Migration Working Papers*, No. 256, OECD Publishing, Paris, https://dx.doi.org/10.1787/7c895724-en. [7]

Manyika, J. et al. (2017), *A Future That Works: Automation, Employment, and Productivity*, McKinsey Global Institute, New York. [18]

Martínez-Plumed, F. et al. (2021), "Research community dynamics behind popular AI benchmarks", *Nature Machine Intelligence*, Vol. 3/7, pp. 581-589, http://dx.doi.org/10.1038/s42256-021-00339-6. [25]

Nakazato, S. and M. Squicciarini (2021), "Artificial intelligence companies, goods and services: A trademark-based analysis", *OECD Science, Technology and Industry Working Papers*, No. 2021/06, OECD Publishing, Paris, https://dx.doi.org/10.1787/2db2d7f4-en. [8]

Nedelkoska, L. and G. Quintini (2018), "Automation, skills use and training", *OECD Social, Employment and Migration Working Papers*, No. 202, OECD Publishing, Paris, https://dx.doi.org/10.1787/2e2f4eea-en. [6]

OECD (2021), "Why accelerate the development and deployment of robots?", in *OECD Science, Technology and Innovation Outlook 2021: Times of Crisis and Opportunity*, OECD Publishing, Paris, https://dx.doi.org/10.1787/0901069e-en. [14]

OECD (2020), *The OECD Artificial Intelligence Policy Observatory - OECD.AI*, https://oecd.ai/ (accessed on 20 January 2021). [2]

OECD (2019), *Artificial Intelligence in Society*, OECD Publishing, Paris, https://dx.doi.org/10.1787/eedfee77-en. [4]

OECD (2019), *Recommendation of the Council on Artificial Intelligence*, https://legalinstruments.oecd.org/en/instruments/OECD-LEGAL-0449#supportDocuments (accessed on 20 January 2021). [1]

OECD (2017), *OECD Digital Economy Outlook 2017*, OECD Publishing, Paris, https://dx.doi.org/10.1787/9789264276284-en. [13]

Ohlsson, S. et al. (2016), "Measuring an artificial intelligence system's performance on a Verbal IQ test for young children", *Journal of Experimental & Theoretical Artificial Intelligence*, Vol. 29/4, pp. 679-693, http://dx.doi.org/10.1080/0952813x.2016.1213060. [28]

PwC (2018), *The Macroeconomic Impact of Artificial Intelligence*, PricewaterhouseCoopers, London, https://www.pwc.co.uk/economic-services/assets/macroeconomic-impact-of-ai-technical-report-feb-18.pdf. [19]

PwC (2017), *Will Robots Really Steal our Jobs? The Potential Impact of Automation on the UK and other Major Economies*, PricewaterhouseCoopers, London. [20]

Squicciarini, M. and J. Staccioli (forthcoming), "Labour-saving technologies and employment levels: Are robots really making workers redundant?", *OECD Science, Technology and Industry Working Papers*. [24]

van der Vlies, R. (2020), "Digital strategies in education across OECD countries: Exploring education policies on digital technologies", *OECD Education Working Papers*, No. 226, OECD Publishing, Paris, https://dx.doi.org/10.1787/33dd4c26-en. [10]

Verhagen, A. (forthcoming), "Opportunities and drawbacks of using Artificial Intelligence for training", *OECD Social, Employment and Migration Working Papers*. [11]

Vincent-Lancrin, S. and R. van der Vlies (2020), "Trustworthy artificial intelligence (AI) in education: Promises and challenges", *OECD Education Working Papers*, No. 218, OECD Publishing, Paris, https://dx.doi.org/10.1787/a6c90fa9-en. [9]

Webb, M. (2020), “The impact of artificial intelligence on the labor market”, *SSRN Electronic Journal*, http://dx.doi.org/10.2139/ssrn.3482150. [23]

Notes

[1] This report will not distinguish between the terms “skills”, “abilities”, “capabilities” and “competences”. They will be used interchangeably. Single chapters, however, may adopt more precise definitions, which will be made explicit.

2. Evaluating revolutions in artificial intelligence from a human perspective

Kenneth D. Forbus, Northwestern University

Artificial intelligence (AI) has made considerable progress over the past 40 years, leading to both important applications now in daily use but also a lot of hype in the popular press. This chapter outlines a framework for understanding progress in AI to help understand how to better measure that progress. It describes three revolutions currently under way: *deep learning*, *knowledge graphs* and *reasoning*. It summarises the progress and limitations in each and contrasts them with human capabilities for learning, knowledge and reasoning. The chapter also summarises a fourth revolution to come, *integrated intelligence*, where the goal is to create systems with agency. It discusses some additional implications for measuring AI progress with respect to learning, knowledge, reasoning and agency.

Introduction

There has been considerable progress in artificial intelligence (AI) over the past four decades. Due to vast increases in the availability of computing power and data, this progress has accelerated over the last 20 years. These increases, combined with steady scientific progress, have led to revolutionary advances. In this context, a revolution occurs when scientific progress and its application through technology lead to a radical improvement in existing capabilities, or new capabilities, that have become widely used and provide significant benefits. This notion of revolution factors out stunts and laboratory experiments. This makes it useful for understanding how to measure AI and to think about how it might impact education and work.

By this criterion, there are three AI revolutions in progress: *deep learning*, *knowledge graphs* and *reasoning*. This chapter discusses them, comparing and contrasting with human capabilities. It then describes a fourth AI revolution just beginning, *integrated intelligence*, which builds on the other three. It concludes with further discussion of measuring progress in AI relative to human capabilities.

Artificial intelligence Revolution 1: Deep learning

The revolution everyone has heard about through the popular press is deep learning (LeCun, Bengio and Hinton, 2015[1]). Neural networks organise computation via large collections of simple computational units metaphorically inspired by neurons in brains. The term "deep" refers to the number of layers in the network. Research on neural networks has received varying amounts of attention since the 1950s. Until recently, the lack of computing power and training data meant that only small networks could be explored. Several decades of research on algorithm improvements, combined with graphical processing units originally developed for video games, finally enabled large networks to be trained using massive amounts of data. This has led to systems that performed far better than previous approaches on tasks such as image captioning, facial recognition, speech recognition and automatic natural language translation. These performance improvements have led to deployed applications in these areas and many others that are in daily use around the world.

Although deep learning models are inspired by biological neurons, they do not necessarily learn in human-like ways. Indeed, given the state of knowledge in neuroscience, claims of biological plausibility must be taken sceptically (Bengio et al., 2015[2]; Marcus, 2018[3]). Research on learning in psychology and AI highlights some important differences between deep learning and human learning. Human learning involves multiple kinds of processes and occurs at multiple time-scales. For example, conceptual change involves articulation of knowledge, working through implications of models over weeks, months and years. It can involve radical discontinuities in conceptual understanding. By contrast, skill learning, which can also take years, is often hard to articulate. Skill learning can be described by a power law, which is smooth. Both conceptual change and skill learning are incremental. By contrast, deep learning requires its training data all at once, but otherwise is more smooth and implicit, like skill learning.

The widespread use of deep learning has exposed three limitations. The first is brittleness, where small changes in inputs lead to dramatically different outputs. The second is data efficiency, i.e. the amount of training data required. The third is explainability, i.e. it is hard to understand why errors occur and how to fix them. Each limitation is discussed in turn.

Brittleness

Deep learning models typically approximate a function, e.g. given an input, it produces an output. Inputs that are close to what the model was trained on often yield reasonable results. Unfortunately, inputs from outside the training set can sometimes yield arbitrarily incorrect results. Changing a few pixels, for example, can cause a scene to be classified differently (Goodfellow, Shlens and Szegedy, 2015[4]). Adding a few small patches can cause the system to perceive a stop sign as a speed limit sign (Eykholt et al., 2018[5]). Indeed, images that look like random noise to people are sometimes confidently recognised by deep learning systems as everyday objects (Nguyen, Yosinski and Clune, 2015[6]). Human vision has limitations, as evidenced by well-known visual illusions. However, human visual illusions are nothing like these misperceptions.

Brittleness is a limitation of deep learning models in general, not just in vision. Deep learning models for question-answering and reading comprehension are subject to similar problems. For example, including extra irrelevant information that a human would ignore can mislead such systems (Jia and Liang, 2017[7]). Part of the problem is that these systems are being tested using instruments that do not necessarily measure what they are intended to measure. Reading comprehension measures designed by AI researchers, for instance, almost always use multiple-choice questions because they are simple to score. However, this method enables systems to pick up correlations between terms in the texts and the answers. Indeed, the systems will sometimes do almost as well in picking out answers even when it does not know the questions (Kaushik and Lipton, 2018[8]).

Data efficiency

Data efficiency is analogous to fuel efficiency. It is the amount of data needed to achieve a particular level of performance. The more layers in a deep learning system, the more parameters there are, and the more training data is required. How much data? Consider learning to play the game of Go. AlphaGo (Silver et al., 2016[9]) and AlphaZero (Silver et al., 2018[10]), by combining deep learning with a hand-built search routine optimised for such games, were able to learn to play Go well enough to beat the best human players. The amount of data used was massive. Achieving AlphaGo's exposure to games of Go, in human terms, would require a person to live about 6 800 years (Forbus, Liang and Rabkina, 2017[11]).[1] Similarly, GPT-3, a large-scale statistical language model, was trained on many billions of tokens, far more than people experience in a lifetime. Nevertheless, while it can sometimes generate text that looks locally plausible, it is often globally incoherent and often the statements it produces are simply wrong (Marcus and Davis, 2020[12]).[2]

By contrast, consider teaching a human assistant to do a routine office task or assembly task. One does not need hundreds of training examples, let alone millions. A growing AI research area, *interactive task learning* (Gluck and Laird, 2019[13]), seeks to train systems with the same amounts and kinds of experience that a person would receive. This requires data efficiency but also the ability to communicate with the system using natural modalities (e.g. language, sketching, gesture, speech). Interactive task learning is a step towards integrated intelligences, which are discussed more below.

Explainability

Explainability is important so that people can understand the rationale for decisions that a system makes and to troubleshoot when problems arise. Deep learning systems are opaque, since they rely on massive sets of numbers that typically do not have any clear interpretation. It should be noted that human learning is not always explainable. The field of psychophysics, for example, explores how perception works, including what is built-in versus learnt. Similarly, cognitive psychologists have studied learning procedural knowledge. In so doing, they have generated both a wealth of behavioural results and computational models that capture temporal properties and error patterns in such learning. For evaluating skills, performance measures often suffice.

By contrast, in science, engineering, law and medicine, as well as many other fields, practitioners are required to explain their results. This puts a premium on explicit conceptual knowledge, as discussed further below. Even in these professions, some forms of knowledge are less easily articulated, commonly called *tacit knowledge* (Forbus, 2019[14]). Such knowledge often concerns how situations in the everyday world can be understood in terms of professional knowledge and the construction of a *mental model* (Gentner and Stevens, 1983[15]). Research in qualitative reasoning (Forbus, 2019[14]) suggests this split between formulating and using mental models is common in professional reasoning (Klein, 1999[16]; Engel, 2008[17]).

Deep learning is indeed a valuable technology, despite these limitations. The trajectory of deep learning also highlights an important lesson about new technologies and their potential impacts. Ideas can take considerable time to hone – conceptual advances and algorithmic improvements can take decades. They may only take off when environmental factors are favourable (e.g. massive amounts of data and of computation). Other approaches to learning in AI (including inductive logic programming, analogy and interactive task learning) are earlier in their trajectories. However, they may yet hit that combination of progress and environmental factors to become revolutions on their own.

There is more to AI than learning. These same factors have led to revolutions in two other areas, as described next.

Artificial intelligence Revolution 2: Knowledge graphs

Knowledge is a key ingredient of intelligence. Decades of research on knowledge representation in AI has led to formal methods for capturing many kinds of facts in machine-usable form at industrial scale. Such compendia are called knowledge graphs.[3]

An overview of knowledge graphs

Knowledge graphs provide explicit representations of knowledge, including descriptions of entities in terms of statements about them, relationships among entities and a variety of other conceptual structures, such as arguments, explanations and plans. Knowledge graphs include *ontologies*, i.e. representations that describe types of things in the world (entities) and facts about those entities. Ontologies are typically constructed by hand. Facts are added via a combination of hand-engineering and automatic methods, such as importing information from databases and extracting from texts such as webpages.

Knowledge graphs have been applied in multiple industries. For example, Google's Knowledge Graph uses 70 billion facts describing over a billion entities to improve web search and perform question-answering (Noy et al., 2019[18]). If a user is looking for a restaurant in San Francisco, for instance, the system needs to understand whether they are searching from California or Colombia. Similarly, Amazon uses one knowledge graph of product information in its recommendation system (Dong, 2018[19]) and another of general information for question-answering via Alexa.[4] In its operations, Facebook uses a knowledge graph of people and their relationship with others (Noy et al., 2019[18]). Several scientific communities have created their own ontologies using Semantic Web representations (Allemang, Hendler and Gandon, 2020[20]) to help index and retrieve data, results and publications. Similarly, publishers have created ontologies to help others access their articles and works.[5]

It is interesting to compare today's knowledge graphs with human knowledge. Billions of facts is quite a lot. Certainly, no human being has memorised, say, all of the named locations on Earth plus all of the people mentioned in Wikipedia. Therefore, along some dimensions, knowledge graphs go beyond what people know. On the other hand, people know massive amounts about the physical world, social world and mental worlds of agents. Estimating the amount of knowledge people have is challenging (Forbus, Liang and Rabkina, 2017[11]). What is clear, however, is that humans use at least three types of knowledge that are rarely found in deployed knowledge graphs. These are multimodal grounding of knowledge, episodic knowledge and inferential knowledge.

Multimodal grounding of knowledge

Some human knowledge is grounded in perception and sensorimotor experience. Humans know what a cat looks like in repose versus at play, and how many dishes can be carried safely. This multimodal grounding of knowledge – what the world looks, sounds, feels and tastes like – is almost entirely missing from today's knowledge graphs.[6]

Episodic knowledge

Episodic knowledge is one's personal experience, which serves as a resource in efficient reasoning. For example, human experience suggests that dogs make good pets whereas sharks do not. Humans can decide this quickly, even if it takes a moment to articulate clear reasons for the difference. People also learn from the experiences of others, as transmitted via stories, allowing cultures to accumulate knowledge across space and time.

The family of techniques called *case-based reasoning* (CBR) use forms of analogy to apply knowledge about previous situations to new problems. Some CBR applications have scaled to millions of cases (Jalali and Leake, 2018[21]). However, current large-scale CBR systems use vector representations[7] rather than the relational information used in other areas in CBR and in knowledge graphs. Relational knowledge is crucial to capture the full range of human knowledge. It includes the ability to represent plans, theories, arguments and social relationships.[8] Although people's general knowledge (often called *semantic memory*) is tightly integrated with their episodic knowledge, large-scale relational CBR systems do not currently exist.

Inferential knowledge

Inferential knowledge consists of general knowledge that can be used to derive new consequences. For example, one might know that dogs are commonly kept as pets in some cultures, because dogs are often affectionate towards people and people often keep pets to provide affection.

Inferential knowledge is valuable because it supports reasoning, which enables systems to come up with novel conclusions about new situations and problems. It typically takes the form of rules in a formal language, allowing algorithms to use them automatically and unambiguously. There are a variety of formal languages used in practice, varying considerably in expressive power. Some are strictly logical. Others incorporate probabilities and statistical reasoning. Still others incorporate procedural (i.e. how-to) knowledge. Cyc is the knowledge graph with the most inferential knowledge. It uses a highly expressive logic, enabling it to be used for a variety of applications.[9]

The trajectory of knowledge graphs

The limitations of knowledge graphs summarised above are natural given that industrial, commercial and scientific knowledge graphs are purpose-built. For example, a genomics ontology will not include cultural products, which are a major component of Google and Bing's knowledge graphs, and they in turn do not include detailed representations of genomes. There is a recognition that the field needs to start building *open knowledge networks*.[10] This will allow existing knowledge graphs to be integrated more easily and extended in ways that will benefit everyone.

As knowledge graphs are applied to more tasks, the range of types of knowledge they include will grow. Multimodal grounding of concepts in perception is important for creating robots that can operate in less constrained environments for tasks such as elder care and housekeeping. Personal assistant software is already helping drive development of formal representations of human events and preferences, to better model the particular people they work with.

Understanding context and higher-quality understanding of natural language is leading to more work on representing inferential knowledge. Today's assistants might handle questions like "What is the weather today?" followed by "How about the weekend?" If the initial question was "What is my schedule today?", the follow-up question about the weekend concerns schedules, not the weather. Handling such changes can be done via semantic representations that explicitly represent the computations used in answering questions (Andreas et al., 2020[22]).

Artificial intelligence Revolution 3: Reasoning

Reasoning is a hallmark of human intelligence. There are many kinds of reasoning, but in their essence they involve combining facts to reach new conclusions. This sub-section examines several dimensions of reasoning, common sense versus professional reasoning, some examples of the reasoning revolution, and compares human and AI reasoning capabilities.

Dimensions of reasoning

Depth of reasoning

Depth of reasoning relates to how many steps are needed to perform a task. A question like "How old is the president of Germany?"[11] provides a simple example. Answering the question requires two steps: identifying the president of Germany and then finding out that person's age. The reasoning revolution is due in part to the ability of some specialised reasoners to go far beyond human capabilities in terms of depth.

Types and breadth of knowledge

Since reasoning involves combining information to produce new conclusions, it invariably involves two dimensions of knowledge: *types* and *breadth*. Consider the kinds of question-answering that IBM's original Watson system performed. To beat the best human players at the game of *Jeopardy!*, Watson integrated a massive amount of knowledge: Wikipedia, multiple reference works, volumes of literature and other information (Fan et al., 2012[23]). Watson's knowledge was thus quite broad but only moderately deep. It was capable of evaluating spatial and temporal constraints on answers, for example, and reasoning about types of entities.

Flexibility of reasoning

Humans marshal vast amounts of multiple types of knowledge efficiently and effectively. We are able to come to some conclusions rapidly, even with very little information, and can combine experience with principles to come up with effective plans in novel circumstances.

Common sense reasoning versus professional reasoning

One of the grand challenges for AI is common sense reasoning (Davis and Marcus, 2015[24]). This includes everyday knowledge of the physical world, such as knowing that it is possible to pull with a string but not push with it.[12] Common sense reasoning gets people through the activities of their days – from navigating, cooking and cleaning to interacting with others.

Professional reasoning is grounded in common sense reasoning. New problems come couched in a mix of professional and everyday terms. For example, a doctor must diagnose a patient's symptoms; a judge reviews a legal case to judge; a designer develops a device. In doing so, they translate from their everyday terms and models into their professional terms and models. Thus, a key part of professional reasoning is *model formulation.* This involves construction of a mental model for a situation in professional terms that enables a problem to be solved.[13] The knowledge needed for professional reasoning includes explicit principles taught (e.g. the definition of a tort, the equations of thermodynamics). However, it also includes strategies and tactics to apply principles to new situations. This strategic and tactical knowledge is often tacit, communicated via apprenticeship and reflection on experience.

Examples from the reasoning revolution

While many issues in understanding reasoning remain open scientific questions, several areas have shown revolutionary progress. The most spectacular AI reasoning systems today operate after model formulation has occurred. In designing a new distributed computing scheme, for example, Amazon engineers write a description of their system using a temporal logic (Newcombe et al., 2015[25]). A *model checker*, which is a type of reasoning system, scrutinises the possible combinations of events in the system to search for bugs. These bugs can be subtle, sometimes only happening after a sequence of over 30 events. Yet, given the velocity of modern computation, such bugs are likely to manifest several times a week, making their detection important.

Many applications use automatic model formulation within specific narrow domains. For example, Facebook uses model checking to evaluate the code being added to the 100 million lines of code that runs their services (Distefano et al., 2019[26]). If an engineer has checked in a software patch that could be problematic, it is flagged for review automatically.

Similar superhuman feats of reasoning occur daily by *satisfiability solvers* used in logistics and operations planning. These satisfiability solvers operate over systems of millions of constraints far more quickly and generate better solutions than human planners can produce [e.g. (Simonis, 2001[27])].

In engineering design, combinations of *qualitative and quantitative models* have been used to do several types of commercially important reasoning. These systems perform causal reasoning, using human-like conceptual models that enable them to provide natural explanations that engineers understand. For example, AutoSteve (Price, 2000[28]) performs failure modes and effects analyses to evaluate what could go wrong in designs for automobile electrical systems. At Xerox, such models are used to evaluate potential new modules for high-performance print engines at design time for cost effectiveness. They are also used as the basis for model-based control software. This software reconfigures itself dynamically on-site as the complex electro-mechanical components of high-end printers change due to reconfiguration of physical parts or parts wearing out (Fromherz, Bobrow and de Kleer, 2003[29]).

Table 2.1. Dimensions of reasoning in people versus AI systems

Dimension	People	AI systems
Depth of reasoning	Medium	Very High
Types of knowledge	Many	Few
Breadth of knowledge	Wide	Narrow
Flexibility of reasoning	High	Low

Reasoning in people vs. AI systems

As Table 2.1 illustrates, AI systems and humans have different strengths and weaknesses. AI systems are better at high-depth reasoning than humans. Asking whether people could be trained to do as well as model checkers and satisfiability solvers is like asking whether people can be trained to outrun an automobile or outlift a crane. The firing frequency of neurons ranges roughly between 1-500 Hertz, but brains have massive parallelism, with billions of neurons. Modern computers, by contrast, are much more serial, but operate around 10 million times faster. For tasks with many sequential reasoning steps (i.e. high-depth reasoning), AI systems have an overwhelming advantage.

People still have the edge over AI systems on common sense reasoning, model formulation and many kinds of professional reasoning. These require many types of reasoning, a wide breadth of knowledge and high flexibility. Of these, more aspects of professional reasoning are likely to be the next to be more broadly automated. It is already happening in engineering domains to some degree, due to the mathematical nature of many engineering models.

However, the experiential component to engineering is harder to capture, and that will take extending more AI systems to accumulate episodic memories. Medicine and law seem to rely even more heavily on experience and rules distilled from experience by practitioners. Ultimately, AI systems will be able to accumulate far more experience in these domains than any human being due to our lifespan limitations and the ability of AI to combine experience across many systems.[14] Scaling up analogical reasoning to handle CBR and distillation of rules from such massive collections will be an important technical challenge.

The coming fourth artificial intelligence Revolution: Integrated intelligence

The three revolutions – deep learning, knowledge graphs and reasoning – are far from over. Moreover, they interact synergistically. For example, deep learning has been used to help build knowledge graphs and estimate which subgoals are worth pursuing in reasoning (like choosing good moves in a game). Knowledge graphs are being integrated with deep learning systems to improve their reasoning capabilities. However, there is a key limitation that none of these revolutions addresses. Even when machine learning is used, today's AI systems are still constructed and maintained by humans. This sub-section outlines the *integrated intelligence* revolution to come, the fourth AI revolution.

From pipelines to organisms

People hand-craft the structure of today's AI systems. Even when machine learning is used, people gather and curate the data, inspect the results, and then tune/tweak the data and the learning algorithms. AI systems do not, on the whole, set their own learning goals and decide what data to gather to achieve them. Human scientists and engineers decide on the ontology for systems and what kinds of information are needed in knowledge graphs. If the kinds of tasks change, humans must decide on the need for new knowledge and how to acquire it. Human engineers tune and tweak reasoning systems and problem formulations to wring the most accurate results and effective performance out of systems.

Today's deployed AI systems can be thought of as pipelines – machines whose architecture is purpose-built for a particular range of tasks. For personal assistant AI systems, such as Google Assistant, Amazon's Alexa, Microsoft's Cortana, Samsung's Bixby or SK Telecom's Aria, these pipelines are massive. Human engineers are constantly updating or even replacing models and algorithms as needed to adapt them to changing circumstances. The exact number of human engineers needed for these tasks is held closely by the companies involved. However, depending on the system, it likely ranges from hundreds to thousands of people.

This reliance on human intervention does not scale. For example, one of the visions in the 20-year AI Roadmap for the US (Gil and Selman, 2019[30]) is the idea of personal AI assistants. Unlike today's assistants, these systems will adapt themselves to the work and personal lives of humans to help them, with humans owning all their own data. Such a vision is impossible with the pipeline model – the systems themselves must be able to maintain, adapt and learn on their own. In other words, pipelines are missing *agency*. Building AI systems that, like biological organisms, have agency is the heart of the coming fourth revolution of integrated intelligence.

The road to the fourth revolution

Several AI research threads are building up the science base for this revolution. Cognitive systems research[15] typically involves multiple AI capabilities, as well as exploring how specific capabilities need to be modified or extended to function as part of larger-scale systems. Cognitive architecture research [e.g. (Anderson, 2009[31]; Laird, 2012[32]; Forbus and Hinrichs, 2017[33])] explores computational models of larger-scale phenomena in human cognition. None of these are deployed for daily use at the scale that would constitute a revolution – yet.[16]

Examining the difference between human development and the way AI systems are engineered today provides a lens for identifying the capabilities needed to build organisms. People successfully learn to become members of a culture and profession. This learning is cumulative, with new material building on previously learnt material in ways that none of today's AI technologies can match. People quickly adapt when joining a new group and when dealing with new circumstances. They can be apprentices, which catalyses learning through interactions with others.

Endowing machines with these capabilities involves numerous scientific challenges, most of which are still being formulated. The recent 20-year AI Roadmap for the US (Gil and Selman, 2019[30]) provides a decomposition of many of these issues and suggests milestones to bridge the gaps.[17] It is thus a resource for projecting progress of the field over that period, based on a consensus view developed by over 90 researchers.

A roadmap is not a schedule, of course. Our understanding of the issues and their difficulties are naturally incomplete and progress requires ample resources. Nor does this enable prediction for the start of daily-use beneficial applications that will herald the start of this revolution. Resources for development of deep learning, knowledge graphs and reasoning ramped up as successful applications were fielded, and the same will apply to integrated intelligence.

Additional implications for measuring artificial intelligence progress

Clearly, no single test of AI progress will suffice for measuring its capabilities. This should not be surprising, since psychometricians long ago gave up on defining a single test for measuring human intelligence. AI systems are still well below the capability and sophistication of human minds in most regards. However, the need to measure their progress on human scales demands at least a battery of tests.

These tests will need to be designed by teams of psychometricians and AI researchers. The psychometricians have a better understanding of how to test extremely complex systems (i.e. humans) and to achieve reliable results over time. The AI researchers have a better understanding of the strengths and weaknesses of the technologies.

A battery should incorporate tests for learning, knowledge, reasoning and agency.

Learning

Separate tests will be needed for perceptual, procedural and conceptual learning as each has different expectations and requirements for data efficiency and explainability. Prior work on modelling conceptual change has used tests developed by cognitive scientists. These tests measure the kinds of models a system learns. They also measure whether trajectories of learnt models are similar to those of human learners (Friedman and Forbus, 2010[34]). Assembling a set of such developmental benchmarks could help ensure that intermediate learnt models make sense to the human collaborators of AI systems.

Knowledge

With explicit knowledge graphs, the amount of knowledge in different areas can be measured directly, with effectiveness for reasoning tested by sampling. Some approaches attempt to use language models and other distributed representation systems as knowledge bases. Evaluating such models is far more difficult, given their lack of interpretability.

For inferential knowledge in some domains, such as verification and scheduling, the knowledge is entirely deductive. The consistency of such knowledge can be automatically verified in many cases. As the range of inferential knowledge is expanded, to include more abductive reasoning and common sense reasoning, benchmark domains will be required for testing, since plausible-sounding rules can turn out to be insufficient and/or problematic.

Testing episodic memory is harder. It requires looking at the experiences required for a job and comparing what the system/person has versus what is required (see vocational education and training discussion below).

Reasoning

Reasoning in technical domains, such as engineering, is worth measuring. Performing technical analyses in engineering domains is reasonably well understood now in many cases, but two more difficult frontiers should also be included. The first is reasoning for model formulation, i.e. can a system frame problems for technical analyses when expressed in everyday terms? The second is reasoning for communication, i.e. can the system interact with professionals using the mixture of technical concepts and everyday concepts that people use when working together to solve problems?

Common sense reasoning is especially difficult to measure since it is broad and often tacit. The knowledge may be tacit because it is grounded in experience. For example, one knows that tipped cups spill because it has been seen to happen, not because of a simulation or an explicit rule.

Agency

Testing the ability of AI systems to manage their own operations and learning over extended periods of time – ultimately, decades – will be extraordinarily difficult. Measuring the ability to handle a suite of complex problems, of varying types and progressive in difficulty, may be a reasonable way to approximate the process. As the complexity of AI systems grows, models of trust will likely be developed that look more like collaboration with other organisms (e.g. working dogs, draft horses, human collaborators) than engineering verification and validation. In other words, experience will lead to trust in systems over time. It is not even clear what verification and validation would mean when, for example, training a system via apprenticeship to fit within the practices of an organisation. In its assessments used in vocational education and training developed in Germany, Chapter 9 provides both a useful breakdown of many workplace skills and examples of assessment techniques that might be adaptable for this purpose.

Conclusions

The three AI revolutions of deep learning, knowledge graphs and reasoning are already having considerable impact, and the fourth revolution of integrated intelligence will likely go further. For understanding AI progress in human terms, these four areas (learning, knowledge, reasoning, agency) are coarser than traditional psychometric taxonomies. However, they provide a useful perspective by focusing on sufficiency for tasks, which is directly relevant for considering the impact of AI on the future of work. For this purpose, they are incomplete: language and vision capabilities need to be assessed as well, for example.

Still, a level of analysis in terms of task capabilities seems useful. Psychometric tests were designed to assess people. They have been, and remain, one source of useful measures to understand progress in AI. However, today's AI systems differ in many ways from people, and tomorrow's likely will as well. In some cases, these differences will be limitations to be overcome; in others, they will be sought deliberately to build systems that complement human strengths and weaknesses. Consequently, more task-oriented measures will be needed as well.

Are there other AI revolutions to come? It seems likely. While deep learning has received much of the press, other learning techniques are being scaled up and more broadly applied as well. Therefore, more revolutions in learning may emerge.

What about natural language? Statistical learning and knowledge graphs have been powering progress on natural language, but there are limits to language on its own. For example, approaches that try to build dialogue systems based on large corpora of human interactions often ignore the critical role of context in dialogue. For example, the meaning of "What shall we do next Thursday?" must be determined in the context of the current conversation. This means that, beyond a certain level, progress in natural language is bound up with progress in integrated intelligences, which provide context for understanding.

What about robotics? There, issues of materials and mechanical engineering are key bottlenecks, making it much harder to estimate progress.

References

Allemang, D., J. Hendler and F. Gandon (2020), *Semantic Web for the Working Ontologist: Effective Modeling for Linked Data, RDFS, and OWL*, Third edition, ACM Books, New York. [20]

Anderson, J. (2009), *How can the Human Mind Occur in the Physical Universe?*, Oxford University Press. [31]

Andreas, J. et al. (2020), "Task-oriented dialogue as dataflow synthesis", *Transactions of the Association for Computational Linguistics*, Vol. 8, pp. 556-571, http://dx.doi.org/10.1162/tacl_a_00333. [22]

Bengio, Y. et al. (2015), "Towards biologically plausible deep learning", *arXiv*, Vol. 1502.04156v2, https://arxiv.org/abs/1502.04156. [2]

Davis, E. and G. Marcus (2015), "Commonsense reasoning and commonsense knowledge in artificial intelligence", *Communications of the ACM*, Vol. 58/9, pp. 92-103, https://doi.org/10.1145/2701413. [24]

Distefano, D. et al. (2019), "Scaling static analyses at Facebook", *Communications of the ACM*, Vol. 62/8, pp. 62-70, http://dx.doi.org/10.1145/3338112. [26]

Dong, L. (2018), "Challenges and innovations in building a product knowledge graph", presentation, January, Paul Allen School of Science and Engineering, University of Washington, Seattle, https://db.cs.washington.edu/events/database_day/2018/slides/luna_productgraph.pdf. [19]

Engel, P. (2008), "Tacit knowledge and visual expertise in medical diagnostic reasoning: Implications for medical education", *Medical Teacher*, Vol. 307/7, pp. 184-188, http://dx.doi.org/10.1080/01421590802144260. [17]

Eykholt, K. et al. (2018), "Robust physical-world attacks on deep learning visual classification", *arXiv*, Vol. 08945, http://dx.doi.org/arXiv:1707.08945. [5]

Fan, J. et al. (2012), "Automatic knowledge extraction from documents", *IBM Journal of Research and Development*, Vol. 56/3/4, pp. 5:1-5:10, http://dx.doi.org/10.1147/JRD.2012.2186519. [23]

Forbus, K. (2019), *Qualitative Representations: How People Reason and Learn about the Continuous World*, MIT Press, Cambridge, MA. [14]

Forbus, K. and T. Hinrichs (2017), "Analogy and qualitative representations in the companion cognitive architecture", *AI Magazine*, Vol. 38/4, pp. 34-42, https://doi.org/10.1609/aimag.v38i4.2743. [33]

Forbus, K., C. Liang and I. Rabkina (2017), "Representation and computation in cognitive models", *Topics in Cognitive Science*, Vol. 9/3, pp. 694-798, http://dx.doi.org/DOI:10.1111/tops.12277. [11]

Friedman, S. and K. Forbus (2010), "An integrated systems approach to explanation-based conceptual change", *Proceedings of the Twenty-Fourth AAAI Conference on Artificial Intelligence, Atlanta, GA*, Vol. 24/1, https://ojs.aaai.org/index.php/AAAI/article/view/7572. [34]

Fromherz, M., D. Bobrow and J. de Kleer (2003), "Model-based computing for design and control of reconfigurable systems", *AI Magazine*, Vol. 24/4, https://doi.org/10.1609/aimag.v24i4.1735. [29]

Gentner, D. and A. Stevens (eds.) (1983), *Mental Models*, Lawrence Erlbaum Associates, Hillsdale, NJ. [15]

Gil, Y. and B. Selman (eds.) (2019), "A 20-year community roadmap for artificial intelligence research in the US", *Workshop Report*, Computing Community Consortium (CCC) and the Association for the Advancement of Artificial Intelligence, http://dx.doi.org/arXiv:1908.02624. [30]

Gluck, K. and J. Laird (eds.) (2019), *Interactive Task Learning: Humans, Robots, and Agents Acquiring New Tasks through Natural Interactions*, MIT Press, Cambridge, MA. [13]

Goodfellow, I., J. Shlens and C. Szegedy (2015), "Explaining and harnessing adversarial examples", *Proceedings of International Conference on Learning Representations 2015*, http://dx.doi.org/arXiv:1412.6572. [4]

Jalali, V. and D. Leake (2018), "Harnessing hundreds of millions of cases: Case-based prediction at industrial scale: 26th International Conference, ICCBR 2018, Stockholm, Sweden, July 9-12, 2018, Proceedings", in Cox, M., P. Funk and S. Begum (eds.), *Case-Based Reasoning Research and Development. ICCBR 2018. Lecture Notes in Computer Science*, Springer, Cham, http://dx.doi.org/10.1007/978-3-030-01081-2_11. [21]

Jia, R. and P. Liang (2017), "Adversarial examples for evaluating reading comprehension systems", *arXiv*, Vol. 07328, http://dx.doi.org/arXiv:1707.07328. [7]

Kaushik, D. and Z. Lipton (2018), "How much reading does reading comprehension require? A critical investigation of popular benchmarks", *Proceedings of the 2018 Conference on Empirical Methods in Natural Language Processing*, pp. 5010-5015, http://dx.doi.org/10.18653/v1/D18-1546. [8]

Klein, G. (1999), *Sources of Power*, MIT Press, Cambridge, MA. [16]

Laird, J. (2012), *The SOAR Cognitive Architecture*, MIT Press, Cambridge, MA. [32]

LeCun, Y., Y. Bengio and G. Hinton (2015), "Deep learning", *Nature*, Vol. 512, pp. 436-444, https://doi.org/10.1038/nature14539. [1]

Lenat, D. and P. Durlach (2014), "Reinforcing math knowledge by immersing students in a simulated learning by teaching experience", *International Journal of Artificial Intelligence in Education*, Vol. 24, pp. 216-250, https://doi.org/10.1007/s40593-014-0016-x. [36]

Lenat, D. et al. (2010), "Harnessing Cyc to answer clinical researchers' ad hoc queries", *AI Magazine*, Vol. 31/3, pp. 13-32, http://dx.doi.org/10.1609/aimag.v31i3.2299. [35]

Marcus, G. (2018), "Deep learning: A critical appraisal", *arXiv*, Vol. 00631, http://dx.doi.org/arXiv:1801.00631. [3]

Marcus, G. and E. Davis (2020), "GPT-3, Bloviator: OpenAI's language generator has no idea what it's talking about", 22 August, MIT Technology Review. [12]

Newcombe, C. et al. (2015), "How Amazon web services uses formal methods", *Communications of the ACM*, Vol. 58/4, pp. 66-73, http://dx.doi.org/10.1145/2699417. [25]

Nguyen, A., J. Yosinski and J. Clune (2015), "Deep neural networks are easily fooled: High confidence predictions for unrecognizable images", *IEEE Conference on Computer Vision and Pattern Recognition (CVPR), 2015*, pp. 427-436, http://dx.doi.org/10.1109/CVPR.2015.7298640. [6]

Noy, N. et al. (2019), "Industry-scale knowledge graphs: Lessons and challenges", *acmqueue*, Vol. 17/2, https://queue.acm.org/detail.cfm?id=3332266. [18]

Price, C. (2000), "AutoSteve: Automated electrical design analysis", *ECAI 2000, Proceedings of the 14th European Conference on Artificial Intelligence, Berlin, Germany, 20-25 August 2000*, https://www.dbai.tuwien.ac.at/event/ecai2000-kbsmbe/papers/w31-10.pdf. [28]

Reeves, B. and C. Nass (2003), *The Media Equation: How People Treat Computers, Television, and New Media like Real People and Places*, CSLI Publications, Stanford. [37]

Silver, D. et al. (2016), "Mastering the game of Go with deep neural networks and tree search", *Nature*, Vol. 529, pp. 484-489, https://doi.org/10.1038/nature16961. [9]

Silver, D. et al. (2018), "A general reinforcement learning algorithm that masters chess, shogi, and Go", *Science*, Vol. 362/6419, pp. 1140-1144, http://dx.doi.org/0.1126/science.aar6404. [10]

Simonis, H. (2001), "Building industrial applications with constraint programming", in Comon, H., C. Marche and R. Treinen (eds.), *Constraints in Computational Logics: Theory and Applications*, Springer, Lecture Notes in Computer Science. [27]

Tunstall-Pedoe, W. (2010), "True knowledge: Open-domain question answering using structured knowledge and inference", *AI Magazine*, Vol. 31/3, pp. 80-92, https://doi.org/10.1609/aimag.v31i3.2298. [38]

Notes

[1] AlphaZero is more data-efficient, but the 21 million games played during training would still require a person to live close to 4 800 years.

[2] They performed over 150 tests to come to this conclusion, which are described at https://cs.nyu.edu/faculty/davise/papers/GPT3CompleteTests.html. A common problem with the evaluation of AI systems is what is called the *Eliza effect*, i.e. that people are easily fooled by AI systems. People's tendency to humanise technologies has been amply documented (Reeves and Nass, 2003[37]).

[3] Another common term is *knowledge base*. Typically, these terms are used interchangeably, but there is a subtle difference. Knowledge graphs are always constructed out of graphs, i.e. node and link data-structures. While all knowledge graphs are knowledge bases, some approaches try to use distributed representations, e.g. statistical language models, as knowledge bases.

[4] Amazon acquired True Knowledge (Tunstall-Pedoe, 2010[38]) and has been building on it for general question-answering services.

[5] The BBC, for example, has made theirs available on line at www.bbc.com/ontologies.

[6] While many knowledge graphs contain URLs for images, there is little processing of such information to use it for reasoning. They typically display it to users to show what a product looks like.

[7] Vector representations consist of a one-dimensional sequence of numbers. Vector, matrix and tensor mathematical representations are commonly used in neural network models.

[8] Relational knowledge is expressed in the form of statements involving predicates and arguments. Statements are analogous to a verb and its arguments in a natural language sentence, but the predicates are drawn from a formal vocabulary and their meaning is constrained by rules of inference. Logics and procedural formalisms are two examples of representational systems that can be used to encode relational knowledge.

[9] Published examples include answering queries integrated across multiple databases by medical researchers expressed in natural language (Lenat et al., 2010[35]) and intelligent tutoring for mathematics (Lenat and Durlach, 2014[36]).

[10] A roadmap for an Open Knowledge Network is laid out in Gil and Selman (2019[30]), as part of a broader 20-year roadmap for AI in the United States.

[11] This example is correctly handled by Google, Bing, Alexa and Bixby at this writing, but they do not handle all such simple two step inferences.

[12] This example is due to the late Marvin Minsky.

[13] Research on qualitative reasoning has done the most to formalise model formulation (Forbus, 2019[14]), but there is still much more research needed on this topic.

[14] Imagine AI assistants for doctors, lawyers, and engineers that accumulate experiences as they work with people. Now imagine that experience is being accumulated across assistants within an organisation

or within a profession, combined into a massive library of experience that would go beyond what any person could ever experience.

[15] Typical venues include the journal and conference *Advances in Cognitive Systems* (www.cogsys.org) and the Cognitive Systems track at the AAAI conference.

[16] ACT-R, SOAR and Companions have been used in deployed systems but not yet at a scale or frequency to constitute a revolution in the sense used here. On the other hand, today's deployed systems may lead to more experiments in autonomy. In an apocryphal story, an apprentice too lazy to manually supervise the steam engine invented the governor to do it for him. Similarly, there may be stirrings of autonomy in today's AI pipelines as engineers find ways to let the system take on more of their work, but that is operating at the margins.

[17] In Gil and Selman (2019[30]), the Integrated Intelligence workshop report is mostly concerned with such issues [e.g. expanding the kinds of memories available to AI systems (page 25) and reducing the maintenance footprint of AI systems (page 26)]. The Meaningful Interaction workshop report addresses multiple relevant issues, including the integration of diverse natural modalities for communication and challenges involved in collaboration with people. The Self-Aware Learning workshop report addresses both expanding the kinds of knowledge that learning techniques can handle (page 66) and controlling their own learning (i.e. durable machine learning systems, page 74).

Part II. Taxonomies and tests of human skills

3. Taxonomy of cognitive abilities and measures for assessing artificial intelligence and robotics capabilities

Patrick C. Kyllonen, Educational Testing Service, Princeton, NJ, United States

This chapter reviews taxonomies of human cognitive abilities and measures of those abilities. It recalls the history of key models and frameworks, analysing their strengths and weaknesses. It gives special attention to the second order of frameworks, which comprises approximately nine distinct abilities, including fluid, crystallised, spatial and broad retrieval/creativity abilities. The primary factors associated with these nine abilities are discussed, along with sample tests and test items reflecting the different abilities. It proposes two additional abilities: emotional intelligence and collaboration/communication. Finally, it discusses the prospects and feasibility of using human abilities and their associated tests to evaluate machine intelligence, discussing the advantage of having a justification for the selection of tasks.

Introduction

This chapter reviews taxonomies of human cognitive abilities and measures of those abilities. It recalls the history of key models and frameworks, analysing their strengths and weaknesses as a group and relative to one another. It gives special attention to the second order of frameworks, which comprises approximately nine distinct abilities, including fluid, crystallised, spatial and broad retrieval/creativity abilities. The primary factors associated with these nine abilities are discussed, along with sample tests and test items reflecting the different abilities. It proposes two additional abilities: emotional intelligence and collaboration/communication. Finally, it discusses the prospects and feasibility of using human abilities and their associated tests to evaluate machine intelligence, discussing the advantage of having a justification for the selection of tasks.

Various taxonomies of human skills and abilities

There are numerous taxonomies of human skills and abilities based on various approaches for developing them. This section first explores human cognitive abilities, and psychometric and sampling models. All these models acknowledge the phenomenon of positive manifold. First noted by Spearman (1904[1]; 1927[2]), positive manifold is a label for the universality of positive correlations between performance scores on any pair of cognitive tests. There have been many attempts to determine the cause of positive manifold (e.g. general factor or bonds or network), or, if a general factor, then the nature of the general factor. The section ends with a discussion of executive function, cognitive architectures and general artificial intelligence (AI).

From Spearman to Cattell-Horn-Carroll

Perhaps the oldest and most well-known taxonomies come from the human cognitive abilities literature (or sometimes, factor analytic tradition). This began with Spearman (1904[1]; 1927[2]) who analysed correlations among scores (i.e. tallies of numbers correct from a set of items) from various cognitive tests, primarily samples of school-like tasks. That led to the fluid-crystallised (Gf-Gc) mode (Horn and Cattell, 1966[3]) and the extended Gf-Gc model (Horn and Blankson, 2005[4]; 2012[5]); the three-stratum model (Carroll, 1993[6]); and their synthesis in the Cattell-Horn-Carroll (CHC) model (McGrew and Woodcock, 2001[7]). Annex 3.A compares the three models.

The various models within this framework agree on a strong general factor that accounts for 30-70% of the between-person variance in any test score [called "gf" in the Horn-Cattell model, see Carroll (1993[6])]. Consider, for example, math, verbal, science and problem-solving scores in the OECD Programme for International Student Assessment (PISA). These scores show intercorrelations ranging from r = .8 to r = .9. There is a general fluid versus crystallised distinction; these two factors can be highly correlated, but developmental trajectories differ (fluid ability peaks at an earlier age). There are roughly 8 to 10 group (second order) factors, and 80 or so primary (first order) latent factors that account for covariances among test scores.

Vernon's hierarchical model

There are alternative approaches within the human abilities/factor analytic tradition, or what is sometimes called the psychometric model of intelligence (Hunt, 2011[8]). The hierarchical model of Vernon (1950[9]) is realised in the g-VPR (general factor, verbal-perceptual-memory, image rotation) model (Johnson and Bouchard, 2005[10]; Johnson, te Nijenhuis and Bourchar, 2007[11]). This model does not differ qualitatively from the Carroll (1993[6]) or CHC models. However, it differs in emphases: there is a general factor at a fourth order, three major factors at the third order (verbal, perceptual and image rotation), about nine factors at the second order and numerous primary factors.

Sampling models

Sampling is another alternative to the abilities model. This tradition began with the bond sampling model (Thomson, 1916[12]), where any test samples a set of mental bonds rather than component abilities per se (Tirre, 1994[13]). More recently, the sampling approach is represented in network (van der Maas et al., 2019[14]) and wiring models (Savi et al., 2019[15]).

Executive function

One popular concept relates the general factor to working memory capacity (Conway and Engle, 1996[16]; Kyllonen and Christal, 1990[17]). Working memory capacity can be characterised as executive attention [i.e. one's capacity to control attention, see Engle (2002[18]) and Kane et al. (2001[19])].

This line of findings arguably underlies the executive functioning literature, which has become popular in education circles (Zelazo, Blair and Willoughby, 2016[20]). Executive functioning is defined as "skills related to working memory, inhibitory control and mental flexibility" (Shuey and Kankaraš, 2018[21]). These skills, in turn, are defined as the temporary activation or storage of information while engaged in cognitive processing (Baddeley, 1986[22]; Cowan, 2017[23]); directing or sustaining attention in the face of distractions (Diamond, 2013[24]); and the ability to shift between different mental sets or tasks (Archambeau and Gevers, 2018[25]), respectively.

Cognitive architectures

Other lines of research that fall outside the factor analytic tradition of conventional abilities contribute to a positing or understanding of a human abilities taxonomy. One of these lines of research is computational modelling, or cognitive architectures. These include Adaptive Control of Thought-Rational (ACT-R) (Anderson and Lebiere, 1998[26]; Anderson et al., 2004[27]); Cortical Capacity-Constrained Concurrent Activation-based Production System (4CAPS) (Just and Varma, 2007[28]); Executive Process Interactive Control (EPIC) (Kieras and Meyer, 1997[29]); SOAR (Newell, 1994[30]); and Hypothesis Generation (HyGene) models (Thomas et al., 2008[31]).

These models are designed to simulate human problem solving. As a side effect, their architectures suggest constructs that may be treated as human abilities. For example, ACT-R distinguishes procedural (production rules) and declarative (chunks) memory. It includes specialised perceptual-motor, goal and declarative memory modules, as well as learning processes. HyGene, which is designed for diagnostic reasoning, includes processes of information sampling, derivation of prototypical representations, generation of candidate hypotheses, probability estimation, hypothesis testing and search termination.

Many of these processes and modules map to the lower-order factors in the hierarchical abilities' models. However, they are implemented more precisely with respect to how they function, which is required for a computer simulation.

Artificial intelligence ability taxonomies

Related to cognitive architectures is the more general AI literature. This is not concerned with simulating human cognition but with building intelligent entities more broadly. A popular AI textbook (Russell and Norvig, 2010[32]) includes chapters entitled problem solving; knowledge, reasoning and planning; knowledge representation; probabilistic reasoning; making simple (and complex) decisions; learning; and perception. These chapters include many methods for addressing these topics, such as induction, case-based reasoning and reasoning by analogy, which also map to the abilities identified in the Carroll (1993[6]) model. Identifying the constructs included in AI can inform discussions of human abilities.

Cognitive tests associated with these taxonomies

Each of the literatures reviewed in the previous section is based on empirical research that involves performance of cognitive tests. The most extensive of these is from the cognitive abilities/factor analytic tradition because the associated taxonomies are derived directly from scores on batteries of cognitive tests. This section looks at cognitive tests associated with these taxonomies.

From school tasks to intellectual tasks

Originally, cognitive tests were essentially samples of school tasks (e.g. spelling) along with perceptual (e.g. pitch perception) and memory (e.g. logical, visual, auditory) tasks from the laboratories of experimental psychology (Wissler, 1901[33]; Spearman, 1904[1]).

Over time, new kinds of intellectual tasks were added, such as Thurstone (1938[34]), and World War II led to a considerable expansion of measures. A unit of the US Army (the Army Air Force) developed tests to measure every conceivable mental function (Humphreys, 1947[35]; Damos, 2019[36]). These included tests of verbal and mathematical skills, reasoning, mechanics, judgement, foresight, planning, integration, memory, attention, mental set, perceptual skills, spatial orientation and visualisation, and general information, as well as a set of motion picture tests (Gibson, 1947[37]; Lamkin, Shafer and Gagne, 1947[38]).

Later, Guilford (1950[39]) expanded even this lengthy list to include new measures to fill out his structure of intellect model (Guilford and Hoepfner, 1971[40]). Perhaps the most significant area of expansion was in measures of divergent thinking, or creativity (Guilford, 1950[39]).

Test kits and tool boxes

Educational Testing Service (ETS) produced a kit of cognitive reference tests. They comprised a sample of tests from the most important 46 factors associated with this work (Ekstrom et al., 1976[41]; 1976[42]). These tests, still used widely in research, are available for free or a nominal charge.

Condon and Revelle (2014[43]) and Dworak et al. (2020[44]) produced the International Cognitive Ability Resource (ICAR). This open-source tool measures 19 domains of cognitive ability, including fluid ability (progressive matrices and matrix reasoning, propositional reasoning, figural analogies, letter and number series, abstract reasoning), emotional ability (emotion recognition), mathematical ability (arithmetic), verbal ability (verbal reasoning), creativity (compound remote associates), face-detection (aka the Mooney Test), perceptual ability (melodic discrimination, a perceptual maze task), and judgement (a situational judgement task). This is not as systematic as the ETS resource. However, given the open-source nature of the ICAR, it may eventually become more comprehensive.

The CHC taxonomy is tightly tied to a commercial instrument: the Woodcock Johnson III Tests of Cognitive Abilities (WJ III) (Schrank, 2011[45]) and Woodcock Johnson IV (WJ IV) (Schrank, Mather and McGrew, 2014[46]). Annex 3.B describes the battery factors and tests.

Executive functioning research is associated with tests that measure working memory capacity, inhibitory attentional control, and cognitive and attentional flexibility. Jewsbury, Bowden and Strauss (2016[47]) show how the CHC model can accommodate these measures. Numerous publications describe working memory measures [e.g. Kyllonen and Christal (1990[17]) and Wilhelm, Hilldebrandt and Oberauer (2013[48])].

The US National Institutes of Health (NIH) provides the NIH toolbox, which comprises a set of 100 stand-alone measures to assess cognition, emotion, motor ability and sensation. NIH cognition measures (Zelazo et al., 2013[49]) include attention and executive function (Flanker Inhibitory Control and Attention Test), working memory (list sorting), executive function (Dimensional Change Card Sort), along with episodic memory, language and processing speed measures. These are separated by age range (3-6, 7-17 and 18+).

The cognitive architecture literature has been primarily driven by laboratory tasks from experimental cognitive psychology. Such measures are most often designed to test specific aspects of cognitive theories. They predominantly measure response time (e.g. fact retrieval, lexical decision) and memory recall (e.g. free recall, recognition memory). They tend to be simpler than tests in the psychometric tradition, as they are typically designed for narrower purposes.

The AI literature is voluminous and therefore difficult to characterise. It is equally difficult to characterise the kinds of cognitive measures associated with the literature.

Criteria for establishing taxonomy and suitable tests

An ideal taxonomy for this project would provide a list of human abilities, identified through a methodology or methodologies that enable a strong scientific justification. Such a list would be comprehensive but parsimonious. In this way, there would be minimal conceptual or empirical overlap between abilities. The definition of the ability would also need to be demonstrable or transparent. These, and other, principles are elaborated below.

Comprehensiveness could prove difficult to demonstrate capability with respect to all the abilities proposed. Parsimony would minimise burden in any application exercise, such as rating jobs. Demonstrability or transparency can be measured with processing requirements that are clear and easily understood. Empirically there must be a strong connection between the test and the construct it is intended to measure. Other considerations for determining the suitability of particular tests are their correlation with the factor of interest; amount of time (or number of items) needed to achieve a reliable score; and susceptibility of the test to (contamination with) other factors (test impurity).

Cattell-Horn-Carroll framework

This section reviews one of the lines of literature in more depth: the human cognitive abilities literature from the factor analytic tradition.

The factor analytic tradition in depth

The factor analytic tradition begins with Spearman (1904[1]; 1927[2]). His analyses of cognitive tests found that a general latent factor accounted quite adequately for the correlations among test scores. However, each test score additionally had to contain test-specific variance.

Thurstone (1938[34]) administered a broader sample of tests to a larger group of college students. Through the development of multiple-factor analysis and the concept of simple structure, Thurstone (1934[50]) showed evidence for a set of narrower group factors (verbal comprehension, word fluency, number facility, spatial visualisation, associative memory, perceptual speed, reasoning). He referred to these as primary mental abilities.

Others showed the Spearman and Thurstone findings were compatible through a hierarchical mode (Undheim, 1981[51]; Gustafsson, 1984[52]) or the similar bifactor model (Holzinger and Swineford, 1937[53]; Holzinger and Harman, 1938[54]; Schmid and Leiman, 1957[55]). In other words, performance on a test could be a function of general, group and specific latent factors simultaneously.

Horn and Cattell (1966[3]) proposed two general factors [fluid (Gf) and crystallised (Gc) ability], along with a set of correlated group factors or broad abilities. This line of research (Horn and Blankson, 2005[4]; Horn and Blankson, 2012[5]) culminated in 80 first-order primary mental abilities and 8 second-order abilities. These are Gc; Gf; short-term memory (Gsm) [later, short-term apprehension and retrieval (SAR)]; long-term memory (Gsl) [later, fluency of retrieval from long-term storage (TSR)]; processing speed (Gs); visual processing (Gv); auditory processing (Ga); and quantitative knowledge (Gq).

Carroll (1993[6]) analysed 460 datasets comprising test correlation matrices accumulated over almost a century of research. He reanalysed them using a version of the Schmidt-Leiman procedure and synthesised findings primarily based on informed but subjective judgements of content (and process) overlap. He proposed a three-stratum model (Carroll's "stratum" is synonymous with the more common term of "order") with a general factor at the apex and eight second-stratum factors.

The second-stratum factors were fluid intelligence, crystallised intelligence, general memory and learning, broad visual perception broad auditory perception, broad retrieval ability, broad cognitive speediness and processing speed. Each of the second-stratum factors covered 4 to 14 first-stratum factors. For example, the second-order fluid intelligence covered the first-order (primary) factors: general sequential reasoning, induction, quantitative reasoning and speed of reasoning. These, in turn, were determined by the correlations among the manifest scores from various tests of those factors.

The CHC model was proposed as a synthesis of the Carroll (1993[6]) and Horn and Cattell (1966[3]) models (McGrew and Woodcock, 2001[7]). It has subsequently been revised and expanded regularly with the incorporation of new research findings [e.g. Schneider and McGrew (2018[56])]. However, these remain three distinct frameworks or models. Carroll (2003[57]) updated his model, and Horn did as well (Horn and Blankson, 2012[5]) to accommodate new findings. Still, it is useful to treat them or their synthesis as a common framework. They differ in some details (Carroll, 2003[57]) but are based on mostly common data and methods.

Critiques and modifications of the CHC framework

The CHC model has become a popular framework for the representation of human abilities, partly or perhaps primarily due to its application in school psychology for student cognitive diagnosis. Nevertheless, several important critiques have recently appeared. These include a special issue of *Applied Measurement in Education* (Beaujean and Benson, 2019[58]; Canivez and Youngstrom, 2019[59]; Geisinger, 2019[60]; McGill and Dombrowski, 2019[61]; Wasserman, 2019[62]).

These critiques identify five limitations: over expansiveness; emphasis of group factors over individual factors; mental speed; treatment of quantitative factor; and its combination of two disparate factors. The issues are summarised below.

Over expansiveness

Like many abilities frameworks (Carroll, 1993[6]; Carroll, 2003[57]; Horn and Blankson, 2012[5]), CHC includes too many abilities with scant justification for their inclusion. More replication would be desirable, using a variety of tests (not just those in the WJ III and WJ IV commercial batteries). Users (e.g. teachers, school psychologists) like having many abilities to test to obtain a more complete picture of a student. However, there is a growing awareness that reliability is crucial to distinguish between tests or factors (Haberman, Sinharay and Puhan, 2011[63]). In addition, profile scores (e.g. a set of scores from several tests or factors) are often not justified due to the importance of the general factor. This critique suggests a smaller number of factors than are typically reported are scientifically justified.

General vs. group factors

The general factor can often be shown to be more important in accounting for test score variance than the group (lower order or lower stratum) factor. However, CHC has mostly denied the general factor. It prefers to emphasise the group factors, which are empirically shown to be highly overlapping.

Mental speed

CHC does not treat mental speed in a way consistent with the recent literature on cognitive processing speed [e.g. Kyllonen and Zu (2016[64])]. New psychometric approaches suggest a rethinking of mental speed with respect to abilities models.

Quantitative factor

In the Carroll (1993[6]) framework, and in the CHC, quantitative ability is a lower-order factor of fluid intelligence. However, Wasserman (2019[62]) points to mathematics prodigies as an indicator that quantitative ability might deserve a higher ranking.

Combining disparate functions

Both Carroll (1993[6]) and CHC frameworks combine knowledge retrieval and idea production in a single long-term memory retrieval factor. However, these are disparate functions. Idea production is thought to be the essence of creativity, whereas knowledge retrieval is considered to be a non-creative process.

Despite these criticisms, the CHC model and its constituents [e.g. Carroll (1993[6]) and Horn and Cattell, (1966[3])] may provide enough of a basis for a justifiable taxonomy of human cognitive abilities. In other words, it may satisfy the criteria of being comprehensive, reasonably succinct and transparent in principle.

g-VPR as an alternative to the CHC framework

Other human abilities frameworks are worth considering in addition to the CHC model. The general plus verbal, perceptual and image rotation (g-VPR) model (Johnson and Bouchard, 2005[10]; Johnson, te Nijenhuis and Bourchar, 2007[11]) has been shown to provide a better account of the test score data than the CHC model.

Some prominent researchers such as Hunt (2011[8]) have suggested g-VPR as a viable alternative to the Carroll (1993[6]) or CHC models. However, showing a slightly superior fit for a few datasets is probably not a sufficient reason for claiming the g-VPR framework as a viable alternative. Even Johnson (2018[65]), one of the architects of g-VPR, has argued their model had not "'carved nature at its joints'" in any battery any better than Carroll had. This is because factor analysis spits back at us only what we put into it, and we have no tasks that uniquely measure any one particular ability or skill..." (p. 24).

What are the most important abilities?

As Carroll (2003[57]) noted in the title of one of his last papers, "Current evidence supports *g* and about ten broad factors." There is considerable agreement across the three major CHC frameworks about those broad factors (see Annex A), although some make distinctions. The major categories are the nine-colour coded distinctions. In addition, one general factor is not listed (because it is at the third stratum). The 80 or so primary (first order) factors are listed in Annex 3.B and Annex 3.C.

General factor and fluid intelligence

The general factor is either identical or close to identical to fluid ability (gf). There is a strong overlap between executive function ability, working memory, attention and gf (Wilhelm, Hilldebrandt and Oberauer, 2013[48]). Most of this research was conducted after Carroll (1993[6]). Still, the primary gf measures are reasoning measures, both quantitative and non-quantitative.

Good examples listed in Annex Figure 3.D.1 are from Carroll's (1993[6]) primary (first order) factors of inductive reasoning, deductive reasoning and quantitative reasoning.

The first primary gf factor of inductive reasoning includes sets (classification tasks, "odd man out" tasks), series (e.g. number, letter, figure series) and matrices tasks. In Raven's progressive matrices (Kyllonen et al., 2019[66]), for example, the goal is to induce a rule or set of rules describing an arrangement of a set of elements then to apply the rule(s) to identify or categorise new elements.

The second primary gf factor of deductive reasoning includes tests for syllogistic reasoning and diagramming relationships using Euler diagrams, as shown in Annex Figure 3.D.1. This example and the other listed illustrate how inductive and deductive reasoning tasks can be implemented in verbal, numerical and spatial content.

The third primary gf factor of quantitative reasoning is illustrated with the Necessary Arithmetic Operations test, which asks respondents to select the operations needed to solve an arithmetic word problem.

All these example tasks (and others listed in Annex 3.D) are singled out because they are good representations of some key primary factors associated with second-order factors. Further discussion of the reasoning factor, the varieties of reasoning and evidence from diverse research traditions can be found in Kyllonen (2020[67]).

Abductive reasoning

Abductive reasoning involves deriving an explanation for a finding or set of facts. Consider the following example taken from a retired form of the GRE: because the process of freezing food consumes energy, many people keep their electric freezers half empty, using them only to store commercially frozen foods. Yet freezers that are half empty often consume more energy than if kept fully stocked.

The example then proposes five possible explanations for the apparent discrepancy. This might be solved with deductive reasoning. However, it follows the form of an abductive reasoning problem in that a phenomenon is presented in search of a cause or explanation. The example presents possible explanations, but abductive reasoning could also involve an open-ended item. In that case, a person would have to retrieve relevant information to come up with an explanation. Consequently, this kind of problem overlaps to some extent with ones in the *broad retrieval ability* category, below.

Crystallised intelligence

Crystallised intelligence, in principle, represents acculturated knowledge but in practice overlaps highly with "verbal ability" (Carroll, 1993[6]). Some of the best example tasks are reading comprehension tests, vocabulary items (open-ended or multiple choice) and cloze tests. A cloze test presents a sentence or paragraph with missing words that need to be provided. This requires knowledge of the topic, vocabulary, grammar rules and the like. Crystallised and fluid intelligence tasks appear to be distinct, but empirically, fluid and crystallised intelligence are highly correlated in individuals. One explanation is that students use reasoning processes in developing verbal knowledge (Marshalek, 1981[68]). Annex Figure 3.D.2 lists examples of tasks.

Broad visual perception

Broad visual perception is commonly called spatial ability. It involves the perception, memory, mental transformation and reasoning about presented or imagined spatial materials. Guilford's blocks test provides an example of imagined spatial materials. Respondents imagine painting a wooden block red, dividing it into 3 x 3 x 3 blocks, then determining the number of blocks with exactly one side painted red. Example items from the most prototypical spatial ability tests covering the most prominent spatial ability primary factors appear in Annex Figure 3.D.3. The factors (and test examples) are spatial visualisation (mental paper folding), closure flexibility (the copying test) and perceptual speed (a picture matching test). Lohman (1979[69]) is a still useful review of this literature.

Broad retrieval ability

Broad retrieval ability is Carroll's label for a set of factors that involve creativity and mental fluency. Prototypical fluency tasks are ones that involve rapidly generating lists of responses that follow a set of rules. This could be generating all the words that begin with "S" and end with "N", or four letter words that do so; an example word fluency item is shown in Annex Figure 3.D.4. An analogous process is figural fluency, such as moving toothpicks around to create a form (see example in Annex Figure 3.D.4). Creativity measures are fluency tests that involve more complex ideas. For example, the consequences test from Christensen, Merrifield and Guilford (1953[70]) (Annex Figure 3.D.4) asks respondents to respond with as many plausible and non-repeating responses as they can in a short interval to prompts such as "What would happen if we didn't have to eat?" or "How can traffic congestion problems be curtailed?"

General memory ability

Carroll (1993[6]) found evidence for a general memory ability factor based on performance on short-term and long-term memory tasks that have been studied in the verbal learning tradition in experimental psychology. These include memory span, associative memory and free recall, as well as a separate visual memory first-order dimension. There was also evidence for a loose learning ability factor.

Memory and learning are obviously important human abilities, but this factor has not been shown to relate uniquely to educational outcomes in the way fluid and crystallised ability have. The factor may represent performance on the peculiar sort of arbitrary memory tasks that psychologists have devised, but not the ability invoked in typical educational learning situations.

Another peculiarity is that simple forward memory span (repeating a string of 7 to 9 digits) seems to invoke an ability different from backward memory span (repeating the string backwards) (Reynolds, 1997[71]). The latter operates more like a working memory test, requiring simultaneous storage and processing (Baddeley, 1986[22]).

Working memory is also highly correlated with fluid intelligence, as reviewed in a previous section. Consequently, it may not be useful to include memory ability factor in a test of AI. Technology may lessen the requirement for memorising arbitrary strings of words and symbols, which is another reason to exclude memory ability from an AI test.

Broad auditory perception

Carroll (1993[6]) found evidence for a distinct broad auditory perception factor, called "listening and hearing" in the Horn-Cattell model, and "auditory processing" in the CHC. These involve speech-sound and general sound discrimination, memory for sound patterns, musical discrimination and the like. These are important human abilities but are more perceptual in nature. They do not seem as pertinent to testing AI as abilities from the other categories.

Psychomotor ability

Psychomotor abilities are important in many jobs and other human activities, such as playing sports and games. Carroll considered this literature outside the scope of his focus on cognitive abilities, but psychomotor ability is represented in the CHC model. Fleishman (1954[72]) provided a taxonomy and set of psychomotor tasks.

Some decades later, Fleishman and Quaintance (1984[73]) and Chaiken, Kyllonen and Tirre (2000[74]) made further comments on psychomotor ability. They suggested a general psychomotor factor that could account for most of the psychomotor tasks. It can be measured with tasks such as multi-limb co-ordination and tracking tasks, such as pursuit motor co-ordination.

Processing speed

Processing speed is an important component of human cognition. Carroll (1993[6]) suggests there was an independent second-order speed factor (i.e. two independent speed factors). The nature of a processing speed factor is a complex topic within cognitive psychology and within the human abilities' literature. This complexity is due to speed-accuracy trade-off, willingness or proclivity to abandon unproductive solution attempts and time management issues generally. In fact, Carroll's two speed factors could be due to interactions among these factors (Kyllonen and Zu, 2016[64]).

It is difficult to imagine how tasks designed to measure a processing speed factor in the human abilities' literature could be used productively to measure AI abilities. A prototypical task is simply an easy version of a fluid or crystallised intelligence test (e.g. an easy vocabulary synonym judgement test). The primary dependent variable is the time it takes to retrieve the answer or solve the simple problem. Thus, little additional information is likely to be obtained by trying to determine machine capabilities on tasks sampled from the set of tests designed to measure human processing speed.

Olfactory, tactile and kinaesthetic abilities

This set of sensory abilities was also considered outside the realm of human cognitive abilities in Carroll (1993[6]). However, these abilities are represented within the CHC framework. This inclusion reflects research attempting to document these abilities within the context of human cognitive abilities. Stankov (2019[75]) summarises the research programme. However, like some of the other factors, this work seems to be outside the central focus of this study, which is primarily based on human cognitive abilities.

Additional abilities

Besides the abilities covered in the previous section, several ability factors could be noted: emotional intelligence, and collaboration and communication.

Emotional intelligence

Emotional intelligence only emerged as a concept with Mayer and Salovey (1993[76]). Therefore, it was not part of thinking about human cognitive abilities at the time of Carroll (1993[6]). Since then, there has been considerable research on the topic.

The literature distinguishes between ratings and performance measures; only the latter would be considered relevant for the *purposes* of testing AI abilities. MacCann et al. (2014[77]) administered an emotional intelligence test battery along with a battery of CHC-type cognitive ability tests (e.g. fluid, crystallised, spatial ability, broad retrieval). They identified first- and second-order emotional intelligence factors based on a set of emotional intelligence measures (two tests of each for emotion perception, emotion understanding and emotion management).

Earlier research MacCann and Roberts (2008[78]) examined the Situational Test of Emotional Understanding (STEU) and the Situational Test of Emotion Management (STEM). STEU presented items such as the following: *Hasad tries to use his new mobile phone. He has always been able to work out how to use different appliances, but he cannot get the phone to function. Hasad is most likely to feel? a) distressed, b) confused, c) surprised, d) relieved or e) frustrated.* The related STEM measure comprises similar types of items. From the standpoint of this study, this factor does represent something distinct from say, crystallised intelligence: it includes a component on reasoning about emotions.

There are other kinds of emotional intelligence measures. These include determining the emotional state of someone photographed (Baron-Cohen et al., 2001[79]; Olderbak et al., 2021[80]) or of someone expressing emotion through language (Scherer and Scherer, 2011[81]; Hellwig, Roberts and Schulze, 2020[82]). The empathic agent paradigm asks test takers to study how another person depicted in vignettes tends to act in situations, and then to apply that knowledge to predict how that person will react in a new situation (Hellwig, Roberts and Schulze, 2020[82]). All these measures make clear that a second-order emotional intelligence factor is an important human cognitive abilities factor distinct from the others discussed here.

Collaboration and communication

Woolley et al. (2010[83]) found evidence for "collective intelligence", meaning that some teams of individuals performed better than other teams across a diverse set of team tasks. Team tasks included brainstorming, planning a shopping trip, group typing, group matrix reasoning and group moral reasoning. This "team effect" was independent of individual abilities on the team (e.g. as indicated by how they performed the task alone). Instead, collective intelligence seemed related to members' emotional intelligence – their ability to read their teammates' emotions, goals and intentions.

The future economy will likely put a premium on teamwork, collaboration and communication (Deming, 2017[84]). Thus, it would seem important to determine the possibility of assessing teamwork skills. PISA 2015 also included a collaborative problem-solving measure (OECD, 2017[85]). In reviewing small groups research (a branch of social psychology), Larson (2010[86]) concluded that some tasks exhibited *synergy.* This is defined as the situation in which a team outperforms the best member of the team, or at least does as well as the best member.

Tasks exhibiting synergy include a letters-to-numbers problem-solving task and a hidden-profile decision-making task. In the latter, different team members are provided overlapping but distinct knowledge about choices. Successful team performance depends on members sharing and considering their common and unique knowledge to arrive at a group decision. It is not clear if a scenario could be set up to evaluate, say, a machine's ability to collaborate, but tasks drawn from this category suggest at least possibilities to consider.

Feasibility of a human abilities framework for assessing artificial intelligence and robotics

This section explores the viability of a human abilities framework to assess AI and robotics.

*The psychometric tradition: CHC and O*NET*

An abilities framework in the psychometric tradition (such as the CHC) has already proven viable. The US Department of Labor rates job requirements with respect to abilities similar to the kinds of abilities listed in the CHC framework through the Occupational Network, or O*NET (National Center for O*NET Development, n.d.[87]); (National Research Council, 2010[88]); (Peterson et al., 1999[89]).

O*NET, an occupational analysis system in the United States, collects ratings on job demands annually. Ratings fall into a variety of categories. These include tasks, generalised work activities, knowledge, education and training, work styles and work contexts).

Significantly, for assessing AI and robotics, ratings are collected on the ability involvement (importance and level) for over 950 occupations (Fleisher and Tsacoumis, 2012[90]). It surveys 52 abilities, while eight occupational analysts provide ratings for every occupation. The abilities are grouped into the categories of cognitive, psychomotor, physical and sensory-perceptual.

The framework is based on Fleishman, Costanza and Marshall-Mies (1999[91]) and Fleishman and Quaintance (1984[73]), but the cognitive part is largely consistent with the CHC framework. Cognitive abilities include oral and written comprehension and expression, fluency of ideas, originality, problem sensitivity, deductive and inductive reasoning, information ordering, category flexibility, mathematical reasoning, number facility, memorisation, speed of closure, flexibility of closure, perceptual speed, spatial orientation, visualisation, selective attention and time sharing. In addition, O*NET surveys perceptual and motor factors such as reaction time, auditory attention and speech recognition. It regularly publishes job descriptions with respect to their standings on these factors.

The current O*NET system does not collect judgements related to emotional intelligence or to collaboration ability, but such ratings could potentially be included. Ability ratings using an abilities framework can be collected for occupational requirements and importance. Thus, abilities could be potentially useful constructs on which to collect ratings regarding machine capabilities.

Three useful concepts to consider for machine intelligence

There are significant differences between human abilities and machine abilities. However, the language and set of concepts have evolved over a century of abilities testing. These may still be useful in considering issues in machine intelligence. Many of these concepts are captured in the Standards for Educational and Psychological Testing (AERA, APA and NCME, 2014[92]). Three – "construct irrelevant variance", "teaching to the test" and "construct underrepresentation" – are discussed below.

Construct irrelevant variance

Construct irrelevant variance is a test fairness issue. It refers to a test intended to measure a construct, such as mathematics. However, performance can be affected by other constructs, such as the ability to understand a diagram or language abilities. If performance is affected by factors that the test is not intended to measure, then a test cannot be considered a fair measure of the construct. This is a major fairness concern motivating accommodations for individuals who might have difficulties with aspects of the test that are not the target of assessment. Consider, for example, sight-impaired individuals.

Teaching to the test

Teaching to the test refers to the notion of instruction related to incidental test features that are not features shared generally with respect to the broader construct the test is intended to measure. Teaching to the test is likely to enhance performance on a test without enhancing the level of the construct the test is designed to assess. Psychometrically, this situation is revealed to the extent that one's performance on a particular test is not consistent with performance on other related tests, a situation sometimes referred to as model misfit.

Construct underrepresentation

Construct underrepresentation refers to a situation in which the set of tests to measure a construct does not reflect the full range of attributes or skills in the construct definition. Here, test performance might only indicate the level of the underlying trait or ability possessed by the individual. However, the test only captures a part of the larger construct. For example, a vocabulary test is a useful indicator of general verbal ability. However, verbal ability should reflect a broader set of skills than vocabular, such as paragraph comprehension or responding to general knowledge questions.

Conclusions

An abilities framework such as the hierarchical model (Carroll, 1993[6]) or the CHC (Schneider and McGrew, 2018[56]) are useful frameworks for evaluating human abilities and are likely to be useful for evaluating machine abilities as well. There is general agreement among various models at some of the major human performance distinctions. The second-stratum factors (Carroll, 1993[6]) and their equivalents in the CHC and Horn-Cattell (1966[3]) models are a useful level for evaluation of the type envisaged for AI and robotics. These might be supplemented by two additional factors – emotional intelligence and collaboration/communication ability. On the other hand, some included second-order factors, such as processing speed, psychomotor ability and sensory abilities, might be less central for understanding machine intelligence in the context of a project designed to evaluate likely future work requirements.

A solid body of evidence can be used both to identify measures of the various second-stratum factors and to evaluate the appropriateness of those measures as good indicators of those factors. Good measures from the standpoint of human abilities have several qualities. First, they produce a reliable assessment of ability in individuals. Second, their scores are highly correlated with the factor of interest (i.e. high factor loadings in a factor analysis). Third, they have high average correlations with scores from other measures of the factor (i.e. they have high centrality with respect to the construct of interest).

High factor loadings and high centrality are also related to the concept of transferability of skills. Two tasks that are highly correlated should share common skills. Conversely, lower correlations and lower factor loadings indicate less commonality with respect to skill requirements. They suggest lower transfer relations between the tasks from a training perspective.

There are many differences between machine and human intelligence. However, the evolved language used to describe tests and their relationships to the abilities intended to be measured is useful for describing issues in machine intelligence. Concepts such as reliability, validity, fairness, measurement invariance, construct representativeness and others may help clarify issues in evaluating machine intelligence in the same way they have for measuring human intelligence.

References

AERA, APA and NCME (2014), *Standards for Educational and Psychological Testing*, American Educational Research Association, the American Psychological Association and the National Council on Measurement in Education, Washington, DC. [92]

Anderson, J. et al. (2004), "An integrated theory of the mind", *Psychological Review*, Vol. 111/4, pp. 1036-1060, https://doi.org/10.1037/0033-295X.111.4.1036. [27]

Anderson, J. and C. Lebiere (1998), *The Atomic Components of Thought*, Lawrence Erlbaum Associates Publishers, Mahwah, NJ. [26]

Archambeau, K. and W. Gevers (2018), "(How) Are executive functions actually related to arithmetic abilities?", in Henik, A. and W. Fias (eds.), *Heterogeneity of Function in Numerical Cognition*, Academic Press, Cambridge, MA, https://doi.org/10.1016/C2016-0-00729-5. [25]

Baddeley, A. (1986), *Working Memory*, Oxford University Press, Oxford. [22]

Baron-Cohen, S. et al. (2001), "The "Reading the Mind in the Eyes" test revised version: A study with normal adults, and adults with Asperger syndrome or high-functioning autism", *Journal of Child Psychology and Psychiatry*, Vol. 42, pp. 241-251, http://dx.doi.org/10.1111/1469-7610.00715. [79]

Beaujean, A. and N. Benson (2019), "The one and the many: Enduring legacies of Spearman and Thurstone on intelligence test score interpretation", *Applied Measurement in Education*, Vol. 32/3, pp. 198-215, https://doi.org/10.1080/08957347.2019.1619560. [58]

Canivez, G. and E. Youngstrom (2019), "Challenges to the Cattell-Horn-Carroll theory: Empirical, clinical and policy implications", *Applied Measurement in Education*, Vol. 32/3, pp. 232-248, https://doi.org/10.1080/08957347.2019.1619562. [59]

Carroll, J. (2003), "The higher-stratum structure of cognitive abilities: Current evidence supports g and about ten broad factors", in Nyborg, H. (ed.), *The Scientific Study of General Intelligence: Tribute to Arthur R. Jensen*, Elsevier Science/Pergamon Press, Oxford. [57]

Carroll, J. (1993), *Human Cognitive Abilities: A Survey of Factor-analytic Studies*, Cambridge University Press, New York. [6]

Chaiken, S., P. Kyllonen and W. Tirre (2000), "Organization and components of psychomotor ability", *Cognitive Psychology*, Vol. 40/3, pp. 198-226, https://doi.org/10.1006/cogp.1999.0729. [74]

Christensen, P., P. Merrifield and J. Guilford (1953), *Consequences form A-1*, Sheridan Supply, Beverly Hills, CA. [70]

Condon, D. and W. Revelle (2014), "The International Cognitive Ability Resource: Development and initial validation of a public-domain measure", *Intelligence*, Vol. 43/March-April, pp. 52-64, http://dx.doi.org/10.1016/j.intell.2014.01.004. [43]

Conway, A. and R. Engle (1996), "Individual differences in WM capacity: More evidence for a general capacity theory", *Memory*, Vol. 4, pp. 577-590. [16]

Cowan, N. (2017), "The many faces of working memory and short-term storage", *Psychonomic Bulletin & Review*, Vol. 24, pp. 1158-1170, https://doi.org/10.3758/s13423-016-1191-6. [23]

Damos, D. (2019), *Technical Review and Analysis of the Army Air Force Aviation Psychology Program Research Reports*, Air Force Personnel Center, Randolph, TX, http://dx.doi.org/10.13140/RG.2.2.35387.36641. [36]

Deming, D. (2017), "The growing importance of social skills in the labor market", *Quarterly Journal of Economics*, Vol. 132/4, pp. 1593-1640, http://dx.doi.org/10.3386/w21473. [84]

Diamond, A. (2013), "Executive functions", *Annual Review of Psychology*, Vol. 64, pp. 135-168, https://doi.org/10.1146/annurev-psych-113011-143750. [24]

Dworak, E. et al. (2020), "Using the international cognitive ability resource as an open-source tool to explore individual differences in cognitive ability", *Personality and Individual Differences*, Vol. 169/109906, https://doi.org/10.1016/j.paid.2020.109906. [44]

Ekstrom, R. et al. (1976), *Manual for Kit of Factor-referenced Cognitive Tests*, Educational Testing Service, Princeton, NJ. [42]

Ekstrom, R. et al. (1976), *Kit of Factor-referenced Cognitive Tests*, Educational Testing Service, Princeton, NJ. [41]

Elliott, S. (2017), *Computers and the Future of Skill Demand*, Educational Research and Innovation, OECD Publishing, Paris, https://dx.doi.org/10.1787/9789264284395-en. [96]

Engle, R. (2002), "Working memory capacity as executive attention", *Current Directions in Psychological Science*, Vol. 11/1, pp. 19-23, https://doi.org/10.1111/1467-8721.00160. [18]

Fleisher, M. and S. Tsacoumis (2012), "O*NET analyst occupational abilities ratings: Analysis cycle 12 results", *HumRRO Research Report*, National Center for O*NET Development, Raleigh, NC. [90]

Fleishman, E. (1954), "Dimensional analysis of psychomotor abilities", *Journal of Experimental Psychology*, Vol. 48/6, pp. 437-454, https://doi.org/10.1037/h0058244. [72]

Fleishman, E., D. Costanza and J. Marshall-Mies (1999), "Abilities", in Peterson, N. et al. (eds.), *An Occupational Information System for the 21st Century: The Development of O*NET*, American Psychological Association, Washington, DC. [91]

Fleishman, E. and M. Quaintance (1984), *Taxonomies of Human Performance: The Description of Human Tasks*, Academic Press, New York. [73]

Frey, C. and M. Osborne (2017), "The future of employment: How susceptible are jobs to computerization?", *Technological Forecasting and Social Change*, Vol. 114/January, pp. 254-280, https://doi.org/10.1016/j.techfore.2016.08.019. [94]

Geisinger, K. (2019), "Empirical considerations on intelligence testing and models of intelligence: Updates for educational measurement professionals", *Applied Measurement in Education*, Vol. 32/3, pp. 193-197, https://doi.org/10.1080/08957347.2019.1619564. [60]

Gibson, J. (ed.) (1947), "Aptitude tests", *Motion Picture Testing and Research Report No. 7*, U.S. Government Printing Office, Washington, DC. [38]

Gibson, J. (ed.) (1947), "Motion picture testing and research", *Research Report*, No. 7, US Government Printing Office, Washington, DC. [37]

Guilford, J. (1950), "Creativity", *American Psychologist*, Vol. 5/9, pp. 444-454. [39]

Guilford, J. and R. Hoepfner (1971), *The Analysis of Intelligence*, McGraw-Hill Book Co., New York. [40]

Guilford, J. and J. Lacey (1947), *Printed Classification Tests Report 5*, U.S. Government Printing Office, Washington, DC. [95]

Gustafsson, J. (1984), "A unifying model for the structure of intellectual abilities", *Intelligence*, Vol. 8/3, pp. 179-203, https://doi.org/10.1016/0160-2896(84)90008-4. [52]

Haberman, S., S. Sinharay and G. Puhan (2011), "Reporting subscores for institutions", *British Journal of Mathematical and Statistical Psychology*, Vol. 62/1, pp. 70-95, https://doi.org/10.1348/000711007X248875. [63]

Hellwig, S., R. Roberts and R. Schulze (2020), "A new approach to assessing emotional understanding", *Psychological Assessment*, Vol. 32/7, pp. 649-662, http://dx.doi.org/10.1037/pas0000822. [82]

Holzinger, K. and H. Harman (1938), "Comparison of two factorial analyses", *Psychometrika*, Vol. 3, pp. 45-60. [54]

Holzinger, K. and F. Swineford (1937), "The bi-factor method", *Psychometrika*, Vol. 2, pp. 41-54. [53]

Horn, J. and A. Blankson (2012), "Foundations for better understanding of cognitive abilities", in Flanagan, D. and P. Harrison (eds.), *Contemporary Intellectual Assessment: Theories, Tests and Issues, 3rd ed.*, The Guilford Press, New York. [5]

Horn, J. and N. Blankson (2005), "Foundations for better understanding of cognitive abilities", in Flanagan, D. and P. Harrison (eds.), *Contemporary Intellectual Assessment: Theories, Tests, and Issues, 2nd ed.*, The Guilford Press, New York. [4]

Horn, J. and R. Cattell (1966), "Refinement and test of the theory of fluid and crystallized general intelligences", *Journal of Educational Psychology*, Vol. 57/5, pp. 253-270, https://doi.org/10.1037/h0023816. [3]

Humphreys, L. (1947), "Tests of intellect and information", in Guilford, J. and J. Lacey (eds.), *Printed Classification Tests Report*, Government Printing Office, Washington, DC. [35]

Hunt, E. (2011), *Human Intelligence*, Cambridge University Press, New York. [8]

Jewsbury, P., S. Bowden and M. Strauss (2016), "Integrating the switching, inhibition, and updating model of executive function with the Cattell-Horn-Carroll model", *Journal of Experimental Psychology: General*, Vol. 145/2, pp. 220-245, http://dx.doi.org/10.1037/xge0000119. [47]

Johnson, W. (2018), "A tempest in a ladle: The debate about the roles of general and specific abilities in predicting important outcomes", *Journal of Intelligence*, Vol. 6/2, p. 24, http://dx.doi.org/10.3390/jintelligence6020024. [65]

Johnson, W. and T. Bouchard (2005), "The structure of human intelligence: It is verbal, perceptual, and image rotation (VPR), not fluid and crystallized", *Intelligence*, Vol. 33/4, pp. 393-416, https://doi.org/10.1016/j.intell.2004.12.002. [10]

Johnson, W., J. te Nijenhuis and T. Bourchar (2007), "Replication of the hierarchical visual-perceptual-image rotation model in de Wolff and Buiten's (1963) battery of 46 tests of mental ability", *Intelligence*, Vol. 35, pp. 69-81, https://doi.org/10.1016/j.in. [11]

Just, M. and S. Varma (2007), "The organization of thinking: What functional brain imaging reveals about the neuroarchitecture of complex cognition", *Cognitive, Affective, & Behavioral Neuroscience*, Vol. 7/3, pp. 153-191, https://doi.org/10.3758/CABN.7.3.153. [28]

Kane, M. et al. (2001), "A controlled-attention view of WM capacity", *Journal of Experimental Psychology: General*, Vol. 130, pp. 169-183, http://dx.doi.org/10.1037//0096-3445.130.2.169. [19]

Kieras, D. and D. Meyer (1997), "An overview of the EPIC architecture for cognition and performance with application to human-computer interaction", *Human-Computer Interaction*, Vol. 12, pp. 391-438, https://doi.org/10.1207/s15327051hci1204_4. [29]

Kyllonen, P. (2020), *Reasoning Abilities*, Oxford Research Encyclopedia of Education, Oxford. [67]

Kyllonen, P. and R. Christal (1990), "Reasoning ability is (little more than) working-memory capacity?!", *Intelligence*, Vol. 14/4, pp. 389-433, https://doi.org/10.1016/S0160-2896(05)80012-1. [17]

Kyllonen, P. et al. (2019), "General fluid/inductive reasoning battery for a high-ability population", *Behavior Research Methods*, Vol. 51/2, pp. 502-522. [66]

Kyllonen, P. and J. Zu (2016), "Use of response time for measuring cognitive ability", *Journal of Intelligence*, Vol. 4/4, p. 14, https://doi.org/10.3390/jintelligence4040014. [64]

Landauer, T. (1986), "How much do people remember? Some estimates of the quantity of learned information in long-term memory", *Cognitive Science*, Vol. 10/4, pp. 477-493, https://doi.org/10.1207/s15516709cog1004_4. [93]

Larson, J. (2010), *In Search of Synergy in Small Group Performance*, Psychology Press, London. [86]

Lohman, D. (1979), "Spatial ability: A review and reanalysis of the correlational literature", *Technical Report*, No. 8, DTIC AD A075972, Stanford Aptitude Research Project, School of Education, Stanford University, CA, https://apps.dtic.mil/sti/p. [69]

MacCann, C. et al. (2014), "Emotional intelligence is a second-stratum factor of intelligence: Evidence from hierarchical and bifactor models", *Emotion*, Vol. 14/2, pp. 358-374, https://doi.org/10.1037/a0034755. [77]

MacCann, C. and R. Roberts (2008), "New paradigms for assessing emotional intelligence: Theory and data", *Emotion*, Vol. 8, pp. 540-551, http://dx.doi.org/10.1037/a0012746. [78]

Marshalek, B. (1981), "Trait and process aspects of vocabulary knowledge and verbal ability", *Technical Report*, No. 15, DTIC AD A102757, Stanford Aptitude Research Project, School of Education, Stanford, University, CA, https://apps.dtic.mil/dtic/tr. [68]

Mayer, J. and P. Salovey (1993), "The intelligence of emotional intelligence", *Intelligence*, Vol. 17/4, pp. 433-442, http://dx.doi.org/10.1016/0160-2896(93)90010-3. [76]

McGill, R. and S. Dombrowski (2019), "Critically reflecting on the origins, evolution, and impact of the Cattell-Horn-Carroll (CHC) model", *Applied Measurement in Education*, Vol. 32/3, pp. 216-231, http://dx.doi.org/10.1080/08957347.2019.1619561. [61]

McGrew, K. and R. Woodcock (2001), *Technical Manual: Woodcock-Johnson III*, Riverside Publishing, Itasca, IL. [7]

National Center for O*NET Development (n.d.), *O*NET Online*, website, https://www.onetonline.org (accessed on 1 December 2020). [87]

National Research Council (2010), *A Database for a Changing Economy: Review of the Occupational Information Network (O*NET)*, National Academies Press, Washington, DC. [88]

Newell, A. (1994), *Unified Theories of Cognition*, Harvard University Press, Cambridge, MA. [30]

OECD (2017), *PISA 2015 Results (Volume V): Collaborative Problem Solving*, PISA, OECD Publishing, Paris, https://dx.doi.org/10.1787/9789264285521-en. [85]

Olderbak, S. et al. (2021), "Reliability generalization of tasks and recommendations for assessing the ability to perceive facial expressions of emotion", *Psychological Assessment, Advance online publication*, https://doi.org/10.1037/pas0001030. [80]

Peterson, N. et al. (1999), *An Occupational Information System for the 21st Century: The Development of O*NET*, American Psychological Association, Washington, DC. [89]

Reynolds, C. (1997), "Forward and backward memory span should not be combined for clinical analysis", *Archives of Clinical Neuropsychology*, Vol. 12/1, pp. 29-40, https://doi.org/10.1016/S0887-6177(96)00015-7. [71]

Russell, S. and P. Norvig (2010), *Artificial Intelligence: A Modern Approach, 3rd edition*, Pearson, London. [32]

Savi, A. et al. (2019), "The wiring of intelligence", *Perspectives on Psychological Science*, Vol. 14/6, pp. 1034-1061, https://doi.org/10.1177%2F1745691619866447. [15]

Scherer, K. and U. Scherer (2011), "Assessing the ability to recognize facial and vocal expressions of emotion: Construction and validation of the emotion recognition index", *Journal of Nonverbal Behavior*, Vol. 35/4, pp. 305-326, https://doi.org/10.1007/s10919-011-0115-4. [81]

Schmid, J. and J. Leiman (1957), "The development of hierarchical factor solutions", *Psychometrika*, Vol. 23, pp. 53-61, https://doi.org/10.1007/BF02289209. [55]

Schneider, W. and K. McGrew (2018), "The Cattell–Horn–Carroll theory of cognitive abilities", in Flanagan, D. and E. McDonough (eds.), *Contemporary Intellectual Assessment: Theories, Tests and Issues*, The Guilford Press, New York. [56]

Schrank, F. (2011), "Woodcock-Johnson III tests of cognitive abilities", in Davis, A. (ed.), *Handbook of Pediatric Neuropsychology*, Springer Publishing Company, New York. [45]

Schrank, F., N. Mather and K. McGrew (2014), *Woodcock-Johnson IV Tests of Achievement*, Riverside, Rolling Meadows, IL. [46]

Shuey, E. and M. Kankaraš (2018), "The Power and Promise of Early Learning", *OECD Education Working Papers*, No. 186, OECD Publishing, Paris, https://dx.doi.org/10.1787/f9b2e53f-en. [21]

Spearman, C. (1927), *The Abilities of Man*, MacMillan, Basingstoke, UK. [2]

Spearman, C. (1904), "'General intelligence' objectively determined and measured", *American Journal of Psychology*, Vol. 15, pp. 201-293, https://doi.org/10.2307/1412107. [1]

Stankov, L. (2019), "Diminished 'g': Fluid and crystallized intelligence and cognitive abilities linked to sensory modalities", in McFarland, D. (ed.), *General and Specific Mental Abilities*, Cambridge Scholars Publishing, Cambridge, UK. [75]

Thomas, R. et al. (2008), "Diagnostic hypothesis generation and human judgment", *Psychological Review*, Vol. 115/1, pp. 155-185, https://doi.org/10.1037/0033-295X.115.1.155. [31]

Thomson, G. (1916), “A hierarchy without a general factor”, *British Journal of Psychology*, Vol. 8, pp. 271-281, http://dx.doi.org/10.1111/j.2044-8295.1916.tb00133.x. [12]

Thurstone, L. (1938), *Primary Mental Abilities*, University of Chicago Press. [34]

Thurstone, L. (1934), “The vectors of the mind”, *Psychological Review*, Vol. 41, pp. 1-32, https://doi.org/10.1037/h0075959. [50]

Tirre, W. (1994), “Bond sampling theory of human abilities”, in Sternberg, R. (ed.), *Encyclopedia of Human Intelligence*, Macmillan, New York. [13]

Undheim, J. (1981), “On Intelligence III: Examining developmental implications of Catell’s broad ability theory and of an alternative neo-Spearman model”, *Scandinavian Journal of Psychology*, Vol. 22/4, pp. 243-249, https://doi.org/10.1111/j.1467-9450.1981.tb00400.x. [51]

van der Maas, H. et al. (2019), “The network approach to general intelligence”, in *General and Specific Mental Abilities*, Cambridge Scholars Publishing, Cambridge, UK. [14]

Vernon, P. (1950), *The Structure of Human Abilities*, Methuen, London. [9]

Wasserman, J. (2019), “Deconstructing CHC”, *Applied Measurement in Education*, Vol. 32/3, pp. 249-268, https://doi.org/10.1080/08957347.2019.1619563. [62]

Wilhelm, O., A. Hilldebrandt and K. Oberauer (2013), “What is working memory capacity, and how can we measure it?”, *Frontiers in Psychology*, Vol. 4, p. 433, http://dx.doi.org/10.3389/fpsyg.2013.00433. [48]

Wissler, C. (1901), “The correlation of mental and physical tests”, *The Psychological Review: Monograph Supplements*, Vol. 3/6, pp. i-62, https://doi.org/10.1037/h0092995. [33]

Woolley, A. et al. (2010), “Evidence for a collective intelligence factor in the performance of human groups”, *Science*, Vol. 330, p. 686, http://dx.doi.org/10.1126/science. 1193147. [83]

Zelazo, P., C. Blair and M. Willoughby (2016), *Executive Function: Implications for Education (NCER 2017-2000)*, National Center for Education Research, Institute of Education Sciences, U.S. Department of Education, Washington, DC, http://ies.ed.gov/. [20]

Zelazo, P. et al. (2013), “II. NIH toolbox cognition battery (CB): Measuring executive function and attention”, *Monographs of the Society for Research in Child Development*, Vol. 78/4, pp. 16-33, http://dx.doi.org/10.1111/mono.12032. [49]

Annex 3.A. Comparison of second-order factors in three hierarchical abilities models

Annex Table 3.A.1. Second-order factors in three hierarchical abilities models

Carroll's (1993) 3-stratum model	Horn-Cattell's Gf-Gc model (1966, 2021)	CHC Model
2F Fluid intelligence	Gf Reasoning under novel conditions[2]	Gf Fluid reasoning[4]
	Gq Quantitative mathematical	Gq Quantitative knowledge
2C Crystallized intelligence[5]	Gc Acculturational knowledge[3]	Gc Comprehension knowledge[4]
		Gkn Domain-specific knowledge
		Grw Reading and writing
2Y General Memory/Learning	SAR/Gsm Short-term apprehension, retrieval[2]	Gsm Short-term-memory[4]
2V Broad Visual Perception	Gv Visualization and spatial orientation	Gv Visual processing[4]
2U Broad Auditory Perception	Ga Listening and hearing	Ga Auditory processing[4]
2R Broad Retrieval Ability	TSR/Glm Long-term storage and retrieval[3]	Glr Long-term storage and retrieval[4]
2S Broad Cognitive Speediness	Gs Speed of thinking	Gs Processing speed[4]
2T Processing Speed		Gt Reaction and decision speed
		Gps Psychomotor speed
		Gp Psychomotor abilities
		Go Olfactory abilities
		Gh tactile abilities
		Gk Kinesthetic abilities

Note: [1] (Horn and Blankson, 2012[5]); [2] Decline with age; [3] Do not decline with age; [4] Appears in WJ III/WJ IV commercial tests; [5] 2H combines 2F & 2C.
Source: Adapted from (Carroll, 1993[6]) (Horn and Cattell, 1966[3]) (Schneider and McGrew, 2018[56]).

Annex 3.B. The WJ III and WJ IV set of factors and tests

Annex Table 3.B.1. WJ III and WJ IV set of factors and tests

Test name	Factor name	Sub-Factor	Description of task requirements
Numerical Reasoning	Gf	Quantitative Reasoning	Determine numerical sequences and a two-dimensional numerical pattern.
Concept Formation	Gf	Induction	Identify rules that make up geometric figures after being exposed to concepts.
Analysis Synthesis	Gf	General Sequential Reasoning	Analyse the structure of an incomplete logic puzzle and solve the missing parts.
Block Rotation	Gv	Mental Rotation, Visualisation	Choose geometric designs that match another design which have been physically rotated to a different position.
Spatial Relations	Gv	Spatial Relations	Select the component parts of whole shape.
Picture Recognition	Gv	Visual Memory	Study five images, remember them and recognise them in a larger set of other arranged images.
Visual Matching	Gs	Perceptual Speed	Quickly find and circle two identical numbers in a row of six numbers in 3 minutes.
Decision Speed	Gs	Mental Comparison Speed	Quickly analyse a row of images and mark two images that are the most closely related in 3 minutes.
Cross out	Gs	Perceptual Speed & Rate of Test Taking	Mark drawings that are identical to the first drawing in the row in 3 minutes.
Rapid Picture Naming	Gs	Naming Facility	Quickly name a series of pictures as fast as possible.
Retrieval Fluency	Glr	Ideational Fluency	State as many words from specified categories as possible in 1 minute.
Visual Auditory Learning: Delayed	Glr	Associative Memory	Recall and relearn (after a 30-minute to 8-day delay) symbols presented in.
Visual Auditory Learning	Glr	Associative Memory	Translate visual symbols after being given orally presented words that are associated with them.
Memory For Names	Glr	Associative Memory	Remember an increasingly large number of names of novel cartoon characters.
Memory For Names: Delayed	Glr	Associative Memory	Recall and relearn (after a 30-minute to 8-day delay) names of novel cartoon.
Sound Blending	Ga	Phonetic Coding Synthesis	Listen to a series of individual syllables, individual phonemes, or both that form words and name the complete words.
Incomplete Words	Ga	Phonetic Coding Analysis	Listen to words with one or more phonemes missing and name the complete words.
Sound Patterns	Ga	Speech-Sound Discrimination	Indicate whether pairs of complex sound patterns are the same or different. The patterns may differ in pitch, rhythm, or sound content.
Auditory Working Memory	Gsm	Working Memory	Listen to a mixed series of words and digits and then to rearrange them by first saying the words in order and then the numbers.
Numbers Reversed	Gsm	Working Memory	Repeat a series of random numbers backward
Memory For Words	Gsm	Memory Span	Repeat lists of unrelated words in the correct sequence
Memory For Sentences	Gsm	Memory Span	Repeat complete sentences.
Picture Vocabulary	Gc	Lexical Knowledge	Name familiar and unfamiliar pictured objects.
Verbal Comprehension	Gc	Language Development & Lexical Knowledge	Name familiar and unfamiliar pictured objects and then say words similar in meaning to word presented, say words that are opposites in meaning to the word presented, and complete phrases with words that complete analogies.

General Information	Gc	General Information	Provide characteristics of objects by responding to questions, such as "Where would you find...?" and " What you would do with...?".
Academic Knowledge	Gc	General Information	Provide information about biological and physical sciences, history, geography, government, economics, art, music and literature.
Oral Comprehension	Gc	Listening Ability	Listen to a short passage and orally supply the word missing at the end of the passage.
Story Recall	Gc	Listening Ability	Listen to a short passage and describe the details.
Verbal Attention (WJIV only)	Gsm	Working memory capacity	Listen to a series of numbers and animal words mixed together and answer questions regarding the sequence.
Number Series (WJIV only)	Gf	Quantitative reasoning	Participants have to identify the correct number in a series of numbers that correctly completed the series. Ex. (2, 4, ?, 8, 10,...)
Letter-Pattern Matching (WJIV only)	Gs	Perceptual speed	Quickly find and circle identical letters and patterns.
Visualisation (WJIV only)	Gv	Mental rotation, Visualisation	Identify two sets of 2D pieces that form a specific shape; also identify two sets of 3D rotated blocks that match another shape.
Phonological Processing (WJIV only)	Ga	Phonetic coding, Word fluency	Name words that begin with a certain sound; also use parts of words to create new ones.
Nonword Repetition (WJIV only)	Ga	Phonetic coding	Listen to a nonsense word and repeat the word exactly.
Segmentation (WJIV only)	Ga	Phonetic coding	Listen to words and break them into syllables and phonemes.

Note: *Appears in WJ IV, not WJ III.
Source: Adapted from (Schrank, 2011[45]); (Schrank, Mather and McGrew, 2014[46]).

Annex 3.C. Factor hierarchy in Carroll's three-stratum model of human cognitive abilities

Annex Table 3.C.1. Factor hierarchy in Carroll's three-stratum model

Stratum III	Stratum II	Stratum I
general intelligence	fluid intelligence	general sequential reasoning induction quantitative reasoning speed of reasoning
	crystallised intelligence	language development lexical knowledge learning ability phonetic coding communication ability oral production and fluency (seven more)
	general memory & learning	memory span associative memory meaningful memory free recall memory visual memory learning abilities broad visual perception
	broad visual perception	visualisation spatial relations coding speed flexibility of closure perceptual speed spatial scanning (six more)
	broad auditory perception	speech-sound discrimination general sound discrimination resistance to auditory stimulus distortion temporal tracking memory for sound patterns musical discrimination and judgement (five more)
	broad retrieval ability	ideational fluency associational fluency expressional fluency naming fluency word fluency originality/creativity (three more)
	broad cognitive speediness	rate of test taking numerical facility perceptual speed
	processing speed	simple reaction time choice reaction time semantic processing time mental comparison speed

Annex 3.D. Sample items

Annex Figure 3.D.1. Example fluid intelligence test items

Figure sets: A test of the Induction factor within the Fluid intelligence domain

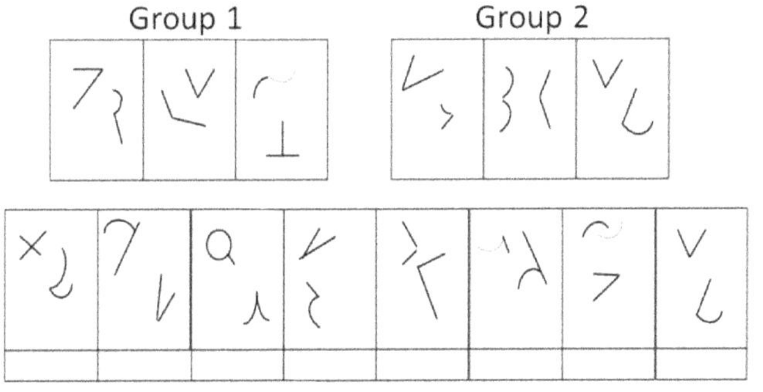

Note: Other Induction test examples include figural, verbal, or numerical sets, series, and matrices tests. The task is to classify the 8 items below into Groups 1 or 2 by inducing the rule from the exemplars above.

Diagramming Relationships: A test of the Sequential (Deductive) Reasoning factor within the Fluid intelligence domain

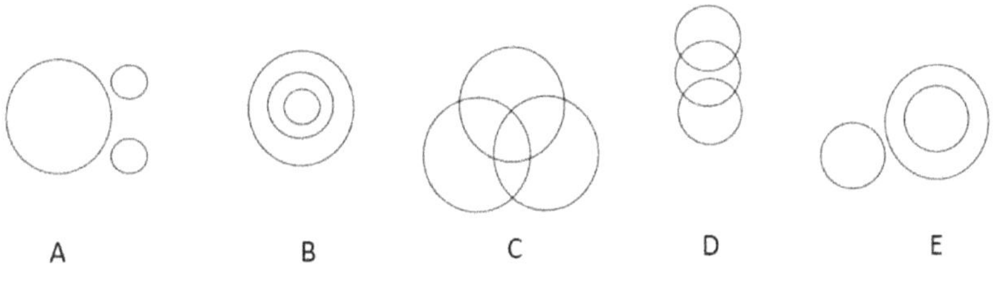

Pines, trees, stones

A B C D E

Note: Other examples include logical deductions. The task is to choose the Euler diagram that reflects the relationships among the listed entities.

Necessary Arithmetic Operations: A test of the Quantitative Reasoning factor within the Fluid intelligence domain

A cyclist in an international bicycle race covered an average of 9 miles every 20 minutes. If she maintained the same average speed, how long did it take her to cycle the remaining 84 miles of the race?

1 – divide and multiply
2 – subtract and divide
3 – add and subtract
4 – divide and add

Note: Other examples include mathematics word problems. The task is to indicate which operations would be required to solve the problem.
Source: Ekstrom et al. (1976[41]).

Annex Figure 3.D.2. Example crystallised intelligence test items

Reading Comprehension: A test of Reading Comprehension within the Crystallized intelligence domain

The metal porch swing virtually sizzled on the old wooden front porch today. But we sat there anyway. Gramma wouldn't hear of anything else. I suggested a walk through the forest, hoping to entertain a breeze or two and to take advantage of the shade. Gramma shook her head. You were supposed to sit on the porch after supper, and that's what we were going to do.

The author implies that

1 – Gramma cooked supper.
2 – Gramma didn't like the forest.
3 – Gramma didn't change her routine.
4 – Gramma couldn't hear very well.

Note: The task is to select the best characterization of the passage from the choices given.

Vocabulary: A test of Lexical Knowledge within the Crystallized intelligence domain

Inclement

1 – balmy
2 – happy
3 – righteous
4 – severe
5 – apprehensive

Note: The task is to identify the closest synonym to the target word.

Cloze: A test of Cloze Ability within the Crystallized intelligence domain

Several different ______ for estimating the approximate functional content of adult human memory have been _____.

Note: The task is to fill in the blanks through inferencing.
Source: First two panels, (Ekstrom et al., 1976[41]); third panel, (Landauer, 1986[93]).

Annex Figure 3.D.3. Example broad visual perception test items

Paper Folding: A test of Spatial Visualization within the Broad Visual Perception domain

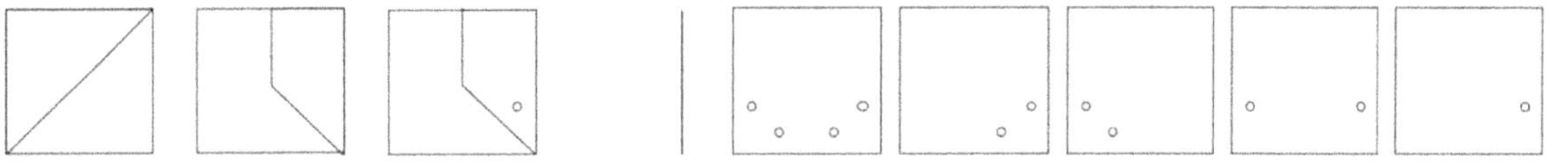

Note: The task is to select the unfolded diagram from the options on the right based on the pattern of folding and punched holes in the depiction on the left.

Copying: A test of Closure Flexibility within the Broad Visual Perception domain

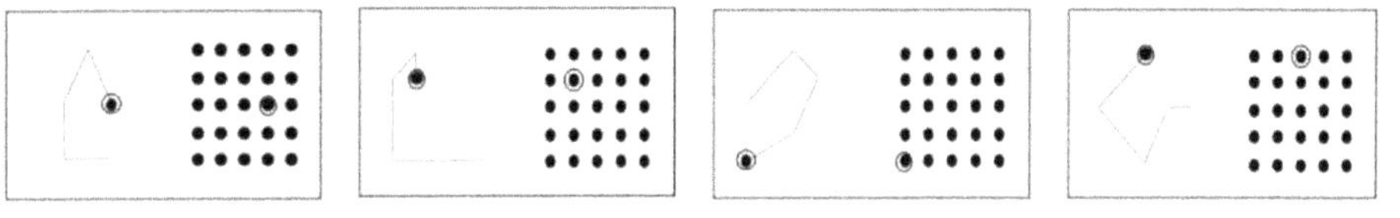

Note: The task is to copy the image on the left by connecting the appropriate dots on the right.

Identical Pictures: A test of Perceptual Speed within the Broad Visual Perception domain

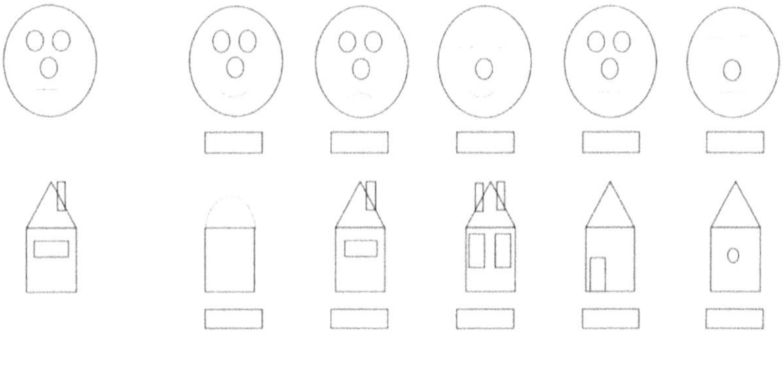

Note: The task is to select the picture on the right that matches the target picture on the left.
Source: Ekstrom et al. (1976[41]).

Annex Figure 3.D.4. Example broad retrieval ability test items

Word Beginnings and Endings: A test of Word Fluency within the Broad Retrieval Ability domain

sun
spin
stain
solution

Now try thinking of some more words beginning with S and ending with N. Write them on the lines below. Names of people or places are not allowed.

Note: The task is to generate as many words as possible within a time limit that meet the constraints given.

Matchsticks: A test of Figural Flexibility within the Broad Retrieval Ability domain

Make as many different solutions as possible, up to five, for each item. Use a different rule for each solution

1. Take away 4 toothpicks. Leave 6 squares.

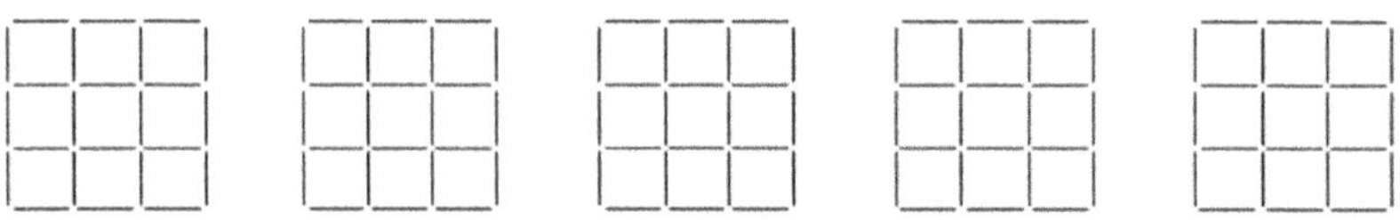

Note: The task is to generate as many solutions as possible within a time limit that meet the constraints given.

Consequences: A test of Creativity within the Broad Retrieval Ability domain

What would happen if the entire United States turned overnight into an arid dessert? Write down as many possibilities as you can think of. You have 3 minutes.

Note: The task is to generate as many implications as possible within a time limit. Scores are based on the number of unique, on topic responses given in 3 minutes.
Source: First two panels, (Ekstrom et al., 1976[41]); third panel, (Christensen, Merrifield and Guilford, 1953[70]).

4. Testing cognitive functions in children: A clinical perspective

Sylvie Chokron, Institut de Neuropsychologie, Neurovision & Neurocognition, Hôpital-Fondation Adolphe de Rothschild; Integrative Neuroscience and Cognition Center, CNRS UMR 8002 & Université de Paris.

Neuropsychological evaluation aims at describing a child's weaknesses and strengths to set an appropriate rehabilitation and education programme, as well as to better educate the school and family about the child's needs. This chapter presents the most commonly used neuropsychological tests, as well as their limits and caveats in the understanding of the cognitive profile of children. It looks at selected cognitive and intelligence tests used in neuropsychological practice, including WWPSI-IV, K-ABC and NEPSY. It also looks at specific tests for attention, memory, visual and spatial cognition, visuo-motor co-ordination and executive functions. The chapter ends with a discussion about intra- and inter-variability during neuropsychological evaluation and dissociation within an intelligence test; intelligence, neuropsychological assessment and learning abilities; and sources of performance differences between children and robots.

Introduction

Neuropsychology represents the linkage between behaviour (including cognitive function) and the brain substrate. Neuropsychological assessment provides standardised, objective and reliable measures of diverse aspects of human behaviour. This allows for the specification of each individual's unique profile (Ivnik, Smith and Gerhan, 2001[1]).

Brain-behaviour relationships in a developing child are both qualitatively and quantitatively different than those in adults. Neuropsychological evaluation in children can thus be seen as measuring the result of the interaction between development, and in particular, brain maturation, environmental stimulations, education, effect of brain lesion (in brain-damaged children), brain plasticity and proposed rehabilitation.

Referral to a child neuropsychologist is appropriate for diverse clinical, research and/or academic reasons. Whatever the context, tests results are considered along with a careful history-taking, precise clinical observations and respect for the socio-cultural context to best describe the child's cognitive function. This is particularly important given that neuropsychological evaluation may have some predictive value for a child. It will, for example, often serve as a basis for academic or vocational decisions and therapeutic programmes.

In clinical practice, the neuropsychological evaluation has several aims. First, it describes the child weaknesses and strengths to set an appropriate rehabilitation and education programme. Second, it aims to educate the school and family about the child's needs.

Testing models have different strategies. Indeed, the kind of test to be used among children referred for an evaluation is still debated. Should it be fixed battery, flexible battery, process approach or personal core battery?

In clinical practice, some flexibility is often required between the models. A given test and the way it is administered will influence the evaluation. A fixed battery, for example, seems useful in obtaining normative and research data. However, an inflexible fixed battery approach often proved insensitive and inappropriate to children with instrumental deficits (verbal, motor, attentional or perceptual). In addition, using a fixed battery might be far from the clinical complaints of the child or from the family or school questioning about the child's difficulties.

These flexible batteries are thus built by adding some own personal core batteries, or some tests to a fixed battery based on individual needs. Tailoring the evaluation can better evaluate domains or subdomains omitted in the rigidly structured battery. In this way, the neuropsychological evaluation can include more qualitative data to nuance the objective, quantitative data obtained through fixed batteries.

Of course, the child must be understood in a larger context. The tests rely on standardised, objective and reliable measures of behaviour. However, neuropsychologists have to compare these measures to the psychological, educational and socio-cultural context of the examined child. This helps understand to which extent the test results are imputable to brain-based mechanisms or to other environmental factors.

The neuropsychologist usually performs a profile analysis at the end of the evaluation to define a child's neurocognitive strengths and weaknesses. The analysis of quantitative features with respect to appropriate normative data is then integrated with qualitative observations about the child's individual style, temperament and environment.

What is observed during the evaluation could largely differ from the actual cognitive level of the child. The child's cognition may be influenced by various factors. These range from personality traits such as intellectual inhibition, anxiety, fear and lack of self-confidence to biological factors such as fatigue and hunger. This intra-variability is inherent to human testing and is not expected in testing artificial intelligence (AI) devices or robots. From this point of view, psychometric tests are more adapted to robots than to humans.

Regarding the framework used for examining data, different approaches to interpreting neuropsychological test data use cross-sectional data: absolute scores, difference scores, profile variability and change scores. Absolute scores refer to a single score from each test that might best differentiate each diagnostic group. Difference scores compare performance on tests sensitive to neurocognitive dysfunction with that on tests resistant to these effects. Profile variability assumes that impairment will affect performance in a different way across a range of tests. Change scores refer to longitudinal data obtained at test-retest intervals (Ivnik, Smith and Gerhan, 2001[1]).

Cognitive evaluation: A Neuropsychological approach

Before describing the most commonly used neuropsychological tests around the world, the major advantages and limitations of intelligence tests are presented.

Advantages of intelligence tests

The ubiquitous IQ test provides the experienced clinician with a well-researched statistical basis for interpretation of individual subtest and factor index scores. Often, a quick appraisal of subtest scaled scores will offer the clinician pertinent cues about potential strengths and weaknesses, despite a successful performance. Although IQ tests are not designed to diagnose neuropsychological deficits, low performance on coding and picture completion, for example, will alert towards an attentional deficit. Conversely, lowered scores in block design subtest with intact verbal scores may point to praxis or visuo-spatial deficit. In addition, intelligence tests often consist of multiple parts that are co-normed, allowing the child's performance on one subtest to be compared directly with performance on another.

When the IQ is above what is expected at his/her age, this well-standardised measure will exclude any neuropsychological deficiency. However, when the IQ is below expectations, the question of the origin of this lowered score arises. A lowered IQ can be associated to various conditions ranging from intellectual deficiency to lack of motivation, cultural or even physical reasons (e.g. sleep debt, virus).

Comparing the performance of a child and a robot on IQ tests raises the question of the underlying cause of the performance. In any case, many factors can potentially influence the performance of a human subject on IQ tests. Thus, they cannot be considered as a reflection of only a subject's intellectual efficiency.

Limitations of intelligence tests

Despite their prevalence in psychological testing, several limitations are associated with the administration of intelligence tests in child neuropsychology practice.

First, IQ tests had not been standardised among pathological populations. It is thus difficult to propose such tests to children with instrumental (verbal, motor, perceptual) or attentional deficits.

Along those lines, the results to IQ tests might be insufficiently sensitive or specific for neuropsychological evaluation. In particular, psychometric tests say nothing about the nature and the aetiology of the deficit causing the lowered IQ. This is probably because intelligence tests are not instruments validated with respect to brain function. Rather, they simply raise a suspicion that must be tested with other measures. For example, the interpretation of a significant split between verbal and performance IQ can be difficult (almost impossible) to interpret only on the basis of the intelligence test. Indeed, the performance scale might be lowered due to motor, spatial, visual or even non-cognitive factors, making discussion on IQ superfluous. This is a major concern in children whose motor or visuo-spatial deficit is still undiagnosed prior to the intelligence test administration.

Thus, whereas IQ tests represent a broad sweep of cognitive performance, neuropsychological evaluation attempts to provide a finer delineation of meaningful elements about how a child perceives, integrates and expresses information. In this way, the neuropsychological evaluation aims at examining the child's discrete behaviour across a wide variety of domains, especially with tests validated with respect to brain function.

Although intelligence tests are often used in children with learning disabilities, they are not "neuropsychological instruments" nor were they constructed for such use. In addition, an IQ score as a summary score does not reveal the basic processes that contributed to, or negatively impacted, the child's functioning. Finally, the duration of standardised intelligence tests in neuropsychological evaluation of children who may suffer from attentional, perceptual or motor disorders is another disadvantage.

Despite all these limitations, an intelligence test is routinely administered to a child referred for a neuropsychological evaluation when such testing has not been recently obtained. In many industrialised countries, the referral source, the socio-medical institutions and the school system often expect its administration.

The cognitive tests most often used in neuropsychology are presented below. In addition, more specific tests evaluating precise cognitive functions are highlighted; they represent an alternative to IQ tests in children with instrumental or learning disabilities.

Selected cognitive and intelligence tests used in neuropsychological practice

This section focuses on the main cognitive tests used in neuropsychological practice. It begins with complete evaluations testing the whole cognitive function before examining other tests that evaluate more specific areas (see Table 4.1). In these tests, neuropsychology considers the following taxonomy of abilities: perception, attention, memory, executive functions, language, verbal and visual reasoning, and calculation. It then looks at how these different abilities are measured.

Table 4.1. Examples of complete and specific tests used in neuropsychological practice

Test	Function	Age
WWPSI-IV	Whole intellectual efficiency	2-7
WISC-V	Whole intellectual efficiency	6-16
K-ABC	Whole intellectual efficiency	3-12
NEPSY	Whole intellectual efficiency	5-16
TEA-ch	Attention	6-12
CMS	Memory	5-16
Rey Figure	Visuo-spatial	3-6 / 6-21
EVA	Neuro-visual	4-6
Purdue Pegboard	Visuo-motor co-ordination	6-10

Neuropsychological evaluation: Complete tests

WWPSI-IV

Tests for younger children

The following five subtests are proposed to younger children (ages 2-6 to 3-11):

- Receptive vocabulary

This subtest measures a child's ability to identify correct responses to spoken words (e.g. identifying a picture of a "fish" that represents the word "fish" spoken by the examiner). However, most of the receptive or expressive vocabulary subtests use visual pictures that must be designated or named by the children.

If the child is unable to visually explore, perceive, analyse or recognise the visual items, it will be impossible for him/her to perform the task. If this happens, it most often leads to a suspicion of a verbal deficit because the task is considered as involving language. However, it is first a visual task that can be failed because of a visual or spatial deficit.

- Information

This subtest measures general cultural knowledge, long-term memory and acquired facts. This subtest is largely linked to culture, education and school knowledge.

- Block design

This subtest measures an individual's ability to analyse and synthesise an abstract design and reproduce that design from coloured plastic blocks. It involves visuo-spatial analysis, simultaneous processing, visual-motor co-ordination, dexterity and non-verbal concept formation.

- Object assembly

As for the block design, this subtest measures an individual's ability to analyse and synthesise an abstract design and reproduce that design from coloured plastic blocks.

- Picture naming

This subtest assesses an individual's ability to name pictorial stimuli. This subtest involves verbal capacities but also visual capacities. It can thus be seen as a language test but also as a visual recognition test.

Tests for older children

The 14 following subtests are proposed to older children (Ages 4-0 to 7-7):

- Block design

This subtest has the same principles as the one for younger children.

- Similarities

This subtest is supposed to measure verbal reasoning. However, like other subtests, it also involves visual, spatial and attentional cognition. Two similar but different objects or concepts are presented. The student is asked to tell how they are alike or different (e.g. what do the ear and eye have in common?)

- Picture concepts

Students are asked to look at two (or three) rows of pictured objects and indicate (by pointing) the single picture from each row that shares a characteristic in common with the single picture(s) from the other row(s). This subtest aims to evaluate categorical and abstract reasoning. However, it first measures visual, spatial and attentional processing.

- Coding

In this subtest, children must learn associations between non-verbal shapes and to complete (draw) the corresponding shape in an empty box as quickly as possible. Non-verbal associative and short-term (working) memory are involved. In addition, this subtest also involves other abilities such as visual analysis, fine motor dexterity, speed, accuracy and ability to manipulate a pencil. Given this context, so many processes contribute to success at this task that it can be failed for multiple reasons. In the simplest form of the test, children use their ink daubers and stamp the shape that goes with each animal according to the "key" given to them. For example, they would stamp a "circle" when they see a fish, a "star" when they see a cat and a "square" when they see a turtle.

- Vocabulary

This subtest measures several verbal abilities: verbal fluency, concept formation, word knowledge and word usage. It is thus influenced by prior knowledge and, in this way, by culture and education.

- Matrix reasoning

In this subtest, the subject must perceive, analyse and understand the logical rule between shapes. As above-mentioned, although this task is supposed to evaluate non-verbal reasoning, it also measures visual, spatial and attentional cognition.

- Comprehension

In this subtest, the child must answer questions based on his or her understanding of general principles and social situations. This subtest measures the students' verbal comprehension but also common sense social knowledge, practical judgement in social situations and social maturity. In addition, it evaluates the student's moral conscience.

- Symbol search

This subtest requires the student to determine whether a target symbol (geometric form) appears among the symbols shown in a search group. This subtest requires perceptual, spatial, attentional and speed abilities. For this reason, children can fail this subtest because of visuo-spatial deficits or slowness.

- Picture completion

This subtest measures a student's ability to recognise familiar items and to identify missing parts (e.g. naming the missing sand in an egg timer). In this way, this subtest requires analysis of fine visual details, as well as visual, attentional and visual memory abilities.

- Information measures

This subtest measures general cultural knowledge and long-term didactic memory. For this reason, this subtest depends on the ability of the student to recall facts and information previously taught in school.

- Word reasoning

This subtest measures verbal abstract reasoning requiring analogical and categorical thinking, as well as verbal concept formation and expression.

- Receptive vocabulary

This subtest requires the child to look at a group of four pictures and point to the one the examiner names aloud. As in most of the verbal subtests, the child needs first to visually analyse and recognise the stimulus. However, this subtest is considered to rely mostly on prior verbal knowledge.

- Object assembly

This subtest embodies the same principles as the subtest for younger children.

- Picture naming

This subtest embodies the same principles as the one for younger children.

Several non-verbal tests, as well as verbal subtests, are influenced by visual perception, spatial cognition and sustained attention. Indeed, among the 14 different subtests, 10 require good abilities in visual and spatial processing. In this way, failure on these subtests can easily result from visuo-spatial or attentional deficits rather than a pure intellectual deficit. In addition, many subtests are time-limited and can thus also be affected by slowness. On the other hand, several verbal tests involve prior word knowledge and in this way are sensitive to culture and education.

K-ABC

The K-ABC (Kaufman and Kaufman, 1983[2]) was normed on 2 000 children, aged 2 years 6 months to 12 years 5 months. This test is based on Luria's theory of simultaneous and successive information processing. Its approach defines intelligence in terms of the child's problem-solving and processing styles. Seven subtests are dedicated to simultaneous processing whereas sequential processing is tested through three subtests. There are also six achievement subtests.

- K-ABC sequential process scale

Hand movements: the child must reproduce a sequence of gestures shown by the examiner.

Immediate memory of numbers: the child must repeat a series of numbers in the order of sequence.

Sequence of words: the child points to object profiles in the order given by the examiner. For added difficulty, children over five years old must also give the colour names before pointing at the objects.

- K-ABC simultaneous process scale

Pattern recognition: the child mentally completes an unfinished drawing and names or describes it to the examiner.

Triangles: triangles, blue on one side and yellow on the other, are presented to the child who assembles them to reproduce a model.

Analogue matrices: the child completes incomplete matrices with pictures.

Spatial memory: the child memorises the space occupied by drawings randomly arranged on a page.

Series of pictures: the child puts back in chronological order pictures that form a little story.

The resolution of the items by the tested child is not limited by time. To help assess cultural or linguistic minorities, as well as children with auditory or verbal deficits, a non-verbal scale of six subtests was included that involves pantomime presentation of instructions. This test battery also includes a series of tests in which the examiner gives instructions by gestures.

The correlation between the Wechsler scale test results (Raiford, 2018[3]), which is the main intelligence scale for children, and the K-ABC test results is quite good. However, in the cases of learning disabilities, the K-ABC generally performs better than the WISC-R. The K-ABC would seem to be better suited than the other tests to analyse the reasons for academic failure. This is especially the case when the failures are unexpected or when the children come from a socio-cultural minority background. The K-ABC would allow a better perception of where the child's difficulties lie, especially in cases of autism spectrum disorders and dyslexia.

NEPSY

The NEPSY was composed of tests created for children from 3-12 years old (Brooks, Sherman and Strauss, 2009[4]). It allows neuropsychologists to understand the cognitive, behavioural and academic problems of young children. At the same time, it allows them to detect and define the deficit in brain-damaged children.

As for the K-ABC, following Luria's approach, the NEPSY allows to identify the primary deficits that may underlie various learning disorders. The NEPSY was originally developed in Finland for young children. It was subsequently extended and normed on 1 000 English-speaking children, 100 at each age level, from 3 to 12 years.

The NEPSY focuses on five broad functional domains: attention and executive functions; language; sensorimotor functions; visuo-spatial treatments; and memory and learning. Each domain is tested with specific subtests at each age (see Table 4.2).

For each domain, there are core tests. In addition, supplementary tests allow a possible deficit to be examined in greater depth when the results of the core tests are low. Additional marks allow a particular deficit to be specified. Finally, qualitative observations are made (e.g. on the child's behaviour or strategy).

Table 4.2. Subtests proposed at each age to test the five domains with the NEPSY

Functional domains (NEPSY)	Subtests Children 3-4 years old	Subtests Children 5-12 years old
Attention and executive functions	• visual attention • statue	• auditory attention/response set • visual attention • tower
Language	• body part naming • phonological processing • comprehension of instructions	• speeded naming • phonological processing • comprehension of instructions
Sensorimotor functions	• imitating hand positions • visuo-motor precision	• imitating hand positions • visuo-motor precision • fingertip tapping
Visuo-spatial treatments	• design copying • block construction	• arrows and design copying • block construction
Memory and learning	• narrative memory • sentence repetition	• narrative memory • memory for faces • memory for names

Neuropsychological evaluation: Specific tests

This section looks at different kinds of tests for attention, memory, visual and spatial cognition, visuo-motor co-ordination and executive functions.

Attention

Vast amounts of information reach humans at every moment, both from the outside world (visual, auditory, tactile, etc.) and also from the body (internal temperature, heartbeat, pains or position of the body in space). The central nervous system cannot process all this information simultaneously. Therefore, it selects the most relevant information to be processed at each moment by the brain (Chokron, 2010[5]).

Among the information that comes to us, some is recurrent, usual and "predictable". Other information will "capture our attention" because it is new, particularly interesting, unexpected, etc. The role of attention is to select and privilege the information to be processed on the basis of its novelty, relevance or the constraints of the moment and motivation. A degree of attentional control is necessary for any successful task performance.

Sustained attention or vigilance corresponds to a state of alertness that allows us to be receptive to the presented information. Under normal waking conditions, humans possess sustained attention abilities that allow us to interact effectively with the outside world. Of course, there are great variations within and between individuals in the way they implement this ability. Sustained attention can thus vary from the normal state (without brain damage) depending on the time of day, physical condition, psychological state, external conditions or motivation (Chokron, 2010[5]).

In addition, there is *selective attention*, which, as its name suggests, is based on the notion of information selection. *Selective spatial attention* refers to the ability to select information in some portions of the external space. *Divided attention* refers to the ability to respond to more than one task or event simultaneously. Finally, *alternating attention or mental shifting* refers to the ability to maintain mental flexibility to shift from one task requirement to another when these have different cognitive requirements.

The ability to orient attention develops gradually from birth. Internally driven ability to scan the environment is actively established by five or six years of age. The ability to focus attention on a sensory source or on a task is expected to be established by the age of seven, and sustained attention abilities develop until adolescence (Helland and Asbjornsen, 2000[6]).

In clinical neuropsychology practice, attention is a process or domain that is assessed as one component contributing to the child's overall neurocognitive competence. Fatigue affects attention, as well as several factors such as motivation, affective state, hunger, etc. Therefore, the failure of a cognitive test (whatever its nature) could be due to a decrease in attentional capacities. Attention is not mediated by a single brain region or by the brain as a whole. Instead, it is carried out by discrete anatomical networks, within which specific computations are assigned to different brain areas (Posner and Petersen, 1990[7]).

A number of attention types, or subdomains, are described in the literature. Commonly, one refers to terms such as focused attention, switching or mental set shifting, and divided attention.

TEA-Ch: Test of everyday attention for children

The Test of Everyday Attention for Children (TEA-Ch) (Manly et al., 1999[8]) is a children's version of the adult eight-subtest Test of Everyday Attention (TEA) (Robertson et al., 1995[9]). TEA-Ch measures different components of attention (selective attention, attentional control, sustained attention) through the nine subtests described below.

- Selective attention

Sky search: this subtest requires that the child filters information to detect relevant information and reject or inhibit distracting information. Specifically, the child must seek pairs of "spaceship" stimuli and rapidly circle all occurrences amid competing non-paired stimuli.

Map mission: in this subtest, the children are given a printed A3 laminated city map. They have to circle as many as possible visual targets (among 80 targets) and ignore distractors.

- Attentional control

Creature counting: this subtest measures attentional control and switching that requires executive functions such as working memory and mental flexibility to count stimuli according to visual cues to either count up or count down.

Opposite worlds: this is a timed measure of attentional control and switching requiring the child to read sequenced chains of numbers as they appear (same world condition) or to inhibit the prepotent response and respond with an alternate number (i.e. 1 for 2, 2 for 1 different world conditions). This subtest makes the stimulus (a digit) and the response (the word "one" or "two") association explicit. This subtest is thus like the conflicting response requirement for the Stroop test.

- Sustained attention

Score!: this subtest presents ten item-counting measures. In each item, between 9 and 15 identical tones of 345 ms are presented, separated by silent interstimulus intervals of variable duration (between 500 and 5 000 ms). Children are asked to silently count the tones (without assistance of fingers) and to give the total at the end.

Code transmission: this subtest is an auditory vigilance-level measure. The child has to listen to a taped 12-minute recording of single-digit numbers presented at 2s intervals. The child has to immediately announce the digit presented just before "55" when they hear the number "55". The score given is the number of digits correctly announced by the participant. There are 40 target presentations. This subtest is a variation of an n-back task.

- Sustained, divided attention and response inhibition

Sky search DT: in this subtest, children have to circle paired spaceship stimuli (as in the sky search task), while also keeping a count of auditory tones until all target stimuli are circled.

Score!DT: this subtest measures sustained auditory attention requiring the child to listen to and count tape recorded tones, while also listening for an animal named by the announcer in a news broadcast.

Walk/Don't walk: this measures sustained attention and response inhibition. A child learns tones that allow progression (go) or require inhibition (no-go) and then makes a mark accordingly. The speed of tone presentation increases as the task progresses. The child must avoid making a mark in the no-go condition.

Memory

Many batteries have been developed largely by including downward extension of tasks from adult memory tests rather than by constructing tests with developmental principles in mind. The most used batteries are the Rivermead Behavioural Memory Test for children aged five-ten years old, the Test of Memory and Learning, and the NEPSY learning and memory subtests, as well as the Children's Memory Scale (CMS) described below.

The CMS is an extension of the Wechsler Memory Scale series for adults (Lichtenberger and Kaufman, 2004[10]). The CMS was designed as part of a standard psychological or neuropsychological evaluation to provide a comprehensive assessment of learning and memory in children and adolescents of ages 5-16 years.

Summary scores of the CMS include verbal immediate, verbal delayed, visual immediate, visual delayed, general memory, learning delayed recognition and attention/concentration indexes. Immediate, delayed recall and delayed recognition scores are converted to scaled scores. Core subtests comprise stories and word tests (for verbal memory indexes); dot locations and faces (for visual/non-memory indexes); and numbers and sequences (for attention/concentration indexes) (Cohen, 2011[11]).

Visual and spatial cognition

Most of the subtests proposed in the different batteries involve visual and spatial cognition for two reasons.

First, it is crucial to evaluate visuo-spatial cognition in children to disentangle a perceptual deficit from an intellectual deficit. Importantly, visuo-spatial deficits can bias not only non-verbal evaluation but also verbal subtests. For example, vocabulary is often evaluated in children through picture naming, which requires visual perception, analysis and understanding.

Second, cortical visual impairments are frequent in at-risk children (born preterm or in a neurological context as with perinatal asphyxia). While visual acuity is often screened, visual function (i.e. visual and spatial cognition) is largely neglected. For this reason, children often receive a diagnosis of intellectual or co-ordination or motor disorder whereas they suffer from a visual deficit that affects cognitive tasks involving vision, or motor or co-ordination tasks.

Specific batteries such as the Developmental Test of Visual Perception (DTVP) (Hammill, Pearson and Voress, 2013[12]) or the Evaluation of Visuo-attentional abilities (EVA) (Cavézian et al., 2010[13]; Chokron, 2015[14]) are available to evaluate visuo-spatial and attentional capacities in children. In the DTVP, for example, the subject has to retrieve and point to the shapes embedded in the figure presented on the top.

In the EVA battery, the subject is first presented with the figure on the top and, then ten seconds later, must find the represented figure among the six propositions.

Visuo-motor co-ordination

Visuo-motor co-ordination is tested as visual cognition is required in a number of tasks. It can be assessed in children to evaluate the presence, nature and impact of a visuo-motor co-ordination disorder on other cognitive tasks.

The Purdue Pegboard, developed in 1948, has been used most extensively in personnel selection for jobs that require fine and gross motor dexterity (Podell, 2011[15]). The test measures the gross motor dexterity of hands, fingers and arms, as well as the fine motor dexterity of fingertips. In addition to being employed in personnel selection, the Purdue Pegboard test has also been used in neuropsychological assessments. It has been found to be sensitive to the presence of brain damage (especially frontal or parietal) or of visuo-motor co-ordination deficit.

The Purdue Pegboard is a board featuring two rows of 25 holes each. At the top of the board are four cups. The pins (pegs) are kept in the outer left and right cups, and the collars and washers are kept in the middle cups. There are four subtests of the Purdue Pegboard, done with the dominant hand, the non-dominant hand and with both hands. The subject has 30 seconds to place as many pins as possible in the holes, starting at the top of the right row in the right-hand test and at the top of the left row in the left-hand test. In the third subtest for both hands, the subject fills both rows, starting at the top, for the same amount of time. In the fourth subtest, the subject is asked to use both hands to construct "assemblies" of a pin, a washer, a collar and another washer for 60 seconds.

Executive functions

Besides global cognitive evaluation, it might be necessary to measure specific cognitive processes such as executive functions. Executive function has been defined in various ways [see Eslinger (1996[16]) for a review]. Executive function maintains an appropriate set to achieve a future goal (Luria, 1973[17]). For Baddeley (1986[18]), executive function refers to those mechanisms by which performance is optimised in situations requiring the simultaneous operation of a number of different cognitive processes. For Welsh, Pennington and Groissier (1991[19]), it involves mostly strategic planning, impulse control and organised search, as well as flexibility of thought and action. For Denckla (1989[20]), it requires the ability to plan and sequence complex behaviours, simultaneously attend to multiple sources of information, grasp the gist of a complex situation, resist distraction and interference, inhibit inappropriate responses and sustain behaviour for prolonged periods.

In this way, executive functions are thus higher functions that integrate others that are more basic, such as perception, attention and memory. These higher functions include the abilities to anticipate, establish goals, plan, monitor results and use feedback (Stuss and Benson, 1986[21]). Executive function also refers to regulatory control (Nigg, 2000[22]) and to a set of processes that guide, direct and manage cognitive, emotional and behavioural functions, especially during active, novel problem solving (Gioia et al., 2000[23]).

According to Baron (2018[24]) executive function can be seen as the "metacognitive capacities that allow an individual to perceive stimuli from his or her environment, respond adaptively, flexibly change direction, anticipate future goals, consider consequences and respond in an integrated or common sense way, utilising all these capacities to serve a common purposive goal."

Subdomains of executive functions

Executive function is thus heterogeneous and includes both broad and specific behaviours. Indeed, executive function has become an umbrella term that encompasses a number of subdomains, some more consistently endorsed than others.

In summary, executive function refers to a set of subdomains such as set shifting, hypothesis generation, concept formation, abstract reasoning, planning, organisation, goal setting, fluency, working memory, inhibition, self-monitoring, initiative, self-control, mental flexibility, attentional control, anticipation, estimation, behavioural regulation, common sense and creativity.

Inhibition and neuropsychological evaluation

As its role in executive function, inhibition mediates response selection in planning and problem-solving tasks (Levin et al., 2001[25]). To respond accurately to a question, one has to inhibit inaccurate responses, as well as incorrect reasoning. During the two last decades, inhibition has received a lot of interest as investigators attempt to parcel out contributions to effective or impaired inhibitory function. A variety of forms of inhibition are described (Nigg, 2000[22]), such as cognitive (intellectual) inhibition, interference control and motor or oculomotor inhibition. Intellectual inhibition may bias the results of cognitive evaluation even in gifted children (with IQ largely above average).

There are substantial data indicating that response inhibition is mediated by frontal cerebral regions (Stuss and Benson, 1986[21]; Mega and Cummings, 1994[26]). Patients with frontal dysfunction may exhibit too much or not enough inhibition depending on the lesion location and the type of task proposed. The developmental trajectory on inhibition tasks appears linked to prefrontal maturation (Levin et al., 2001[25]). Anterior regions of the frontal cortex continue to develop throughout childhood and into adolescence, thus influencing the level of inhibition.

Because frontal lobe is developing in children, unlike in adults, a clinician must consider that various strategies and/or neural pathways might operate at different maturational stages. Often the qualitative observations of the child's performance will add critical insight into inhibitory strength or weakness. As a result, behavioural observations and error analysis become particularly useful. For example, repetition errors suggest a failure to successfully self-monitor. Perseverative errors further suggest difficulty inhibiting previous response patterns and shifting to a new response set.

Inhibition can also be evaluated in attentional tests where, for example, subjects have to inhibit distractors in visual search tasks or in divided attention tasks. The Wisconsin Card Sorting Test (WSCT) is commonly used to measure inhibition and executive function (Kolakowsky-Hayner, 2011[27]). This test assesses judgement, reasoning, hypothesis generation, initiation, flexibility and inhibition. In the standard administration of the WCST, four stimulus cards are placed in front of the child. Two sets of 64 response cards become the child's deck. The child must match each consecutive response card to the examiner's stimulus cards according to the examiner's (unstated) principle or rule. However, the principle keeps changing at a designated time. The child must thus discover the principle and adjust the sorting accordingly. The child will propose an answer and be told if it is right or wrong. The criterion is six complete correct sorts or until all 128 cards are attempted.

Discussion

The demand for neuropsychological evaluation varies considerably according to many needs. What are the deficits observed in the context of a neurological pathology? What are the effects of treatment? What is the nature of a learning disability? What are the risks of extreme prematurity or what are the effects of a psychological disorder? What are the child's needs in terms of academic adaptation or management? Is there a neuropsychological explanation to a specific behaviour?

Depending on the origin of the request and the needs, the neuropsychological evaluation takes the form of a screening intervention, a complete check-up with a view to a diagnosis or a more in-depth analysis with a view to a rehabilitation or a pedagogical project. The neuropsychological evaluation must establish not only the weak and strong points of cognitive functioning but also give the best possible account of the difficulties encountered in daily life. For this reason, questionnaires are also often used with parents and teachers to assess socio-adaptive behaviour, emotional and behavioural disorders and quality of life. Indeed, these factors may well influence or interact with the cognitive abilities being assessed. At the same time, more ecological tools using stimulations have proven their relevance in child neuropsychology.

For a long time, the neuropsychological evaluation process was confronted with a lack of tools. Today, the offer in terms of tests is considerable. Nevertheless, this does not guarantee a valid assessment of the child's cognitive profile.

The next section discusses these tests, the interpretation biases that may result and functions that remain impossible to evaluate.

Intra and inter-variability during neuropsychological evaluation and dissociation within a test

The performance of a robot, machine or program is considered stable and invariable over time. However, neuropsychological tests have shown significant variability in the performance of human adults and children. This could occur between different subtests of a battery or within the same subtest over time.

Measuring natural intra-individual variability in neuropsychological tests is thus crucial. It allows the evaluator to avoid systematically interpreting these variations as an improvement or deterioration in performance. This variability could be associated with a recovery or a worsening of the disorders in the case of a brain injury.

Schretlen et al. (2003[28]) investigated the normal range of intra-individual variation in neuropsychological tests in adults. The authors derived 32 z-transformed scores from 15 tests administered to 197 adult participants in a study of normal ageing. The difference between each person's highest and lowest scores was computed to assess his or her maximum discrepancy (MD). The results show that 66% of participants produced MD values that exceeded three SDs. Eliminating each person's highest and lowest test scores decreased their MDs, but 27% of participants still produced MD values exceeding three. Although conducted in adults, this study revealed that marked intra-individual variability is common in normal participants, which of course, is not expected in machines.

Along the same lines, Zabel et al. (2009[29]) examined the test-retest reliability of selected variables from the computerised continuous performance test (CPT). Participants were 39 healthy children aged 6-18 without intellectual impairment. The authors found that test-retest reliability was modest for CPT scores. This study suggests a considerable degree of normal variability in attentional scores over extended test-retest intervals in healthy children. These findings suggest a need for caution when interpreting test score changes in neurologically unstable clinical populations. They also underline that one cannot expect stable performance during a test (especially involving attentional resources) over time in humans unlike with robots and machines.

Taken together, these studies emphasise the difficulty to directly compare human to machine performance in neuropsychological, psychometric or attentional tests. In this way, machine performance could only be compared to a range of human performance.

Difficulty in establishing the origin of failure in a neuropsychological test

Despite the use of standardised tests in child neuropsychology, it remains difficult to precisely establish the cognitive origin of a deficit. For example, a subject can fail a spatial reasoning subtest due to an attentional, visual, spatial or reasoning deficit. In a similar way, the naming task, most often considered as a verbal task (lexical evocation) can be failed because of a visual recognition problem. Indeed, in the WWPSI, the vast majority of the proposed subtests involve visual perception or visuo-spatial analysis. They can therefore be failed due to a perceptual rather than intellectual disorder.

Only the clinical sense of the neuropsychologist can enable him/her to propose a set of tests involving the different processes involved in the failed subtest. By analysing the associations and dissociations between the performance in these different tests, the neuropsychologist will be able to rule on the cognitive process at stake. This point can be illustrated through a simple example of dissociation between performances on similar tasks that differ only in the sensory modality involved. A child may be able to write a word when it is dictated orally but be unable to copy it when it is presented visually. Is there a disorder of perception and/or visual analysis that makes the copying task impossible?

In a similar way, the performance of a battery of perceptual, language and memory tests for several hours in a row, without difficulty, makes it unlikely there is a disabling sustained attentional disorder. Conversely, a decrease in performance as a test is taken argues in favour of a problem sustaining attention over time. Because of the multitude of cognitive processes involved in such complex tasks, a failure on a given task, analysed in isolation, does not say anything about the underlying deficient processes.

In addition, regarding machines, the use of psychometric tasks will require a good understanding of the way the task is executed to compare robot and human performance. Different strategies used by robots and humans, for example, could explain the discrepancy between their performance. When asked to name an object presented visually, humans cannot avoid activating its function, although this is not useful for the naming task. For this reason, a tool will take longer to be named than a flower because humans cannot prevent themselves to activate its function (although this is totally irrelevant to the task). Of course, a machine will never do that. For this reason, it will remain difficult to compare the performance of machines and humans. Machines can thus outperform humans because they have less access to related knowledge.

Intelligence, neuropsychological assessment and learning abilities

Many authors have attempted to correlate IQ with children's academic success or adults' career success. The results in this area are contradictory. No clear link can be established between IQ score, grade level and subsequent success.

High potential children can also have relatively disabling learning and academic difficulties. These disorders can sometimes be so marked that they can mask the diagnosis of intellectual precociousness. Neuropsychologists will often see children with severe academic difficulties for whom a psychometric assessment is requested to eliminate an intellectual disability. Yet these children may obtain an IQ score much higher than the average. These children can sometimes present a heterogeneous profile on psychometric tests with an over-investment of verbal skills to the detriment of visual-spatial skills.

The origin of these disorders is not clearly understood. However, it is often hypothesised that these children need tasks of sufficient complexity to recruit their attention and enable them to complete a task. Thus, in a completely counter-intuitive way, a high IQ may interfere with the performance of overly simple tasks for which the subject does not recruit his or her attention and therefore scores below average. Moreover, this is observed in some subjects with more difficulty recalling a series of numbers in the same order (right-side up) than in reverse (reverse order).

Similarly, there is no strong argument in favour of a correlation between IQ and academic achievement, except perhaps for subjects whose performance is well below their age group. Some studies do show such correlations, but a critique is nonetheless warranted.

IQ tests were originally developed with the constraint of correlating with grade level. Not surprisingly, some subtests such as the arithmetic test correlate well with academic performance. IQ and academic achievement are both products of socio-cultural level and family environment. There is no formal evidence that IQ is exceptionally high among geniuses. The reverse is also true: a low IQ does not necessarily go hand in hand with a total lack of intelligence. This is especially true since failure on IQ tests may not be due to an intellectual deficit. Rather, it could be due to a deficit in the perceptual, motor and language skills required for IQ tests.

In this way, if comparing children to robots through IQ tests, the question arises of what exactly is being compared. IQ tests, which are supposed to test intellectual efficiency, are unlikely to actually measure intellectual ability. Rather, these tests measure a subject's ability to respond to tests that are supposed to measure intelligence (Chokron, 2014[30]). This ability may be subject to a large number of factors far removed from intellectual ability. One can thus expect that a robot, a machine or an algorithm, can answer a question in a more adapted, invariable and systematic way if it has been well programmed for that. Does this mean it will be more intelligent than the human subject?

Sources of performance differences between children and robots

For the past few decades, tests have been over-used both in children for school orientation decisions and for adults in the context of professional recruitment. This is all the more surprising since IQ does not absolutely predict academic or professional success. It is sometimes even the opposite since a non-negligible number of intellectually precocious children can present significant learning disorders.

Furthermore, training in one of the particular subtests of the IQ scales, such as number memory, induces an increase in performance for that subtest. However, it does not improve performance in the other subtests or in the same subtest carried out with other items (letters instead of numbers, for example). Therefore, how could one imagine the IQ score could predict an employee's skills in a specific position that requires the performance of tasks completely different from those required during the test?

These results are confirmed in various studies among different populations such as young street vendors in Brazil (Carraher, Carraher and Schliemann, 1985[31]). In this study, the participants demonstrate remarkable mental arithmetic skills. Yet, when subjected to the arithmetic subtest of psychometric tests, their performance proves to be much lower than expected for their age group. Performance on the same problem proposed in an abstract way during a test and in a concrete way in an ecological situation generally has little correlation.

This type of study underlines the need to consider two aspects of tasks when assessing cognitive abilities. First, there are the processes involved in the task to be performed. Second, there is the ecological character of the task and its proximity to tasks the subject performs in everyday life. This also refers to the distinction between a "psychometric test" and an ecological task. This must be considered when comparing subjects to each other or to machines. Along those lines, the notion of task seems more suitable to evaluate human performance than the one of test.

The comparison of cognitive performance between human subjects and robots also refers to other important distinctions. Human subjects show a variability in their performance over time, as well as between different tasks in a battery that is not expected in robots. Moreover, robots, unlike human subjects, are not expected to be hampered by their affective state, their attentional or motivational level.

Most intellectual efficiency tests evaluate conscious processes, whereas a large part of cognitive processes is unconscious or implicit (Schacter, 1992[32]). Intellectual efficiency would inevitably be different without the good functioning of these conscious and unconscious processes. However, it remains difficult to model these unconscious processes. In a comparison of intellectual performance between robots and humans, it might be interesting to consider how the distinction between conscious and unconscious processes can be operationalised.

Conclusion

Child neuropsychology has made tremendous progress over the past 50 years. As the profession has grown, there has been a better understanding of cognitive and brain development and a growing interest in developmental issues. Tests for children are becoming increasingly specific and standardised rather than simple adaptations of tests for adults. In addition, research in child psychology is improving our knowledge of cognitive processes during development. It also makes it possible to develop new tests that correspond more closely to the dynamics of the child's cognitive and cerebral development. All these elements contribute to the development of child neuropsychology.

Nevertheless, in spite of this considerable progress, much research is needed to compensate for persistent weaknesses in child neuropsychological testing. Indeed, intelligence tests also contribute to these weaknesses. This point is raised not to undermine the role and value of neuropsychological or cognitive assessment. Rather, it seeks to raise awareness of the limitations of testing.

These limitations could be addressed both in clinical practice and research. In clinical practice, they can be addressed through the clinical meaning and judgement, knowledge and experience of the well-trained and responsible neuropsychologist in both normal and abnormal child populations. Meanwhile, in comparing human to machines' intellectual performance, researchers can reflect on how much problem-solving processes may differ in and between humans depending on a multitude of factors that do not affect robots in the same way.

In the case of a direct comparison between the performance of human subjects and of machines on psychometric tests, researchers would need to compare the subjects during tasks aiming at measuring basic functions. These comprise perceptual, attentional or memory tasks. They should involve as little affect as possible and use abstract stimuli to avoid biases in human subjects due to the use of semantic data that are totally inappropriate for the task. In addition, tasks should be performed within a limited period of time (ideally a few minutes for each subtest) to avoid any sustained attentional difficulties in humans that are absent in robots.

Moreover, to avoid any problem of motivation or investment in the human subjects, researchers should choose voluntary subjects. The subjects need to invest themselves totally in the requested task to give the best of themselves. The proposed task may bring into play subjective judgements, involving affects, emotions and choice. It may also consider previous experience or data stored in memory that could be related to the resolution of the presented task. The more this occurs, the more humans will likely outperform machines. Conversely, machines can be expected to outperform humans as tasks become simpler, more automated, more abstract and even more repetitive. Of course, selecting only certain items and not the entire test battery to compare performance of machines to humans will require a reference population of human subjects. It will be impossible to use the test norms obtained when standardising the entire battery.

References

Baddeley, A. (1986), *Working Memory*, Oxford University Press, Oxford. [18]

Baron, I. (2018), *Neuropsychological Evaluation of the Child : Domains, Methods, and Case Studies*, Oxford University Press, Oxford. [24]

Brooks, B., M. Sherman and E. Strauss (2009), "NEPSY-II: A developmental neuropsychological assessment, Second edition", *Child Neuropsychology*, Vol. 16/1, pp. 80-101, http://dx.doi.org/10.1080/09297040903146966. [4]

Carraher, T., D. Carraher and A. Schliemann (1985), "Mathematics in the streets and in schools", *British Journal of Developmental Psychology*, Vol. 3/1, pp. 21-29, https://doi.org/10.1111/j.2044-835X.1985.tb00951.x. [31]

Cavézian, C. et al. (2010), "Assessment of visuo-attentional abilities in young children with or without visual disorder: Toward a systematic screening in the general population", *Research in Developmental Disabilities*, Vol. 31/5, pp. 1102-1108, http://dx.doi.org/10.1016/j.ridd.2010.03.006. [13]

Chokron, S. (2015), "Evaluation of visuo-spatial abilities (EVA) : A simple and rapid battery to screen for CVI in young children", in Lueck, A. and G. Dutton (eds.), *Impairment of Vision due to Disorders of the Visual Brain in Childhood: A Practical Approach*, American Foundation for the Blind Press, Arlington, VA. [14]

Chokron, S. (2014), *Peut-on mesurer l'intelligence?*, Editions le Pommier, Paris. [30]

Chokron, S. (2010), *Pourquoi et comment fait-on attention ?*, Editions le Pommier, Paris. [5]

Cohen, M. (2011), "Children's memory scale", in Kreutzer, J., J. DeLuca and B. Caplan (eds.), *Encyclopedia of Clinical Neuropsychology*, Springer, New York. [11]

Denckla, M. (1989), "Executive function, the overlap zone between attention deficit hyperactivity disorder and learning disabilities", *International Pediatrics*, Vol. 4/2, pp. 155-160. [20]

Eslinger, P. (1996), "Conceptualizing, describing and measuring components of executive function", in Lyon, R. and N. Krasnegor (eds.), *Attention, Memory and Executive Function*, Paul H Brookes, Baltimore, MD. [16]

Gioia, G. et al. (2000), *Behaviour Rating Inventory of Executive Function*, Psychological Assessment Resources, Inc., Odessa, FL. [23]

Hammill, D., N. Pearson and J. Voress (eds.) (2013), *Development Test of Visual Perception – Third Edition (DTVP – 3)*, Pearson, Toronto. [12]

Helland, T. and A. Asbjornsen (2000), "Executive function in dyslexia", *Child Neuropsychology*, Vol. 6/1, pp. 37-48, http://dx.doi.org/10.1076/0929-7049(200003)6:1;1-B;FT037. [6]

Ivnik, R., G. Smith and J. Gerhan (2001), "Understanding the diagnostic capabilities of cognitive tests", *The Clinical Neuropsychologist*, Vol. 15/1, pp. 114-124, http://dx.doi.org/10.1076/clin.15.1.114.1904. [1]

Kaufman, A. and N. Kaufman (1983), *Kaufman Assessment Battery for Children Interpretive Manual*, American Guidance Service, Circle Pines, MN. [2]

Kolakowsky-Hayner, S. (2011), "Wisconsin card sorting test", in Kreutzer, J., J. DeLuca and B. Caplan (eds.), *Encyclopedia of Clinical Neuropsychology*, Springer, New York. [27]

Levin, H. et al. (2001), "Porteus maze performance following traumatic brain injury in children", *Neuropsychology*, Vol. 15/4, pp. 557-567, http://dx.doi.org/10.1037//0894-4105.15.4.55. [25]

Lichtenberger, E. and A. Kaufman (eds.) (2004), *Essentials of WPPSI-III Assessment*, John Wiley & Sons Inc., Hobken, NJ. [10]

Luria, A. (1973), *The Working Brain. An Introduction to Neuropsychology*, Penguin, Harmondsworth, UK. [17]

Manly, T. et al. (1999), *The Test of Everyday Attention for Children Manual*, Thames Valley Test Co Ltd. [8]

Mega, M. and J. Cummings (1994), "Frontal subcortical circuits and neuropsychiatric disorders", *Journal of Neuropsychiatry and Clinical Neuroscience*, Vol. 6/4, pp. 358-370, http://dx.doi.org/10.1176/jnp.6.4.358. [26]

Nigg, J. (2000), "On inhibition/dishinibition in developmental psychopathology: Views from cognitive and personality psychology and a working inhibition taxonomy", *Psychological Bulletin*, Vol. 126/2, pp. 220-246, http://dx.doi.org/10.1037/0033-2909.126.2.220. [22]

Podell, K. (2011), "Purdue pegboard", in Kreutzer, J., J. DeLuca and B. Caplan (eds.), *Encyclopedia of Clinical Neuropsychology*, Springer, New York. [15]

Posner, M. and S. Petersen (1990), "The attention system of the human brain", *Annual Review of Neuroscience*, Vol. 13, pp. 25-42, http://dx.doi.org/10.1146/annurev.ne.13.030190.000325. [7]

Raiford, S. (2018), "The Wechsler Intelligence Scale for Children—Fifth Edition Integrated.", in Flanagan, D. and E. McDonough (eds.), *Contemporary intellectual assessment: Theories, tests, and issues, 4th ed.*, The Guilford Press, New York, NY, US. [3]

Robertson, I. et al. (1995), *Test of Everyday Attention*, Thames Valley Test Company, Ltd., Bury, St. Edmonds, UK. [9]

Schacter, D. (1992), "Implicit knowledge: New perspectives on unconscious processes", *Proceedings of the National Academy of Sciences U.S.A.*, Vol. 89/23, pp. 11113-11117, http://dx.doi.org/10.1073/pnas.89.23.11113. [32]

Schretlen, D. et al. (2003), "Examining the range of normal intraindividual variability in neuropsychological test performance", *Journal of the International Neuropsychology Society*, Vol. 9/6, pp. 864-70, http://dx.doi.org/10.1017/S1355617703960061. [28]

Stuss, D. and D. Benson (1986), *The Frontal Lobes*, Raven Press, New York. [21]

Welsh, M., B. Pennington and D. Groissier (1991), "A normative developmental study of the executive function: A window on prefrontal function in children", *Developmental Neuropsychology*, Vol. 7, pp. 131-149, https://doi.org/10.1080/87565649109540483. [19]

Zabel, T. et al. (2009), "Reliability Concerns in the Repeated Computerized Assessment of Attention in Children", *The Clinical Neuropsychologist*, Vol. 23/7, pp. 1213-1231, http://dx.doi.org/10.1080/13854040902855358. [29]

Annex 4.A. Subtests

Annex Table 4.A.1. NEPSY subtests: Age range and description

Domain	Age range	Description of subtest
Attention and executive functions		
Animal sorting	7-16	Card sorting task assessing concept formation.
Auditory attention and response set	5-16	Auditory selective and sustained attention task involving pointing to stimuli according to consistent and inconsistent examiner cues; it assesses shifting, inhibition and maintaining a new and complex set.
Clocks	7-16	Clock drawing task.
Design fluency	5-12	Non-verbal fluency task during which the child must draw as many unique designs in a given time limit from both structured and unstructured dots arrays.
Inhibition	5-16	Visual task involving three components: naming, inhibition and switching. Naming involves rapid naming of shapes (squares and circles) in the direction of arrows (up or down). Inhibition involves rapid opposite naming of shapes (saying "square" for "circle", saying "circle" for "square") or arrows (saying "up" when they are pointing down or saying "down" when they are pointing up). Switching involves rapidly saying the correct shape or arrow direction if the object is coloured white or saying the opposite shape or arrow direction if the object is coloured black.
Statue	3-6	Response inhibition and motor persistence task by having child maintain a body position while ignoring examiner's sound distractors.
Language		
Body part naming and identification	3-4	Naming task for younger children that involves naming body parts on a structure of a child or on the child's own body, as well as recognition of body part names.
Comprehension of instructions	3-16	Auditory comprehension task that requires the child to point to the correct picture in response to examiner commands with increasing syntactic complexity.
Oromotor sequences	3-12	Oromotor co-ordination task involving repetition of "articulatory sequences" (i.e. tongue twisters).
Phonological processing	3-16	Two part test that requires the child to identify words from word segments and then to create a new word by omitting or substituting word segment or phonemes.
Repetition of nonsense words	5-12	Phonological encoding and decoding task that requires the child to repeat nonsense words within specific semantic and initial letter categories.
Memory and learning		
List memory	7-12	Test of verbal learning and memory that involves learning over five trials, an interference list, immediate recall and delayed recall for a list of 15 words.
List memory delayed	7-12	
Memory for designs	3-16	List of visuo-spatial memory for novel visual designs that involves learning, immediate and delayed recall of the position of designs on two-dimensional grids.
Memory for designs delayed	5-16	
Memory for faces	5-16	Face recall task involving recall of a series of photographs of children's faces.
Memory for faces delayed	5-16	
Memory for names	5-16	Name-face association task that involves repeated exposure trials to a set of cards on which are children's faces; the child is required to learn and recall the name associated with each face.
Memory for names delayed	5-16	
Narrative memory	3-16	Story recall task that involves the examiner reading a story to the child, followed by immediate free recall, immediate cued recall and immediate recognition (for ages three–ten only).
Sentence repetition	3-6	Sentence repetition task where sentences are aurally presented to the child. The child recites the sentences to the examiner immediately after they are presented.
Word list interference	7-16	Recall task that involves two aurally presented series of words that are each repeated after presentation. The child is then asked to recall both series of words in order.

Sensorimotor		
Fingertip tapping	5-16	Tapping task that assesses motor speed and finger dexterity.
Imitating hand positions	3-12	Hand position task that Involves the child copying complex hand/finger position demonstrated by the examiner.
Manual motor sequences	3-12	Rhythmic task involving imitation of hand movement sequences.
Visuo-motor precision	3-12	Paper-and-pencil task involving timed eye-hand co-ordination in which the child is asked to rapidly trace a path on paper without crossing any lines.
Social perception		
Affect recognition	3-16	Affect task where the child is asked to recognise affect (e.g. happy, sad, anger) from photographs of children's faces in various tasks. The task progresses from affect identification to recognition memory for affect.
Theory of mind	3-16	Theory of mind task that involves showing a child pictures and asking questions, reading brief passages about people's experiences and asking questions, comprehension of abstract phrases and matching of facial expression (feeling) to a person's experience.
Visuo-spatial processing		
Arrows	5-16	Judgement of line orientation task involving selection of the arrow(s) from a number of arrows with different orientations that point(s) to the centre of a target.
Block construction	3-16	Block test that requires the child to reproduce three-dimensional block constructions using models or two-dimensional pictures, using unicoloured blocks.
Design copying	3-16	Motor and visual-perceptual test that involves the child copying two-dimensional geometric designs of increasing difficulty on paper.
Geometric puzzles	3-16	Mental rotation, visuo-spatial analysis and attention to detail test in which a child matches two shapes outside of a grid to two of several shapes inside grid.
Picture puzzles	7-16	Test of visual discrimination, spatial localisation and visual scanning that involves identifying the location of smaller parts of a picture on a grid for the larger picture.
Route finding	5-12	Visuo-spatial task involving finding the correct route leading to a target on a map.

5. Understanding and testing social-emotional skills

Filip De Fruyt, Edulab 21, Institute Ayrton Senna, Brazil

This chapter reviews the background and conceptualisation of social-emotional skills and how these are related to consequential outcomes in education and the labour market. It discusses taxonomies on how to structure and operationalise these skills, converging on an overarching model representing five broad groups. In addition, it explores how these skills are connected to various types of work performance. Multiple common methods to assess social-emotional skills are reviewed. These comprise Likert-based and ipsative self- and observer reports, more objective tests and situational judgement tests, as well as behavioural residue indicators.

Introduction

In recent years, the assessment and learning of social-emotional skills in education and policy making have garnered more attention (Kankaraš, 2017[1]; Kankaraš and Suarez-Alvarez, 2019[2]). Language or math achievement have long been traditional indicators of scholastic performance. Today, in addition to those skills, social-emotional skills are considered as both means and end products of education processes.

Learning and training social-emotional skills became an explicit part of educational curricula for several reasons. Social-emotional skills are assumed to (in)directly affect consequential outcomes in the short and long term. These include outcomes such as employability and work performance, health and longevity but also happiness, interpersonal relatedness and civic citizenship (Heckman, Stixrud and Urzua, 2006[3]; John and De Fruyt, 2015[4]; Holbein, 2017[5]; Taylor et al., 2017[6]). In addition, there is also evidence that social-emotional skills facilitate learning processes at school and contribute to academic achievement (Durlak et al., 2011[7]; Sklad et al., 2012[8]; Taylor et al., 2017[6]; Corcoran et al., 2018[9]).

This chapter reviews the background and conceptualisation of social-emotional skills and how these are related to consequential outcomes in education and the labour market. It discusses taxonomies on how to structure and operationalise these skills, converging on an overarching model representing five broad groups. In addition, it explores how these skills are connected to various types of work performance. Multiple common methods to assess social-emotional skills are reviewed. These comprise Likert-based and ipsative self- and observer reports, more objective tests and situational judgement tests, as well as behavioural residue indicators.

Defining social-emotional skills and their conceptual space

There are probably as many definitions of social-emotional skills as there are models structuring these skills. Literature identified social-emotional skills as three types of individual capacities (John and De Fruyt, 2015[4]; Scheerens, van der Werf and de Boer, 2020[10]; John and De Fruyt, 2015[4]). First, they manifest in consistent patterns of thoughts, feelings and behaviours. Second, they can be developed through formal and informal learning experiences. Third, they influence important socio-economic outcomes throughout the individual's life. Such a definition is comprehensive enough to accommodate a wide range of skills, while highlighting their malleability and consequential effects for individuals and society.

The term social-emotional skills is to be preferred over "non-cognitive skills" because many of the skills subsumed under the previous definition also require specific cognitive abilities (De Fruyt, Wille and John, 2015[11]). Collaborative problem solving, for example, entails both cognitive and non-cognitive abilities.

Likewise, the term social-emotional skills is desired over "soft skills". It is unclear why skills that are sometimes hard to learn are called "soft". A better term often used in this debate is "transferable skills". This underscores the idea that these skills are transportable from one context to another, and hence contribute to the individual's adaptation across different life challenges.

In the labour market, organisations usually describe jobs and job vacancies in terms of "competences". Hoekstra and Van Sluijs (2003[12]) define a competence as: "the ability to perform a particular type of task effectively or respond appropriately to a particular type of problem." They conceptualise competences as the result of the interaction between an "expertise" in interaction with a "behavioural repertoire".

In this interaction between expertise and behavioural repertoire, the expertise component is highly (but not exclusively) determined by cognitive ability factors. Conversely, an individual's behavioural repertoire is chiefly determined by personality traits. The observed competence levels, however, are not merely a product of someone's expertise and behavioural repertoire: attention and emotion fluctuations may interfere and impact on the finally manifested competence level.

Understanding socio-emotional skills also requires distinguishing between "how one is behaving typically" (also called a trait approach) relative to "how well one can behave" (or solve a problem). The latter is often represented as a maximal performance kind of construct such as mental abilities.

The distinction between "how well one can behave'" and "how one behaves typically" can be made conceptually. However, it is much harder to distinguish at an operational level. For example, the social-emotional skills measure known as SENNA (Primi et al., 2016[13]) has both self-efficacy ("how well one can behave") and identity ("how one behaves typically") items to assess social-emotional skills.

In sum, the terms social-emotional skills and transferrable skills are to be preferred over "soft" or "non-cognitive" skills. These concepts are synonymous with the term "competences" that is used more frequently on the labour market. Personality traits and cognitive abilities are to be considered as key building blocks of social-emotional skills, competences and transferrable skills.

Towards an integrative taxonomy

There is a broad amalgam of different social-emotional skill taxonomies. Some advocate only a few, while others propose 100 or more skills.

Domains of social-emotional skills

Elias et al. (1997[14]) described six major domains of social and emotional learning. These include recognising and managing emotions, setting and achieving positive goals, appreciating the perspectives of others, establishing and maintaining positive relationships, making responsible decisions and handling interpersonal situations constructively.

Durlak et al. (2011[7]) advocated that social-emotional learning programmes should foster five broad competence sets: self-awareness, self-management, social awareness, relationship skills and responsible decision making.

From a different angle, Saarni (1999[15]; 2011[16]) focused on what a child needs to learn to become an emotionally and socially competent adult. She distinguished eight affect-oriented behavioural, cognitive and regulatory skills that are presumed prerequisites for emotional competence:

- awareness of one's own emotional state
- skills in discerning and understanding the emotions of others
- skill in using the common vocabulary of emotion and expression
- capacity for empathic and sympathetic involvement in others' emotional experiences
- skill in realising that inner emotional states need not correspond to outer expression
- capacity for adaptive coping with aversive or distressing emotions by using self-regulatory strategies that ameliorate the intensity or temporal duration of such emotional states
- awareness that relationships are defined by emotional genuineness of expressive display and reciprocity
- capacity for emotional self-efficacy (i.e. individuals can accept their own emotional experience and view themselves as generally feeling the way they want to feel).

Trilling and Fadel (2009[17]) distinguished among a set of more than 160 different skill terms. These include terms as abnegation and altruism, engagement and enthusiasm, innovation and inquisitiveness, self-discipline and self-control, stability and tranquillity. Conceptually, these authors grouped these terms into the broader themes of what they call the 4Cs: creativity/innovation, critical thinking, communication and collaboration.

Grit: the jingle-jangle fallacy

Although there might be obvious reasons why these frameworks look different, this mixture of models and diversity of vocabularies hampered an integrative and in-depth debate among various stakeholders. The field further suffers from the jingle-jangle fallacy. In this fallacy, similarly named constructs across frameworks refer to different skills (jingle), whereas nearly identical skills are labelled differently (jangle).

The construct of "grit" provides an example of the fallacy. Grit, referring to perseverance and passion for achieving long-term goals, received considerable attention in education over the past decade (Duckworth et al., 2007[18]). Recent behaviour-genetic (Rimfeld et al., 2016[19]) and meta-analytic work (Crede, Tynan and Harms, 2016[20]) showed that grit is similar to the personality trait of "conscientiousness".

Grit has a well-documented history as an important trait for learning within the personality field (Poropat, 2009[21]; Poropat, 2014[22]; Poropat, 2014[23]). Dumfart and Neubauer (2016[24]) showed that no other factors than conscientiousness and intelligence predicted languages and grade point average in eighth graders in Austria. In line with Rimfeld et al. (2016[19]), these authors demonstrated that grit showed no incremental validity beyond conscientiousness.

Restructuring the field of socio-emotional skills

The field of socio-emotional skills needs the kind of restructuring that occurred in the personality field. The challenge to bring order in the amalgam of (overlapping) social-emotional skill terms closely resembles the efforts of personality psychologists to structure the hundreds of personality descriptive terms available.

Such terms were finally summarised in the Big Five personality dimensions (John, 1990[25]) as Neuroticism, Extraversion, Openness to experience, Agreeableness and Conscientiousness. Today, personality psychologists agree these five dimensions form the largest common denominator to describe personality differences observable in various age and cultural groups (De Fruyt and Van Leeuwen, 2014[26]; McCrae and Terracciano, 2005[27]).

Advances on the Big Five framework

The availability of the Big Five empirical framework helped solve the discussion on differentially labelling rather similar constructs and examine the overlap among presumably distinct constructs. This breakthrough considerably advanced the personality field leading to improved knowledge on how traits are best assessed and how personality traits develop across life.

Primi et al. (2016[28]) advanced the discussion, demonstrating that items and scales of frequently used measures to evaluate social-emotional skills learning could be easily mapped within the Big Five scheme. A joint factor analysis of seven measures showed that all their items could be easily structured under the umbrella of the five major dimensions of personality. The measures analysed were the Nowicki-Strickland Locus of Control Scale (Nowicki and Strickland, 1973[29]), Rosenberg Self-Esteem Scale (Rosenberg, 1979[30]), Strengths and Difficulties Questionnaire (Goodman, 1997[31]), Big Five Inventory (John and Kentle, 1991[32]), Self-Efficacy Questionnaire for Children (Muris, 2001[33]), Core Self-evaluations (Judge et al., 2003[34]) and the Grit Scale (Duckworth and Quinn, 2009[35]).

Following analysis of the socio-emotional skills literature, John and De Fruyt (2015[4]) grouped social-emotional skills into five broad domains: collaboration, engaging with others, emotion regulation, task performance and open-mindedness. These domains parallel the psycho-social systems described by John and Srivastava (1999[36]). They are also fully in line with the empirical analyses by Primi et al. (2016[28]) of the content covered in instruments frequently used to assess social-emotional skills.

In the John and De Fruyt (2015[4]) framework, the engaging with others and collaboration fields describe social-emotional skills related to vertical (hierarchical) and horizontal (attaching/bonding) interactions among persons. Conversely, the emotion regulation field groups those skills that help individuals to deal with anxieties and uncertainties, recover from setbacks and control impulses. Both the interpersonal and the emotion regulation fields have been intensively studied in psychology. Finally, open-mindedness and task performance skills are crucial for adaptation and learning, grouping social-emotional skills related to exploration and exploitation, respectively, of the world around us.

These broad domains group more specific skills that are described in Figure 5.1. For example, a domain like collaboration encompasses skills like compassion, respect, trust and harmonious relationship building. Conversely, open-mindedness includes curiosity, creativity, aesthetic sensitivity, appreciating diversity, self-awareness, and autonomy and independence. The domains of the social-emotional skill framework map directly onto the Big Five framework as used by personality psychologists, i.e. collaboration maps to agreeableness, engaging with others maps to extraversion, emotion regulation maps to neuroticism, task performance maps to conscientiousness, and open-mindedness maps to openness to experience.

The Big Five framework has been successfully applied, along these lines, to structure competence models within the human resources field (De Fruyt et al., 2006[37]) or to classify the numerous social-emotional skills listed in the 21st century educational literature (John and De Fruyt, 2015[4]). The model is further helpful to structure needs and requirements of different jobs on the labour market, and connects well with the job descriptions included in Occupational Network (O*NET).

*Combining the Big Five framework with O*NET*

The O*NET database has turned out to be indispensable for practitioners and researchers to connect the worlds of education and the labour market. O*NET contains a detailed and updated set of job descriptions and requirements. This provides an overview of social-emotional skill demand at the level of the labour market.

The O*NET skills and abilities' sections, together with the work styles and values' descriptions for job titles, easily translate into the social-emotional skills subsumed under the Big Five framework. Work styles, for example, are a more "neutral term" in O*NET to refer to personality traits. Conversely, skills and abilities are manifestations of either personality traits or specific cognitive abilities or the interaction between them (Hoekstra and Van Sluijs, 2003[12]).

Relying on O*NET, McCloy et al. (2017[38]) recently identified those characteristics that are critical for effective performance in a broad range of occupations in 79% of O*NET occupations. This resembles 81% of the total US workforce in 2012 and 82% of the projected workforce in 2022. The top 20 skills of their analysis exclusively contained skills from each of the five domains of the social-emotional skills model proposed by John and De Fruyt (2015[4]), supplemented with problem solving. This type of knowledge makes it possible to compute an index of "work readiness" for each individual student relying on his/her scores on this identified skill set.

The proposed model claims to be fairly comprehensive, accommodating most social-emotional skills listed in the literature. However, the framework is probably less suited to classify those skills with a stronger cognitive component. Such cognitive skills would include critical thinking, metacognition, complex problem solving, or surface/deep-level information processing, for example. The more social-cognitive of the skills certainly have associations with the key dimensions of the proposed social-emotional skill model. However, they are also linked to models of cognitive ability (Bartram, 2005[39]).

From personality traits and cognitive abilities to interests and values

Traits and cognitive abilities form the cornerstones of social-emotional skills. However, constructs of other individual differences are important to understand, assess and predict an individual's behaviour and performance at work. Contemporary psychology on individual differences converges on a well-studied set of four groups of constructs: personality traits, cognitive abilities, interests and values.

There are well-researched models for each of these constructs. For personality traits, there are the Big Five/HEXACO models (Goldberg, 1993[40]; Ashton, Lee and Goldberg, 2004[41]). For cognitive abilities, there is the Cattell-Horn-Carroll (CHC) model (McGrew, 2009[42]). For (vocational) interests, there is Holland's RIASEC model (Holland, 1997[43]). Finally, for values, there is the Schwartz value model (Schwartz, 1994[44]).

Vocational interests

Vocational interests can be defined as relatively stable individual differences that affect (choice and performance) behaviour through individuals' preferences for particular work activities and environments (Wille and De Fruyt, 2019[45]). Holland's (1997[43]) vocational interest model describes people's interests and preferences in terms of their resemblance with six core interest domains, i.e. realistic, investigative, artistic, social, enterprising and conventional, that are structured in a hexagonal space.

This resemblance is often expressed in a RIASEC letter code, e.g. SEAICR, with the first letter resembling the person's prime interest field, the second letter her/his second interest theme, etc. The model can be further used to describe and structure environments (e.g. educational majors or jobs), enabling the computation of a fit index between a person's interests and his/her (new) environment. O*NET provides RIASEC letter codes for a broad range of jobs.

Cognitive abilities

Several models have been proposed over the past 100 years to structure cognitive abilities. CHC (McGrew, 2009[42]) is one of the most comprehensive for research and diagnostic specificity. To represent the positive manifold (Spearman, 1927[46]) among various intelligence measures, the model proposes a hierarchy of cognitive abilities. A broad general factor is on top, with more specific types of intelligence structured underneath (www.themindhub.com).

Personal values

Personal values can be defined as: "broad beliefs concerning desirable, trans-situational goals that serve as guiding principles in the individual's life" (Vecchione et al., 2020[47]). The Schwartz (1994[44]) model on values provides a cross-cultural and comprehensive account of people's values, distinguishing among ten core values that people may identify with to various extents.

The values are structured in a circumplex model, in clockwise order from the top: universalism, benevolence, conformity, tradition, security, power, achievement, hedonism, stimulation and self-direction. The quadrants are labelled clockwise from the top as self-transcending, conservation, self-enhancement and openness to change. The model is well-supported and used in large-scale cross-cultural research (e.g. European Social Surveys).

Interplay between the four core groups

The pairwise relationships between these four groups of core constructs have been extensively described, often meta-analytically. They represent distinct, though somewhat related, constructs. Personality and cognitive abilities, for example, correlate poorly. However, interest dimensions are to some extent associated with personality traits (Barrick, Mount and Gupta, 2003[48]) and cognitive abilities (Passler, Beinicke and Hell, 2015[49]). Meanwhile, personality traits are further associated with values (Parks-Leduc, Feldman and Bardi, 2015[50]).

Such background information is key to understanding the complex interplay of ingredients of specific social-emotional skills. For example, a social-emotional skill like "collaboration" requires specific interpersonal personality traits at work *and* verbal abilities. However, it will also be determined and embedded in a context related to the interests (e.g. the field one is working in) and the values that the individual embraces.

Against this background of (small to moderate) interrelationships, a good conceptual framing and understanding is needed of how these constructs hang together and affect work(-related) behaviour. From this perspective, "personality traits" and "abilities" are arguably key ingredients to explain *how* people will perform (in their jobs). Conversely, interests and values are predictors of the field (*what and where*) in which people will use and apply their social-emotional skills. For example, two persons can be alike in terms of their personality traits and abilities. Yet depending on their field of work (different interests and values; e.g. an artist vs. a cardiovascular surgeon), their salaries and options on the labour market may be different.

Understanding work behaviour and performance

Social-emotional skills are considered pivotal today to understand and predict work performance. In other words, they relate to attainment of desired outcomes but also avoidance of unwanted outcomes. Work performance is understood in terms of five main types of performance: task, contextual, adaptive, learning and counterproductive. Task performance (quantity and quality) basically refers to the tasks explicitly listed in a job description; contextual performance refers to how well a person gets along with others and contributes to the team; adaptive performance refers to how well a person deals with change and insecurity; learning performance refers to what and how people learn, what they do to strengthen employability; and counterproductive performance/derailment is an undesirable type of performance.

Social-emotional skills and performance

Social-emotional skills are differentially related to these different types of performances as described below.

- Task performance

Task performance (Renn and Fedor, 2001[51]) is overall predicted by cognitive ability (Salgado et al., 2003[52]) and conscientiousness (Salgado, 1997[53]; Barrick, Mount and Judge, 2001[54]). However, other personality traits may be predictive as well depending on the nature of the job. In sales, for example, interpersonal traits are also important.

- Contextual performance

Contextual performance (Van Scotter and Motowidlo, 1996[55]) is more related to the interpersonal factors of extraversion (explaining the frequency of social interaction) and agreeableness (describing the quality of interpersonal behaviour).

- Adaptive performance

Adaptive performance (Pulakos et al., 2020[56]) relates to flexibility and openness, and how easily change triggers anxiety in the individual. This is related, in turn, to the dimensions of openness to experience and emotional stability (De Fruyt and Rolland, 2013[57]), respectively.

- Learning performance

For an understanding of learning performance, the combination of openness to experience and conscientiousness is critical. The so-called learning circumplex (De Fruyt et al., 2008[58]) is supplemented with cognitive ability (Salgado et al., 2003[52]).

Predicting and understanding performance

Various types of counterproductive work behaviour/derailment are predicted by low agreeableness and conscientiousness (Salgado, 2002[59]), honesty/humility (Pletzer et al., 2019[60]) eventually combined with low emotional stability (De Fruyt et al., 2009[61]), especially for the prediction of derailment. A direct translation of these types of work performance into the skill language of the OECD framework can be found in Table 5.1.

Predicting and understanding different types of work performance involves choosing the right social-emotional skill constructs, as well as analysing the nature of the kind of performance that one is interested in. In some cases, performance can be defined in terms of the best possible solution for a problem, referring to the maximal potential of the individual's skills to accomplish a goal. Other types of performances refer to more daily and habitual expressions. These might include being friendly and respectful, or how one generally deals with stressors and everyday hassles.

From a select-out perspective, one wants to avoid derailment and counterproductive work behaviour in organisations. The prevalence of some of these behaviours might be infrequent, but a single malevolent act (e.g. a pilot making a mistake) can have a devastating impact. Consequently, in addition to selecting the appropriate social-emotional skill constructs, one also has to adequately model the relationships between such constructs and various performance indicators in structural relationship models and equations.

Most contemporary psychology and econometric models are primarily focused on understanding (work) performance differences *between* individuals. This was mainly driven by a focus in selection psychology to select those individuals among job applicants that will best perform in a job. Work performance in this context was mainly understood as a kind of static process where an individual provides a consistent level of performance.

Today, job performance is also considered a dynamic construct, with people demonstrating "performance fluctuation" around a mean.

Debusscher, Hofmans and De Fruyt (2016[62]) examine state neuroticism and task performance using an experience sampling design. They demonstrate that 60.9% of the variance of momentary task performance and 66.7% of the variance of state emotional stability was within-person. Hence, how well people (maximally) master a particular skill may differ from how people behave typically with variation across the day.

Debusscher, Hofmans and De Fruyt (2016[62]) convincingly demonstrate that people do not always perform to their maximal mastery level but rather show fluctuations. Others observe that roughly half of the variance in job performance is *within* the person (Debusscher, Hofmans and De Fruyt, 2014[63]); (Sosnowska et al., 2020[64]). However, it requires specific assessment tools and methodology to capture such variance (Lang et al., 2019[65]; Lievens et al., 2018[66]). AI has great potential to complement and eventually replace ambulatory assessments with ecologically valid information that can be electronically assessed and estimated and updated by algorithms.

Assessing social-emotional skills

This section reviews how social-emotional skills are assessed. It identifies several caveats in terminology, reflects on rating scales, and proposes situational judgement tests as an alternative assessment method. It explores more objective assessments to supplement self-report and rating-scale approaches. Finally, it examines personal living spaces and digital footprints as a source of information on personality traits, cognitive abilities, interests and values.

Correct use of terminology

Tests vs. assessments

The term "test" should be avoided (or correctly used) when it comes to the assessment of social-emotional skills. A "test" refers to examining someone's maximal knowledge or capacity to solve a problem or perform. In a test, one typically assesses a person's specific knowledge, cognitive abilities, memory, attention span or physical fitness. Often, tests include a set of items or exercises with an increasing level of difficulty (e.g. a reasoning task) or effort (e.g. physical tasks), with cognitive tasks usually having a right or wrong answer.

An exam for obtaining a driver's licence, for example, first tests the knowledge of traffic rules, with "pass or fail" as the result. In addition, an examiner wants to know how this person performs typically in day-to-day traffic situations. Thus, the (broader) term "assessment" is preferred to "test", which refers to the examination of these more typical modes of behaving and acting. In assessments, some of the behaviours can be considered "right or wrong". However, they may also point to a continuum of behaviours that are more subtle and gradual in nature (e.g. "anticipatory driving" or "respecting others"). The broader term "assessment" better reflects this complexity.

Constructs vs. assessment methods

Constructs such as social-emotional skills should not be equated with the methods used to assess them. Constructs can be assessed using various methods that may either supplement or complement each other. A person's collaborative skills, for example, can be described via self- or peer reports. However, they can also be assessed in a group exercise rated by independent assessors or via a situational judgement test.

Prediction in psychological measurement further relies on the principles of aggregation and triangulation. Aggregation measures across multiple indicators of a construct, while triangulation measures across different methods to assess this construct. Social-emotional skills, for example, are assessed using multiple items covering the bandwidth of a construct.

One also tries to explain incremental validity by picking up variance in a construct using alternative assessment methods. Peer ratings, for example, can complement self-ratings on assertiveness. Meanwhile, scores on a situational judgement test can supplement self- and peer reports. Schmidt, Oh and Shafer (2016[67]) have conducted a meta-analytic summary of the validity of methods commonly used to predict work and training performance.

Rating scales

Social-emotional skills are most frequently assessed using self-reports on an item set covering the bandwidth of a set of skills. Homogeneous groups of items are presented with Likert scales and anchors referring to either descriptive labels or frequency indicators. A descriptive label might be a five-point scale ranging from "not characteristic at all", "barely characteristic", "more or less characteristic", "characteristic" to "very characteristic". Frequency indicators might include "not at all", "once a month", "a few times a month", "once a week", etc. An example of such Likert-based assessment for skills enclosed in the OECD model can be found in Kankaraš, Feron and Renbarger (2019[68]) (Table 3.5).

In accordance with the aggregation principle, multiple items referring to various nuances of emotion regulation are administered to assess a person's standing on this skill. Often, the item set includes both positively and negatively keyed items (e.g. "I can control my emotions well" versus "My emotions overwhelm me completely"). This allows to correct for acquiescence bias (a response tendency to say "yes" to items) (Primi et al., 2020[69]).

The psychometric characteristics of the scale and demonstration of measurement equivalence are both critically important. The former measures reliability, consistency and structural characteristics, while the latter allows comparisons across groups (e.g. across gender, age, culture). Such Likert-based assessments are sometimes preceded by an anchoring-vignette to be in a position to correct raw scale scores for group-reference bias (Primi et al., 2016[28]; Primi et al., 2016[13]).

"Observer reports" of an individual's social-emotional skills form an alternative and complementary approach to self-ratings. The addition of this type of information adds predictive power to understand various outcomes (Connelly and Ones, 2010[70]). Similar psychometric requirements account for observer reports as for self-ratings. Preferably, the observer is well acquainted with the target individual. For example, the observer could be a parent or teacher rating the social-emotional skills of a child, partners rating each other or a supervisor rating a collaborator or vice versa. However, zero-acquaintance reports also have some small validity. Observer or peer reports are most useful to assess overtly observable behaviours (e.g. assertiveness). They are less suited to assess more internalising forms of behaviour such as emotion regulation skills.

Ipsative assessments of social-emotional skills have been introduced as an alternative for Likert scales. This is because the latter may be subject to different rater biases, including self-presentation bias (i.e. how people tend to portray a more socially desirable picture of themselves). In an ipsative assessment, the person must select from different sets of three to four items. The items in a set, which refer to different social-emotional skills, are illustrated in the triplet of items that follows:

I can calm down myself easily.
I get distracted quickly.
I can easily keep my promises.

The three items are indicators of emotion regulation, concentration and sense of responsibility, respectively. The subject indicates the item that is most like her/him and least like her/him. Although ipsative assessment creates interdependencies among the assessed constructs, more recent psychometric techniques have been developed to deal with this problem (Brown and Maydeu-Olivares, 2013[71]). As a result, ipsative item administration can be used to assess differences between individuals. More research, however, is needed on how to design optimal groups of items, including both positive and negative indicators of a construct so the model can be identified properly.

Situational judgement tests

Situational judgement tests (SJTs) are an alternative method to assess personality or social-emotional skills. In an SJT, persons are presented items describing a short situation, the so-called stem, together with a set of possible reactions. The stem can be a narrative description, but more recent approaches also use videos, animations or avatars to present the situation. Assessees are required to select from the set of responses the one that bests represents how they will behave in that situation, or are requested to rank those responses.

The responses are indicators of specific social-emotional skills. Subject matter experts *a priori* align responses to constructs, or their allocation is determined in a separate study and derived empirically. Assessees are presented several of such SJT items and their position on constructs is computed across their choices or rankings of responses across the different SJTs.

An example of an SJT stem assessing an element of leadership behaviour could be as follows: "You (as the team leader) have scheduled an online meeting with members of your team to discuss an adjustment in the work plan. You see in the chat function of the communication tool that one team member comments disrespectfully about someone from a different department (not participating in the meeting). What do you do?" The assessee can select from:

a) You react immediately and ask to clarify his point.

b) You react during the meeting, and tell him we do not communicate within the organisation with this tone.
c) You wait until the meeting is over and approach him individually on line to tell him that this is not the tone we use to communicate within the organisation.
d) You deny the comment in the chat because you consider it irrelevant for solving the work-planning problem.

SJTs have become popular in assessment and selection psychology due to their face validity and favourable reactions from assessees (because the situations have a realistic and attractive appeal on candidates applying for jobs). There is meta-analytic evidence supporting SJTs' predictive validity, beyond self-ratings on personality inventories (McDaniel et al., 2007[72]).

However, SJTs have large developmental costs. They must often be developed to reflect the kind of (work) situations and behaviours that one wants to understand and predict. In a recent review, Lievens (2017[73]) suggested that the SJT format provides interesting possibilities for assessing both between and within individual differences. These possibilities emerge because the various situational descriptions in an SJT item set provide opportunities to study trait or skill variability (flexibility) across different situations.

The use and especially the design of SJTs, however, was recently questioned. Several studies demonstrated the situation descriptions were unnecessary or only contributed to increased predictive validity for specific criteria (Krumm et al., 2015[74]; Schäpers et al., 2020[75]). For example, Schäpers et al. (2020[75]) demonstrated that SJT validity to predict in-role performance and organisational citizenship behaviour was not substantively affected when removing the stem. Conversely, they found it did have an impact for predicting interpersonal adaptability and efficacy for teamwork. They further showed the impact of dropping the stems on applicant reactions was negligible. Overall, these studies question the heart of the SJT approach. They suggest that SJTs are less situation-dependent than previously thought.

More objective assessments

Duckworth and Yeager (2015[76]) convincingly argued to supplement self-report and rating-scale approaches with more objective forms of assessing socio-emotional skills. They said rating scales may be less effective, especially for summative educational assessment, when stakes of evaluation are high. Both positive and negative self-presentation styles may distort self-descriptions when financial consequences are attached to the outcome of the evaluation. Schools may be rewarded for good but also for poor performance (getting extra funds to remediate). Therefore, intentional distortion of self-descriptions is possible in both directions.

Other threats of the validity of self-reports are teaching-to-the-test bias or increasing awareness of a skill that was the target of an intervention. For example, after acquiring a better understanding of empathy through an intervention, students may adjust (towards the lower end) their self-described skill mastery level. In other words, they may describe themselves as lower on empathy skills after an intervention to improve empathy. Taken at face value, this would lead to the paradoxical conclusion that empathy skill-level decreased due to the intervention.

Several studies have attempted to develop more objective measures to assess specific social-emotional skills. Santacreu, Rubio and Hernandez (2006[77]), for example, described three computerised tests that examine risk tendency: Roulette, Betting Dice and Crossing the Street. In the Roulette test, people have to bet and can win or lose for obtaining a prize. The Betting Dice test is similar to the Roulette test. In the third test, a figure on the computer screen has to cross a street as quickly as possible without causing accidents with satisfactory levels of reliability and validity. Falk et al. (2015[78]) assessed altruism using a donation task where people could contribute a certain amount of money to a good cause.

The Balloon Analogue Risk Task (BART) is an interesting objective measure to assess risk taking (a specific form of lack of impulse control) (Lejuez et al., 2002[79]) or derivative tasks. In BART, participants are shown 16 balloons on a computer screen, one after another. They are asked to pump the balloon that eventually will burst, and will get a dollar for each (extra) pump.

Balloon burst points, however, are randomised. If a balloon pops before they cash their earnings, they lose everything collected with pumping that single balloon. When a balloon explodes, they are presented a fresh balloon and continue until they finish with all 16. The key dependent variable is the amount of money collected at the end of the experiment, which gives them extra chances for gaining a substantive and attractive prize.

Alternatives to BART are so-called lottery tasks, where people have to imagine they won $100 000 with a lottery. The "bank" approaches them for a financial investment with the following strategy: double the investment in two years with a similar chance that half of the money will be lost. Participants are requested to indicate the amount of money they will invest under this condition: a) nothing (i.e. decline the offer), b) $20 000, c) $40 000, d) $60 000, e) $80 000 or f) the full amount of $100 000.

The BART and lottery tasks are interesting examples because they strongly correlate with self-report measures of risk taking (Dohmen et al., 2011[80]). Grover (2018[81]) reported correlations of .73 and .62 between a self-report measure of seven items (with a Cronbach's alpha of .93) with the number of adjusted pumps and the number of explosions in the BART task, respectively. Financial risk taking in the lottery task had a correlation of .79 with the self-report measure.

The previous examples made clear that more objective forms of assessment usually require more advanced ways of test administration. They also eventually require extra time beyond the easily administered paper-and-pencil self-descriptions. In addition, the number of constructs that can be evaluated more objectively is probably time-bound. Social-emotional skill inventories such as SENNA 2.0 (Primi et al., 2016[13]) assess 17 different skills using 153 items and can be easily administered in 25 to 30 minutes. Performance-based assessment of the same 17 constructs will probably require considerable extra time but also specific infrastructure.

The available objective measures cover only a small part of the proposed social-emotional skill framework. The domains of emotion regulation, engaging with others and openness are poorly covered by objective assessment tools. In addition to availability, the construct validity of objective measures needs additional research. More objective measures typically correlate poorly with social-emotional skill self-reports, and it is unclear why this is the case (Ortner and Proyer, 2015[82]). In addition, objective measures of social-emotional skills are frequently task-based with considerable cognitive and motivational load. This puts into question their validity as pure indicators of social-emotional skills.

Behavioural residue and digital footprint

Personal living spaces and digital footprints are a final source of information on people's standing on social-emotional skill building blocks such as personality traits, cognitive abilities, interests or values. An office or bedroom, for example, could represent personal living space. A digital footprint could be represented by social media activity (Facebook profiles, Twitter activity, Internet searches, cell phone use and tracking information).

Gosling et al. (2002[83]) demonstrated that people's offices and bedrooms contained so-called behavioural residue. Observers can make consensual and reliable ratings of people's personality based on bits and pieces of information in these personal spaces.

More behavioural residue can be found in people's use of online social networking sites and their digital footprint. Back et al. (2010[84]) found that Facebook profiles reflect instances of their owner's personality and not self-idealisation. Ratings of the Facebook pages showed correlation coefficients mounting to .39

and .41 with self- and peer-rated personality for extraversion and openness to experience, respectively, when ratings of the Facebook page were averaged across raters. Correlations were .25 and .24 when single observer ratings were considered. Facebook ratings were not correlated with ideal self-ratings, when corrected for actual personality ratings.

Another study by Park et al. (2015[85]) analysed Facebook status messages and developed algorithms associating language use in Facebook posts with self-descriptions of personality. A subsequent validation study showed that Facebook language-based assessments were correlated on average .38 with self-ratings of personality. Similar accuracy results were obtained from analysing individuals' smartphone behaviours, looking at communication and social behaviours, music consumption, app usage, mobility, overall phone activity and, finally, day- and night-time activity (Stachl et al., 2020[86]).

Specific combinations of these passive variables that could be read from an individual's smartphone were useful to predict openness to experience, conscientiousness and extraversion self-ratings, except for Agreeableness. Conversely, only carefreeness and self-consciousness from the emotional stability domain were significantly related with smartphone behaviour indices.

Table 5.1. Work performance indices and associated social-emotional skills

Type of performance	Social-emotional skills (OECD model)	Specific skills and narrative
Task performance	Cognitive ability (CHC model) Task performance	Taking responsibility for tasks, persisting and steering own performance
Contextual performance	Engaging with others Collaboration	Interacting and connecting with others, working in teams; trusting others and being trustworthy; acting respectfully
Adaptive performance	Open-mindedness Emotion regulation	Being open and flexible, welcoming new ideas and people; controlling negative and defensive emotions and reactions
Learning performance	Cognitive ability (CHC model) Open-mindedness Task performance	Having a curious mindset and acting upon that; making an effort, persisting when it gets tough and being ambitious to learn
(prevention of) Derailment	Collaboration and Task performance Emotion regulation	Controlling selfish tendencies and negative emotions that may harm others; following rules and honouring commitments

Figure 5.1. OECD Framework on social-emotional skills

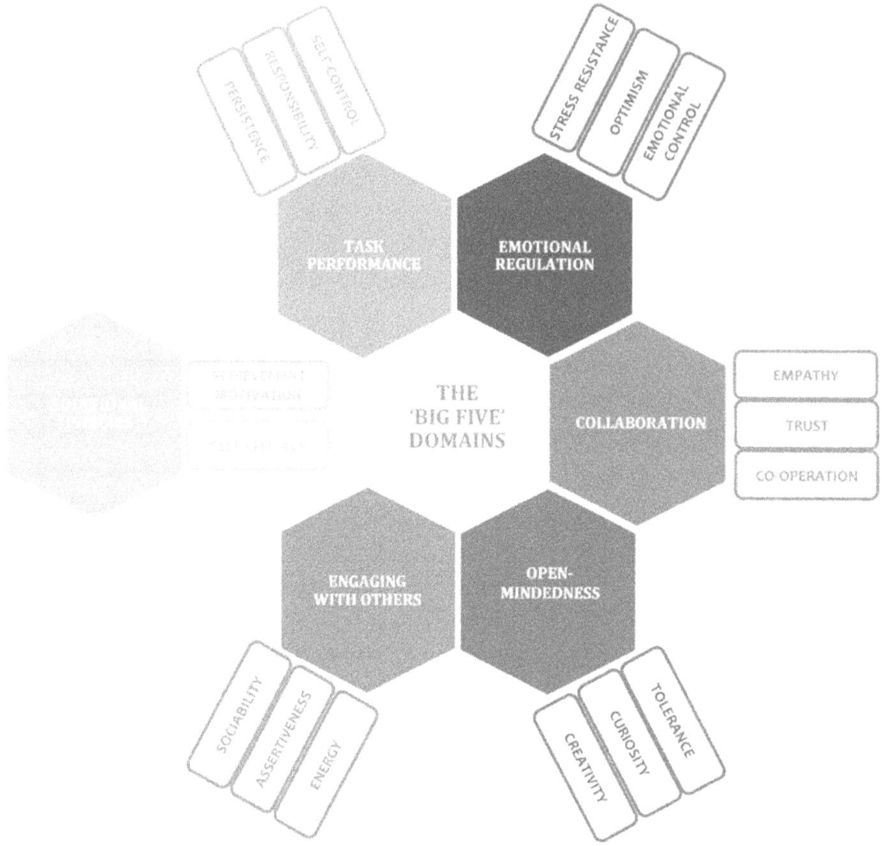

Source: Kankaraš and Suarez-Alvarez, (2019[68]).

References

Ashton, M., K. Lee and L. Goldberg (2004), "A hierarchical analysis of 1,710 English personality-descriptive adjectives", *Journal of Personality and Social Psychology*, Vol. 87/5, pp. 707-721, http://dx.doi.org/10.1037/0022-3514.87.5.707. [41]

Back, M. et al. (2010), "Facebook profiles reflect actual personality, not self-idealization", *Psychological Science*, Vol. 21/3, pp. 372-374, https://doi.org/10.1177/0956797609360756. [84]

Barrick, M., M. Mount and R. Gupta (2003), "Meta-analysis of the relationship between the five-factor model of personality and Holland's occupational types", *Personnel Psychology*, Vol. 56/1, pp. 45-74, http://dx.doi.org/10.1111/j.1744-6570.2003.tb00143.x. [48]

Barrick, M., M. Mount and T. Judge (2001), "Personality and performance at the beginning of the new millennium: What do we know and where do we go next?", *International Journal of Selection and Assessment*, Vol. 9/1-2, pp. 9-30, https://doi.org/10.1111/1468-2389.00160. [54]

Bartram, D. (2005), "The great eight competencies: A criterion-centric approach to validation", *Journal of Applied Psychology*, Vol. 90/6, pp. 1185-1203, http://dx.doi.org/ttps://doi.org/10.1037/0021-9010.90.6.1185. [39]

Brown, A. and A. Maydeu-Olivares (2013), "How IRT can solve problems of ipsative data in forced-choice questionnaires", *Psychological Methods*, Vol. 18/1, pp. 36-52, http://dx.doi.org/10.1037/a0030641. [71]

Connelly, B. and D. Ones (2010), "An other perspective on personality: Meta-analytic integration of observers' accuracy and predictive validity", *Psychological Bulletin*, Vol. 136/6, pp. 1092-1122, http://dx.doi.org/10.1037/a0021212. [70]

Corcoran, R. et al. (2018), "Effective universal school-based social and emotional learning programs for improving academic achievement: A systematic review and meta-analysis of 50 years of research", *Educational Research Review*, Vol. 25, http://dx.doi.org/10.1016/j.edurev.2017.12.001. [9]

Crede, M., M. Tynan and P. Harms (2016), "Much ado about grit: A meta-analytic synthesis of the grit literature", *Journal of Personality and Social Psychology*, Vol. 113/3, pp. 492-511, https://doi.org/10.1037/pspp0000102. [20]

De Fruyt, F. et al. (2006), "Police interview competencies: assessment and associated traits", *European Journal of Personality*, Vol. 20/7, pp. 567-584, http://dx.doi.org/10.1002/per.594. [37]

De Fruyt, F. et al. (2009), "Assessing personality at risk in personnel selection and development", *European Journal of Personality*, Vol. 23/1, pp. 51-69, https://doi.org/10.1002/per.703. [61]

De Fruyt, F. and J. Rolland (eds.) (2013), *Personality for Professionals Inventory: PfPI Handleiding*, Pearson, Amsterdam. [57]

De Fruyt, F. and K. Van Leeuwen (2014), "Advancements in the field of personality development", *Journal of Adolescence*, Vol. 37/5, pp. 763-769, https://doi.org/10.1016/j.adolescence.2014.04.009. [26]

De Fruyt, F. et al. (2008), "Sex differences in school performance as a function of conscientiousness, imagination and the mediating role of problem behaviour", *European Journal of Personality*, Vol. 22/3, pp. 167-184, https://doi.org/10.1002/per.675. [58]

De Fruyt, F., B. Wille and O. John (2015), "Employability in the 21st century: Complex (interactive) problem solving and other essential skills", *Industrial and Organizational Psychology-Perspectives on Science and Practice*, Vol. 8/2, pp. 276-U189, http://dx.doi.org/10.1017/iop.2015.33. [11]

Debusscher, J., J. Hofmans and F. De Fruyt (2016), "From state neuroticism to momentary task performance: A person x situation approach", *European Journal of Work and Organizational Psychology*, Vol. 25/1, pp. 89-104, https://doi.org/10.1080/1359432X.2014.983085. [62]

Debusscher, J., J. Hofmans and F. De Fruyt (2014), "The curvilinear relationship between state neuroticism and momentary task performance", *Plos One*, Vol. 9/9, http://dx.doi.org/10.1371/journal.pone.0106989. [63]

Dohmen, T. et al. (2011), "Individual risk attitudes: Measurement, determinants, and behavioural consequences", *Journal of the European Economic Association*, Vol. 93/3, pp. 522-550, https://doi.org/10.1111/j.1542-4774.2011.01015.x. [80]

Duckworth, A. et al. (2007), "Grit: Perseverance and passion for long-term goals", *Journal of Personality and Social Psychology*, Vol. 92/6, pp. 1087-1101, http://dx.doi.org/10.1037/0022-3514.92.6.1087. [18]

Duckworth, A. and P. Quinn (2009), "Development and validation of the short grit scale (Grit-S)", *Journal of Personality Assessment*, Vol. 91/2, pp. 166-174, https://doi.org/10.1080/00223890802634290. [35]

Duckworth, A. and D. Yeager (2015), "Measurement matters: Assessing personal qualities other than cognitive ability for educational purposes", *Educational Researcher*, Vol. 44/4, pp. 237-251, https://doi.org/10.3102/0013189X15584327. [76]

Dumfart, B. and A. Neubauer (2016), "Conscientiousness Is the most powerful noncognitive predictor of school achievement in adolescents", *Journal of Individual Differences*, Vol. 37/1, pp. 8-15, https://doi.org/10.1027/1614-0001/a000182. [24]

Durlak, J. et al. (2011), "The impact of enhancing students' social and emotional learning: A meta-analysis of school-based universal interventions", *Child Development*, Vol. 82/1, pp. 405-432, https://doi.org/10.1111/j.1467-8624.2010.01564.x. [7]

Elias, M. et al. (eds.) (1997), *Promoting Social and Emotional Learning: Guidelines for Educators*, Association for Supervision and Curriculum Development, Alexandria, VA. [14]

Falk, A. et al. (2015), *The Nature and Predictive Power of Preferences: Global Evidence*, https://EconPapers.repec.org/RePEc:iza:izadps:dp9504. [78]

Goldberg, L. (1993), "The structure of phenotypic personality-traits", *American Psychologist*, Vol. 48/1, pp. 26-34, http://dx.doi.org/doi: 10.1037//0003-066x.48.1.26. [40]

Goodman, R. (1997), "The strengths and difficulties questionnaire: A research note", *Journal of Child Psychology and Psychiatry*, Vol. 38/5, pp. 581-586, http://dx.doi.org/10.1111/j.1469-7610.1997.tb01545.x. [31]

Gosling, S. et al. (2002), "A room with a cue: Personality judgments based on offices and bedrooms", *Journal of Personality and Social Psychology*, Vol. 82/3, pp. 379-398, https://doi.org/10.1037/0022-3514.82.3.379. [83]

Grover, H. (2018), *The Upside to the Dark Side: An Empirical Investigation into the Moderators and Mediators of the Dark Triad and Work Related Outcomes*, University College London, https://discovery.ucl.ac.uk/id/eprint/10063615/. [81]

Heckman, J., J. Stixrud and S. Urzua (2006), "The effects of cognitive and noncognitive abilities on labor market outcomes and social behaviour", *Journal of Labor Economics*, Vol. 24/3, pp. 411-482, http://dx.doi.org/10.3386/w12006. [3]

Hoekstra, H. and E. Van Sluijs (eds.) (2003), *Managing Competencies: Implementing Human Resource Management*, Royal Van Gorcum, Nijmegen. [12]

Holbein, J. (2017), "Childhood skill development and adult political participation", *American Political Science Review*, Vol. 111/3, pp. 572-583, http://dx.doi.org/10.1017/S0003055417000119. [5]

Holland, J. (ed.) (1997), *Making Vocational Choices: A Theory of Vocational Personalities and Work Environments (3rd. ed.).*, Psychological Assessment Resources, Odessa, FL. [43]

John, O. (1990), "The "Big Five" factor taxonomy: Dimensions of personality in the natural language and in questionnaires.", in *Handbook of personality: Theory and research.*, The Guilford Press, New York, NY, US. [25]

John, O. and F. De Fruyt (2015), *Framework for the Longitudinal Study of Social and Emotional Skills in Cities*, OECD Publishing, Paris. [4]

John, O. and R. Kentle (1991), *The "Big Five" Inventory – Versions 4a and 54*, University of California, Berkeley, Institute of Personality and Social Research, Berkeley, CA. [32]

John, O. and S. Srivastava (1999), *The Big Five Trait taxonomy: History, measurement, and theoretical perspectives.*. [36]

Judge, T. et al. (2003), "The core self-evaluations scale: Development of a measure", *Personnel Psychology*, Vol. 56/2, pp. 303-331, https://doi.org/10.1111/j.1744-6570.2003.tb00152.x. [34]

Kankaraš, M. (2017), "Personality matters: Relevance and assessment of personality characteristics", *OECD Education Working Papers*, No. 157, OECD Publishing, Paris, https://doi.org/10.1787/8a294376-en. [1]

Kankaraš, M., E. Feron and R. Renbarger (2019), "Assessing students' social and emotional skills through triangulation of assessment methods", *OECD Education Working Papers*, No. 208, OECD Publishing, Paris, https://doi.org/10.1787/717ad7f2-en. [68]

Kankaraš, M. and J. Suarez-Alvarez (2019), "Assessment Framework of the OECD Study on Social and Emotional Skills", *OECD Education Working Papers*, No. 2017, OECD Publishing, Paris, https://doi.org/10.1787/5007adef-en. [2]

Krumm, S. et al. (2015), "How "situational" is judgment in situational judgment tests?", *Journal of Applied Psychology*, Vol. 100/2, pp. 399-416, http://dx.doi.org/10.1037/a0037674. [74]

Lang, J. et al. (2019), "Assessing meaningful within-person variability in Likert-scale rated personality descriptions: An IRT tree approach", *Psychological Assessment*, Vol. 31/4, pp. 474-487, http://dx.doi.org/10.1037/pas0000600. [65]

Lejuez, C. et al. (2002), "Evaluation of a behavioral measure of risk taking: The Balloon Analogue Risk Task (BART)", *Journal of Experimental Psychology-Applied*, Vol. 8/2, pp. 75-84, http://dx.doi.org/10.1037//1076-898x.8.2.75. [79]

Lievens, F. (2017), "Assessing personality-situation interplay in personnel selection: Toward more integration into personality research", *European Journal of Personality*, Vol. 31/5, pp. 424-440, https://doi.org/10.1002/per.2111. [73]

Lievens, F. et al. (2018), "The predictive power of people's intraindividual variability across situations: Implementing whole trait theory in assessment", *Journal of Applied Psychology*, Vol. 103/7, pp. 753-771, https://doi.org/10.1037/apl0000280. [66]

McCloy, R. et al. (2017), "Identifying universally critical characteristics of O*NET occupations. A prelude to assess workforce readiness", presented at the SIOP Leading Edge Consortium, Atlanta. [38]

McCrae, R. and A. Terracciano (2005), "Universal features of personality traits from the observer's perspective: Data from 50 cultures", *Journal of Personality and Social Psychology*, Vol. 88/3, pp. 547-561, http://dx.doi.org/10.1037/0022-3514.88.3.547. [27]

McDaniel, M. et al. (2007), "Situational judgment tests, response instructions, and validity: A meta-analysis", *Personnel Psychology*, Vol. 60/1, pp. 63-91, https://doi.org/10.1111/j.1744-6570.2007.00065.x. [72]

McGrew, K. (2009), "CHC theory and the human cognitive abilities project: Standing on the shoulders of the giants of psychometric intelligence research", *Intelligence*, Vol. 37/1, pp. 1-10, https://doi.org/10.1016/j.intell.2008.08.004. [42]

Muris, P. (2001), "A brief questionnaire for measuring self-efficacy in youths", *Journal of Psychopathology and Behavioural Assessment*, Vol. 23, pp. 145-149, https://doi.org/10.1023/A:1010961119608. [33]

Nowicki, S. and B. Strickland (1973), "A locus of control scale for children", *Journal of Consulting and Clinical Psychology*, Vol. 40, pp. 148-154, https://doi.org/10.1037/h0033978. [29]

Ortner, T. and R. Proyer (2015), "Objective personality test", in Ortner, T. and F. Vijve (eds.), *Behaviour-based Assessment in Psychology: Going beyond Self-report in the Personality, Affective, Motivation, and Social Domains*, Hogref. [82]

Park, G. et al. (2015), "Automatic personality assessment through social media language", *Journal of Personality and Social Psychology*, Vol. 108/6, p. 934, http://dx.doi.org/10.1037/pspp0000020. [85]

Parks-Leduc, L., G. Feldman and A. Bardi (2015), "Personality traits and personal values: A meta-analysis", *Personality and Social Psychology Review*, Vol. 19/1, pp. 3-29, http://dx.doi.org/10.1177/1088868314538548. [50]

Passler, K., A. Beinicke and B. Hell (2015), "Interests and intelligence: A meta-analysis", *Intelligence*, Vol. 50/May-June, pp. 30-51, https://doi.org/10.1016/j.intell.2015.02.001. [49]

Pletzer, J. et al. (2019), "A meta-analysis of the relations between personality and workplace deviance: Big Five versus HEXACO", *Journal of Vocational Behaviour*, Vol. 112, pp. 369-383, https://doi.org/10.1016/j.jvb.2019.04.004. [60]

Poropat, A. (2014), "A meta-analysis of adult-rated child personality and academic performance in primary education", *British Journal of Educational Psychology*, Vol. 84/2, pp. 239-252, http://dx.doi.org/10.1111/bjep.12019. [22]

Poropat, A. (2014), "Other-rated personality and academic performance: Evidence and implications", *Learning and Individual Differences*, Vol. 34, pp. 24-32, https://doi.org/10.1016/j.lindif.2014.05.013. [23]

Poropat, A. (2009), "A meta-analysis of the five-factor model of personality and academic performance", *Psychological Bulletin*, Vol. 135/2, pp. 322-338, https://doi.org/10.1037/a0014996. [21]

Primi, R. et al. (2020), "True or false? Keying direction and acquiescence influence the validity of socio-emotional skills items in predicting high school achievement", *International Journal of Testing*, Vol. 20/2, pp. 97-121, https://doi.org/10.1080/15305058.2019.1673398. [69]

Primi, R. et al. (2016), "Development of an inventory assessing social and emotional skills in Brazilian youth", *European Journal of Psychological Assessment*, Vol. 32/1, pp. 5-16, https://doi.org/10.1027/1015-5759/a000343. [13]

Primi, R. et al. (2016), "Anchoring vignettes: Can they make adolescent self-reports of social-emotional skills more reliable, discriminant, and criterion-valid?", *European Journal of Psychological Assessment*, Vol. 32/1, https://doi.org/10.1027/1015-5759/a000336. [28]

Pulakos, E. et al. (2020), "Adaptability in the workplace: Development of a taxonomy of adaptive performance", *Journal of Applied Psychology*, Vol. 85/4, pp. 612-624, http://dx.doi.org/10.1037/0021-9010.85.4.612. [56]

Renn, R. and D. Fedor (2001), "Development and field test of a feedback seeking, self-efficacy, and goal setting model of work performance", *Journal of Management*, Vol. 27/5, pp. 563-583, http://dx.doi.org/0.1177/014920630102700504. [51]

Rimfeld, K. et al. (2016), "True grit and genetics: Predicting academic achievement from personality", *Journal of Personality and Social Psychology*, Vol. 111/November, pp. 780-789, http://dx.doi.org/10.1037/pspp0000089. [19]

Rosenberg, M. (1979), *Conceiving the Self*, Basic Books, New York. [30]

Saarni, C. (2011), "Emotional development in childhood", in Tremblay, R. et al. (eds.), *Encyclopedia on Early Childhood Development: Emotions*, Centre of Excellence for Early Childhood Development, Montreal, QC. [16]

Saarni, C. (1999), *The Development of Emotional Competence*, The Guilford Press, New York. [15]

Salgado, J. (2002), "The big five personality dimensions and counterproductive behaviours", *International Journal of Selection and Assessment*, Vol. 101/2, pp. 117-125, https://doi.org/10.1111/1468-2389.00198. [59]

Salgado, J. (1997), "The five factor model of personality and job performance in the European Community", *Journal of Applied Psychology*, Vol. 82/1, pp. 30-43, http://dx.doi.org/10.1037/0021-9010.82.1.30. [53]

Salgado, J. et al. (2003), "International validity generalization of GMA and cognitive abilities: A European community meta-analysis", *Personnel Psychology*, Vol. 56/3, pp. 573-605, http://dx.doi.org//doi.org/10.1111/j.1744-6570.2003.tb00751.x. [52]

Santacreu, J., V. Rubio and J. Hernandez (2006), "The objective assessment of personality: Cattell's T-data revisted and more", *Psychology Science*, Vol. 48/1, pp. 53-68, https://psycnet.apa.org/record/2006-07428-004. [77]

Schäpers, P. et al. (2020), "The role of situations in situational judgment tests: Effects on construct saturation, predictive validity, and applicant perceptions", *Journal of Applied Psychology*, Vol. 105/8, pp. 800-818, https://doi.org/10.1037/apl0000457. [75]

Scheerens, J., G. van der Werf and H. de Boer (2020), *Soft Skills in Education. Putting the Evidence in Perspective*, Springer International Publishing. [10]

Schmidt, F., I. Oh and J. Shaffer (2016), *The Validity and Utility of Selection Methods in Personnel Psychology: Practical and Theoretical Implications of 100 Years of Research Findings*, Department of Management and Organizations, University of Iowa. [67]

Schwartz, S. (1994), "Beyond individualism/collectivism: New cultural dimensions of values", in Kim, U. et al. (eds.), *Individualism and Collectivism: Theory, Method, and Applications*, Sage, Thousand Oaks, CA. [44]

Sklad, M. et al. (2012), "Effectiveness of school-based universal social, emotional, and behavioural programs: Do they enhance students' development in the area of skill, behaviour, and adjustment?", *Psychology in the Schools*, Vol. 49/9, pp. 892-909, https://doi.org/10.1002/pits.21641. [8]

Sosnowska, J. et al. (2020), "New directions in the conceptualization and assessment of personality – A dynamic systems approach", *European Journal of Personality*, Vol. 34/6, pp. 988-998, https://doi.org/10.1002/per.2233. [64]

Spearman, C. (1927), *The Abilities of Man*, MacMillan, Basingstoke, UK. [46]

Stachl, C. et al. (2020), "Predicting personality from patterns of behaviour collected with smartphones", *Proceedings of the National Academy of Sciences of the United States of America*, Vol. 117/30, pp. 17680-17687, https://doi.org/10.1073/pnas.1920484117. [86]

Taylor, R. et al. (2017), "Promoting positive youth development through school-based social and emotional learning interventions: A meta-analysis of follow-up effects", *Child Development*, Vol. 88/4, pp. 1156-1171, http://dx.doi.org/10.1111/cdev.12864. [6]

Trilling, B. and C. Fadel (eds.) (2009), *21st Century Skills: Learning for Life in our Times*, Wiley. [17]

Van Scotter, J. and S. Motowidlo (1996), "Interpersonal facilitation and job dedication as separate facets of contextual performance", *Journal of Applied Psychology*, Vol. 81/5, pp. 525-531, https://doi.org/10.1037/0021-9010.81.5.525. [55]

Vecchione, M. et al. (2020), "Stability and change of basic personal values in early adolescence: A 2-year longitudinal study", *Journal of Personality*, Vol. 88/3, pp. 447-463, http://dx.doi.org/10.1111/jopy.12502. [47]

Wille, B. and F. De Fruyt (2019), "Development of vocational interests in adulthood", in Nye., C. and J. Rounds (eds.), *Vocational Interests in the Workplace: Rethinking Behaviour at Work*, Routledge, New York. [45]

6. Assessing collective intelligence in human groups

Anita Williams Woolley, Carnegie Mellon University

This chapter summarises the research on collective intelligence (CI) in human groups over the last ten years. It describes key factors that lead to development of CI in human teams and discusses the approach taken to measuring it. It also examines different task taxonomies, including the McGrath Group Task Circumplex, and looks at the nature of interdependence in CI. Newer research on enhancing CI is explored, including possible roles for artificial intelligence (AI). The chapter discusses how these approaches could provide useful directions for shaping and evaluating AI, and particularly artificial social intelligence, for enabling teamwork in more complex settings.

Introduction

The ability to collaborate in teams is becoming increasingly important. Across every sector of society and the economy, there is growth in the delivery of emergency and medical services in teams (Hughes et al., 2016[1]), the management of business in teams (Pearce, 2004[2]) and even the development of new knowledge in science and technology in teams (Wuchty, Jones and Uzzi, 2007[3]).

At the same time, problems arise that are difficult for teams to address. These problems are due to complexity associated with geographic dispersion; language and cultural barriers; the large number of people required to co-ordinate; and the broad diversity of expertise, beliefs and motivations needed to be aligned to make progress. The right knowledge and tools to manage those barriers might enable powerful engines of progress for society. However, a workforce would need tools for effective collaboration and support from tools and technology to help manage the associated complexity.

Enhancing collaboration to enable teams to address the toughest problems requires deep understanding of what helps teams perform well. Technology can likely play a role; however, knowing how to create technologies that will enable good teamwork consistently is also needed. This will require understanding both the individual characteristics and abilities that enable effective collaboration, and the group level features and behaviours that support good teamwork across a wide range of problems of varying complexity.

Research on collective intelligence (CI) has focused on conceptualising and measuring the capabilities underlying teamwork. Woolley et al. (2010[4]) used an analogy between the individual intelligence of a person and the CI of a group. Research literature commonly defines individual intelligence with a statistical factor ("g" for general intelligence) that captures and predicts how well a person will perform on a wide range of different tasks. Woolley and colleagues found that, similar to individual intelligence, a single factor predicted over 40% of the variance in performance when groups completed a range of different tasks. They called this factor "collective intelligence."

Research on CI in human groups builds upon and extends traditional research on team performance. The latter is typically focused on elements that enable a team to perform well on a particular task. Conversely, research on CI examines what enables a group to perform tasks that vary in complexity over time. Accordingly, subsequent studies have tested and shown that measures of a group's CI predicts future performance in a variety of settings (Woolley et al., 2010[4]; Engel et al., 2014[5]; Kim et al., 2017[6]; Aggarwal et al., 2019[7]; Glikson et al., 2019[8]).

What are the key factors that cultivate collective intelligence?

Several factors appear to be key drivers of CI in teams. These include individual characteristics, such as social perceptiveness, which enhance the quality of collaboration; group compositional features, such as various forms of diversity; and group process behaviours. These are discussed below.

Individual characteristics

Researchers have explored which individual characteristics can enable group CI. Most personality variables have relatively weak correlations with CI (Engel et al., 2014[5]; Woolley et al., 2010[4]). However, gender composition of the group emerged early as an important characteristic. Early studies showed a significant correlation between the proportion of women in the group and its CI (Woolley et al., 2010[4]; Engel et al., 2014[5]; Kim et al., 2017[6]).

The relationship between having more women in the group and CI is explained by another characteristic: social perceptiveness. Social perceptiveness is the ability to take another's perspective and reason about their intentions, knowledge, beliefs and emotions (Premack and Woodruff, 1978[9]). In studies of CI, groups with higher average social perceptiveness are more collectively intelligent. Women, on average, tend to be more socially perceptive than men. Having more women in a group, then, raises CI because it raises the average level of social perceptiveness. Furthermore, groups with higher social perceptiveness tend to engage more in collaboration processes that enhance CI, which are discussed further below.

Group composition factors

In addition to the influence of specific individual characteristics, research on CI in groups finds several influential compositional factors. First, it appears that cognitive style diversity is important. Aggarwal et al. (2019[7]) evaluated individual member cognitive style using the Object-Spatial Imagery and Verbal Questionnaire (Blazenkova, Kozhevnikov and Motes, 2006[10]). They found that groups with moderate levels of cognitive style diversity reach the highest levels of CI. In related research, Aggarwal and Woolley (2018[11]) found that high levels of cognitive diversity were associated with group difficulties in settling on a shared task strategy. Task performers with different cognitive styles tended to approach the task in different ways. This finding has been illustrated in research in educational settings as well (Blazhenkova and Kozhevnikov, 2020[12]). Thus, to be collectively intelligent, a group needs a variety of perspectives but also a means of integrating them to produce better ideas or decisions and to act upon them.

Box 6.1. Cognitive styles and collective intelligence

Cognitive styles refer to consistencies in an individual's manner of acquiring and processing information (Ausburn and Ausburn, 1978[13]). Research in the 1970s argued for a visual-verbal cognitive style dimension (Richardson, 1977[14]; Paivio, 1979[15]). However, neuroscience research demonstrates that the visual areas of the brain divide into two distinct pathways: the object (ventral) and the spatial (dorsal) pathways (Ungerleider and Mishkin, 1982[16]). Furthermore, researchers have documented a trade-off in the abilities associated with processing in these areas; individuals with above-average object visualisation abilities (such as artists) tend to have below-average spatial visualisation abilities. The inverse is true for those with above-average spatial visualisation abilities (scientists).

These observations led to the identification of two different cognitive styles associated with visualisation (i.e. object visualisation and spatial visualisation) in addition to the verbalisation cognitive style. In the past two decades, research has demonstrated these cognitive styles can be identified in children as young as eight years of age. They are also associated with related abilities and preferences for associated academic subjects (Blazhenkova, Becker and Kozhevnikov, 2011[17]).

A self-report survey instrument, the Object-Spatial Imagery and Verbal Questionnaire (Blazhenkova and Kozhevnikov, 2008[18]) has been developed and validated as a reliable measure of individual cognitive style in adults. A companion version has been developed for children (Blazhenkova, Becker and Kozhevnikov, 2011[17]). Items include statements such as, "I can easily imagine and mentally rotate three-dimensional geometric figures" as an indicator of spatial visualisation. Another statement is "When reading fiction, I usually form a clear and detailed mental picture of a scene or room that has been described" as an indicator of object visualisation.

In teams, diversity in cognitive styles has important implications for collaboration and collective intelligence. Teams with insufficient cognitive style diversity lack the abilities and perspectives to reach the highest levels of collective intelligence (Aggarwal et al., 2019[7]). Conversely, teams with high levels of cognitive style diversity can struggle to collaborate effectively (Woolley et al., 2008[19]; Aggarwal and Woolley, 2013[20]). This occurs because of different approaches taken by individuals with different cognitive styles to organise information and solve problems (Blazhenkova and Kozhevnikov, 2020[12]).

Highly cognitively diverse teams could be supported to collaborate effectively (Woolley et al., 2007[21]; Woolley et al., 2008[19]). However, this requires knowing the cognitive style strengths of all team members and helping them match members to the best task and roles. Artificially intelligent systems could be designed to sense the cognitive styles of individuals and represent information in a format best suited to their strengths. In addition, such systems could make cognitive style differences apparent to team members and aid in "translating" between them to facilitate co-ordination. For example, a strong verbaliser might be trying to convey a project plan through long prose to a team member who is a strong spatial visualiser. Such a system could help the verbaliser translate the prose into a schematic or a diagram. Conversely, the same system could also sense when a team was too homogenous in cognitive style. It would thus prompt them to consider alternative approaches or seek other inputs to broaden their options. In these ways, artificial social intelligence could play an important role in both creating and facilitating diverse perspectives in teams.

There is evidence to suggest that racial diversity is beneficial for CI. Chikersal et al. (2017[22]) found that CI was higher when partners were of different ethnic backgrounds. In that study, the researchers examined facial expression synchrony, or the degree to which partners were mirroring the positivity or negativity of each other's facial expressions during their interaction. They found CI was higher in pairs with greater synchrony in facial expressions. The level of racial diversity of each pair enhanced synchrony in facial expressions. This is likely because people exhibit increased attention to interaction partners who are of a different race. They are more likely to attend to information they supply (Phillips and Loyd, 2006[23]) and process information more carefully in their presence in general (Sommers, 2006[24]). This heightened attention can translate into higher CI levels.

Group process

In addition to the characteristics of the people in the group, interaction processes in which group members engage advance CI. In the last few decades, there has been growing recognition that team cognition plays an essential role in enabling effective teamwork (DeChurch and Mesmer-Magnus, 2010[25]). Similarly, Gupta and Woolley (2021[26]) recently articulated a Transactive Systems Theory of Collective Intelligence. It describes the interconnected memory, attention and reasoning systems that operate at the individual and collective levels to enable the emergence of CI.

These transactive cognitive systems give rise to three observable group processes: the group's level of effort, their task strategy and their use of member knowledge and skill (Hackman, 1987[27]). Effort relates to the total amount of work that members put towards their collective task. Task strategy encompasses a group's decisions regarding aspects of its work. This includes what gets completed first and what tasks to divide among members versus what tasks to do all together, etc. Use of member knowledge and skill captures a group's proficiency at achieving agreement between relative member skills and their contributions to work on a task. Groups that make better use of member talents achieve higher levels of CI. Groups that engage in higher levels of effort, who engage in better task strategies, and who use the specific knowledge and skills of members more effectively develop higher levels of CI. Each of these pathways to CI suggest specific ways in which AI might be developed to enhance CI, as will be discussed further below.

Measuring collective intelligence

The primary approach to measuring CI in teams focuses on capturing group performance on a variety of group tasks that require different modes of interaction for completion. A group's performance scores across a diverse battery of tasks then leads to an inference about the group's CI. This approach is modelled after the traditional psychometric approach to measuring intelligence in individuals.

A measurement battery should incorporate tasks that are sufficiently diverse to draw an inference of CI. Diversity is more important for the collaboration processes needed for completion than for content.

Identifying task types

Figure 6.1. McGrath's Group Task Circumplex

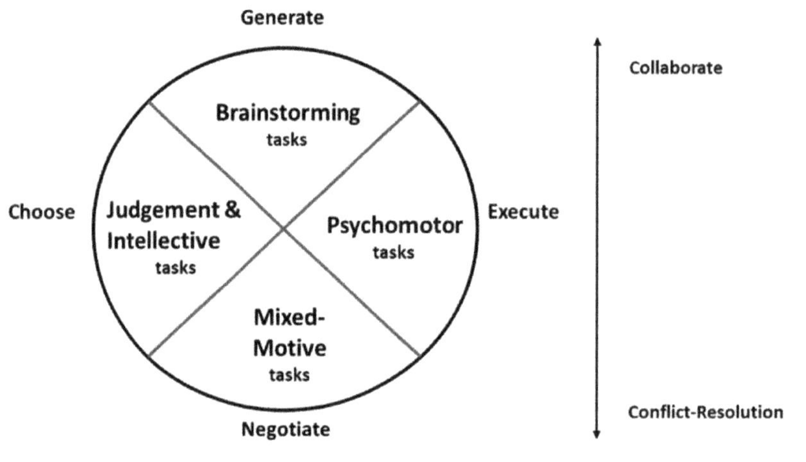

Source: Adapted from McGrath (1984[28]).

Several different task taxonomies have been proposed to capture and describe the essential differences in the types of tasks groups are asked to perform. McGrath's Group Task Circumplex (see Figure 6.1); (McGrath, 1984[28]), for example, articulates four major types of tasks: generate, choose, negotiate and execute. These types vary in terms of the degree to which they require collaboration versus resolution of conflicting preferences. "Generate" tasks require teams to think as broadly and divergently as possible to develop a wide range of ideas. "Choose" tasks, by contrast, require groups to pool information or perspectives to converge on the single best or correct answer. "Negotiate" tasks require the resolution of conflicting objectives for group members to arrive at a solution that works best for the group. "Execute" tasks require the careful co-ordination of physical inputs to accomplish a specified product as quickly and accurately as possible.

In addition to sampling task types, the level of task interdependence needed for the team to accomplish the task successfully should vary in a diverse battery. In the context of teamwork, one major driver of task complexity is the level and number of interdependencies a task involves (Thompson, 1967[29]; Steiner, 1972[30]; Wageman, 1995[31]). The inference of CI in a group is most strongly supported by evidence that the group can handle tasks of varying complexity.

Thompson (1967[29]) conceptualises three levels of interdependence (see Figure 6.2). This framework specifies the ways in which task contributors must combine their efforts to accomplish a task.

"Reciprocal" interdependence is the highest level. Here, the final product cannot be traced back to the inputs of any one member. The solution to a complex problem often results from reciprocal interdependence. Different contributors provide different perspectives or pieces of information that get combined synergistically.

"Sequential" interdependence represents an intermediate level. Here, each member works on and hands off their work to another member, who passes it down the line (e.g. a typical factory assembly line).

Figure 6.2. Interdependence of group members' intelligence

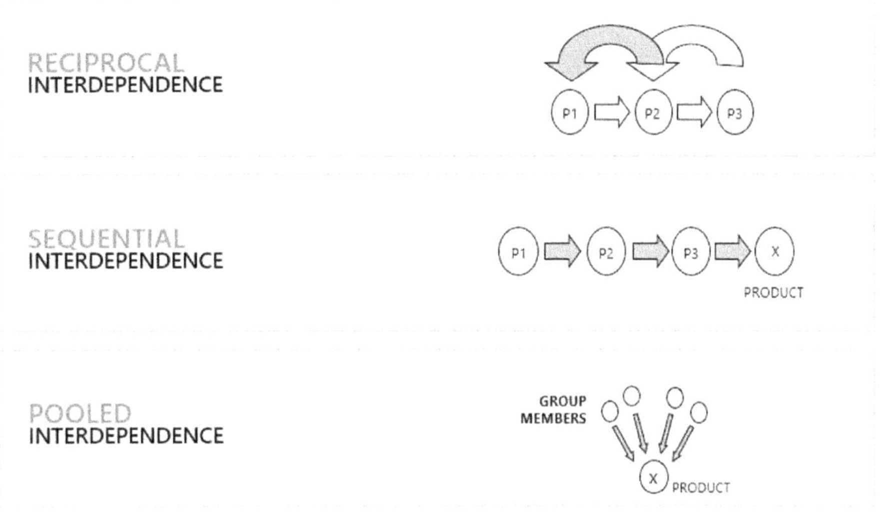

"Pooled" is the lowest level. Here, each member works relatively independently and then contributes their part to a shared product (e.g. independently written book chapters in an edited volume, which are typically assembled with little to no interaction among authors).

The task types identified in the McGrath Group Task Circumplex can vary in their level of interdependence. For instance, "generate" tasks are generally best completed primarily using pooled interdependence; such tasks benefit from a broader range of divergent perspectives; however, there can be problems for which members need to integrate knowledge to generate solutions, which would involve higher levels of interdependence. "Choose" and "execute" tasks can also vary in the interdependence employed. A choose task completed with pooled interdependence could be an election or a crowd prediction of a future event such as a stock price. A choose task completed with sequential interdependence would benefit from multiple reviewers checking and correcting, such as a complex math problem. A choose task completed with reciprocal interdependence might be a hiring decision, where many different individuals need to give input and integrate information.

Box 6.2. Task types, interdependence and the workplace

Most jobs in organisational settings involve a variety of task types and interdependence levels. The number of different task types or interdependence levels vary by job type. For instance, a custodian's job might consist mostly of execute tasks. It could involve a small amount of decision making (i.e. choose) and maybe some occasional planning or creative problem solving (i.e. generate), conducted mostly with pooled or sequential interdependence with co-workers. A management position might incorporate more task types and more variation in interdependence. In these roles, individuals frequently shift across task types (e.g. developing options, making a choice, negotiating with others). They also shift between independent work and highly collaborative, reciprocally interdependent work. Performance in complex jobs, particularly those requiring ongoing learning of new functions, is shown to be the most strongly correlated with measures of individual intelligence and cognitive ability (Schmidt and Hunter, 1998[32]; Murphy, 1989[33]). Complex jobs involve a high level of social skill to facilitate high interdependence. Wages for these kinds of jobs have grown significantly faster over the last four decades than for jobs that require technical skills alone (Deming, 2017[34]).

Developing test items

The Test of Collective Intelligence (TCI) (Riedl et al., 2021[35]; Kim et al., 2017[6]) aims to sample different task types that vary in levels of interdependence. It also includes items that tap into both verbal and non-verbal content. The latter has become a larger priority over time. The inclusion of more non-native English speakers in studies suggested the CI of a group might be underestimated if all test items depended too much on facility with English.

The Platform for Online Group Studies (POGS), a web-based platform, facilitates the administration of the TCI. It enables collaborators to work together from anywhere in the world, seeing the work of their fellow group members in real time. POGS regulates how much time groups work on each task and provides a chat feature to facilitate co-ordination. Administering the tasks on the platform enables the capture of detailed information about each member's activities. This allows for a more accurate calculation of process measures in evaluating how different collaboration behaviours contribute to CI (Riedl et al., 2021[35]).

"Generate" items

Many standard examples of items from the "generate" quadrant of the McGrath Group Task Circumplex resemble standard brainstorming tasks. This is true as well for items on the TCI that tap into this quadrant. Figure 6.3 provides a verbal example; others in the TCI battery are more mathematical (i.e. brainstorm equations that satisfy specified constraints). Groups do best on tasks of this type when members work relatively independently to come up with as broad a range of ideas as possible. This finding is consistent with extant research on group brainstorming [e.g. Paulus and Yang (2000[36])]. Groups are scored based on both the number of unique ideas generated and the creativity of their ideas. Creativity is measured based on how often the same idea appears within the corpus of ideas submitted by groups across all samples.

Figure 6.3. Brainstorming task for the "generate" quadrant of the McGrath Group Task Circumplex

Note: A screenshot from the Platform for Online Group Studies (POGS).

Some "generate" tasks in the TCI vary in the level of interdependence required. For instance, some problems require that ideas later in the list incorporate information from the prior item in the list (e.g. use one of the same numbers in the equation). This then requires groups to work with at least a sequential, if not reciprocal, level of interdependence to generate a large list of ideas effectively.

"Choose" items

Common examples of tasks in the TCI related to the "choose" category resemble problem-solving and decision-making tasks used in a variety of studies of both individual and team performance. Some of these involve problems with demonstrably correct answers, such as unscrambling a word or solving a puzzle. Others involve matters of judgement, such as rating the quality of poems or photographs.

Many "choose" tasks can be completed by groups using pooled or sequential interdependence. For example, members may independently suggest an answer and the group goes with the one favoured by most (pooled). In another scenario, a member attempts to solve the problem and others review it afterward to check their work (sequential). Some "choose" tasks require reciprocal interdependence, such as a Sudoku puzzle designed to permit more than one solution; any one member's choice of number has implications for the numbers other members select in the grid.

Figure 6.4. Suduko task for the "choose" quadrant of the McGrath Group Task Circumplex

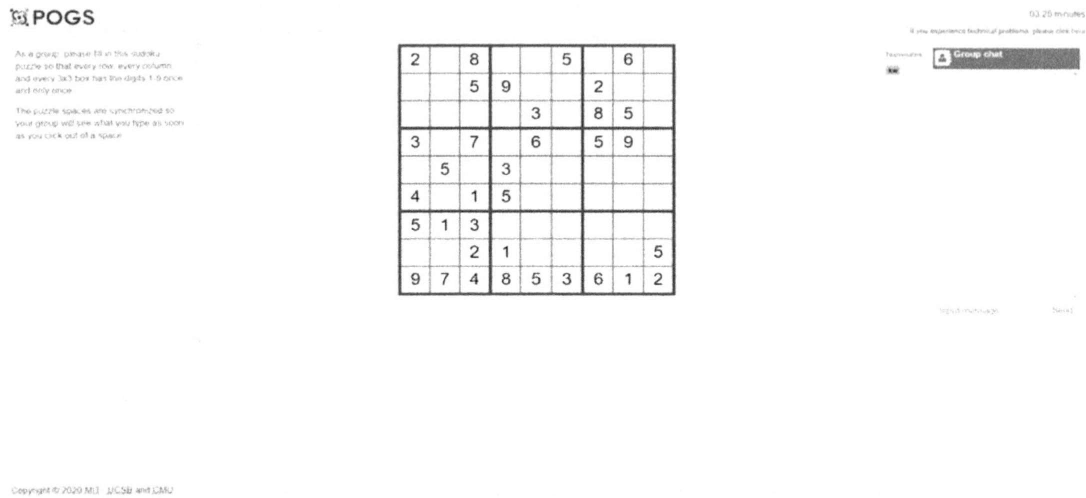

Note: A screenshot from the Platform for Online Group Studies (POGS).

"Negotiate" items

As described above, "negotiate" tasks involve the resolution of conflicting motives. In these cases, an action that would provide better rewards for one member might compromise the outcomes of other members of the group or the group as a whole, and vice versa. A number of behavioural economics games facilitate the operationalisation of such situations, such as the minimum effort game (Anderson, Goeree and Holt, 2001[37]). In this game, each member chooses from a set of options without communicating with other members. Their payoff is determined based on the combination of their choice and the lowest choice of anyone else in the group. If they can trust other members to choose the highest number (i.e. 5), then they can win the maximum points. However, if they choose 5 and anyone chooses 1, they lose points (see Figure 6.5). This type of game creates incentives for co-operation; in some other games, such as "the prisoner's dilemma", individuals are rewarded for defecting.

Figure 6.5. Minimum effort game for the "negotiate" quadrant of the McGrath Group Task Circumplex

"Execute" items

Tasks that fall into the "execute" category typically involve carefully co-ordinated psychomotor movements so the team can quickly and accurately complete a specified task. In the TCI, for example, team members need to retype a given text as quickly as possible into a shared online document. Members need to be careful not to type over each other or to skip over large blocks of text; missing text results in loss of points. In a variation meant to increase interdependence, different members can only type specified words, which forces team members to work together more closely in reproducing the text.

Box 6.3. The psychometric structure of collective intelligence

In their initial studies, Woolley et al. (2010[4]) found strong evidence of a single general factor in the performance scores of teams performing tasks sampled from existing team task taxonomies, such as the McGrath Group Task Circumplex. That factor was replicated in subsequently published studies (Engel et al., 2014[5]; Meslec, Aggarwal and Curşeu, 2016[38]; Kim et al., 2017[6]; Aggarwal et al., 2019[7]). Some researchers have raised questions about the factor structure of collective intelligence (CI), questioning if there is truly a single factor or rather multiple factors (Barlow and Dennis, 2016[39]; Credé and Howardson, 2017[40]). Others have also raised questions about the size of the role of individual intelligence (Bates and Gupta, 2017[41]).

A more recent analysis illustrated the group process was more than twice as important for predicting CI as individual skills. Riedl and colleagues (2021[35]) meta-analytically examined data from 22 different samples, including more than 1 300 teams that worked on a range of tasks such as those above. Their dataset included detailed process information describing how team members worked together on the tasks, as well as what level of skill each member possessed for working on the tasks independently. While they showed group process to be over twice as important as individual skills for predicting CI, they also identified variance in the degree to which team process predicted performance on different individual tasks of different types compared to individual skills. This suggested that studies showing a larger role for individual ability tended to oversample types of tasks that depend more upon individual skill. Furthermore, Riedl et al. (2021[35]) found that individual characteristics such as social perceptiveness were influential because they led teams to engage in the processes most strongly predictive of CI. In addition, it appears that whether a single factor or multiple factors of CI emerge depends on some methodological choices. These include whether one team member or multiple team members can record answers to problems simultaneously, and whether all task types are represented in the battery used to measure CI.

Using artificial intelligence to enhance collective intelligence

Can technology be used to enhance human group CI? Given the important role of group process and individual skill in creating CI, the answer is almost certainly "Yes".

Technological tools could enhance both individual skill and group process in a variety of ways. Gupta and colleagues (2019[42]) have experimented with "digital nudges". Like other types of nudges (Thaler and Sunstein, 2009[43]), digital nudges are subtle changes in a digital environment that make it more likely that people will make certain choices over others – in this case, choices about how to do that results in group processes resembling those known to enhance CI.

In their studies, the researchers deployed "bots" and other tools to prompt teams to consider critical aspects of group process. For example, a bot facilitator would ask the team questions that would lead them to consider whether they were using information about which member was good at what in deciding how to divide a task, or whether they had co-ordinated their task strategy to cover all parts of the task. The researchers found that technological nudges helped encourage some group processes, particularly those related to getting groups to use information about individuals' skills and abilities in structuring work.

In a similar study, Glikson et al. (2019[8]) found that use of a digital nudge that provided feedback about the relative effort of members enhanced group CI. This was especially true in teams with members who were low in conscientious. The nudge effectively served as a "conscience" to help lesser-contributing members realise they needed to increase their effort.

Given the consistent role of characteristics related to social intelligence in advancing group CI, another avenue for enhancing CI might be to help improve the social and communications skills of collaborators. How much these abilities can be enhanced in individuals is an ongoing debate. Some researchers have shown efficacy of interventions in improving individual skills in these areas through training (Kidd and Castano, 2013[44]; Hodzic et al., 2018[45]; Kotsou et al., 2019[46]). Others question that evidence (Panero et al., 2016[47]; Panero et al., 2017[48]; Kidd and Castano, 2017[49]).

In still other areas, researchers are experimenting with technological tools to augment individuals' natural ability to pick up on social cues. This helps individuals interpret otherwise ambiguous situations [e.g. Voss et al. (2016[50])]. Therefore, as technology develops, CI may be enhanced in the context of human-computer systems. Specifically, AI might enable technology to augment the individual skills and group collaboration processes that enable CI to emerge.

Using collective intelligence to evaluate or enhance artificial intelligence

Can the same approach to evaluating CI also be used to assess AI capabilities? Arguably, such an approach could complement perspectives on evaluating the utility of AI. This section explores the role of production and co-ordination technologies, and artificial social intelligence (ASI) in diagnosis and assessment.

Production and co-ordination technologies

Much of the work to date on AI has focused on development of production technologies. These are technologies designed to produce more accurate or higher-quality output more effectively or efficiently than traditional approaches. Decision aids that improve medical diagnoses or autonomous vehicles that can drive more safely than humans are examples of production technologies.

Some argue that developing AI as a co-ordination technology has even more potential for enhancing CI. Such technology operates by co-ordinating and combining the inputs of other contributors. Examples that exist today include shared ride systems such as Uber or Lyft, or social network platforms that connect people with similar interests or complementary needs. As we think about evaluating and increasing the level of intelligence in technology, we might consider the types of tasks and levels of interdependence the technology can facilitate as an important indicator. High levels of intelligence would be evident in technology that could sense that different group members had conflicting goals or preferences and help resolve them or facilitate problem solutions requiring high levels of knowledge integration. Such a capability will only become possible as the field of AI makes progress in developing artificial social intelligence.

Artificial social intelligence

The next frontier in AI is artificial social intelligence (ASI) – the capacity of technology to pick up on the knowledge, beliefs, goals and emotions of users to anticipate their needs or potential response to events. To facilitate high CI, an ASI system would need to perform several kinds of functions that humans with high levels of social and emotional intelligence perform naturally. A high-functioning ASI technology would accomplish this by facilitating collective memory, attention and reasoning processes (Gupta and Woolley, 2021[26]).

ASI enhances collective memory by:

- sensing the skills and abilities of the human participants
- suggesting roles based on ability
- conducting or facilitating information transfer so that all participants know who is doing what and have the relevant information for their task.

ASI enhances collective attention by:

- reallocating work if some members are overloaded while others are underutilised
- helping manage the work cue based on group priorities
- drawing the group together to discuss tasks or problems as needed.

ASI enhances collective reasoning by:

- sensing when individual teammates are feeling frustrated, unmotivated or distracted, and intervening or prodding the team to address it
- sensing when participants are annoyed with or ignoring the ASI system itself and altering its approach accordingly.

Groups in which collective memory, attention and reasoning are functioning at a high level are characterised by several factors. They demonstrate a strong ability to match member skills with tasks and roles; they co-ordinate task strategy so their work gets accomplished efficiently; and their members exhibit uniformly high and consistent levels of effort and commitment (Riedl et al., 2021[35]; Gupta et al., 2019[42]). The intelligence level of an ASI agent or system could be tested by its ability to enhance CI in human groups working across a range of different tasks and levels of interdependence. Specifically, it would be tested on its ability to enhance collective memory, attention and reasoning processes, and how well it led the group to exhibit optimal skill use, task strategy and high collective effort.

Agents with ASI could play a valuable role to facilitate CI in more challenging systems. Teamwork can become more complicated with more members or dispersion of members across different locations. In these contexts, agents with ASI could help facilitate CI in ways that humans alone could not manage. ASI abilities would need to be tested in highly complex environments, where teams are adapting to different types of tasks and levels of interdependence, and the development of collective memory, attention and reasoning is a significant challenge.

References

Aggarwal, I. and A. Woolley (2018), "Team creativity, cognition and cognitive style diversity", *Management Science*, Vol. 65/4, http://dx.doi.org/10.1287/mnsc.2017.3001. [11]

Aggarwal, I. and A. Woolley (2013), "Do you see what I see? The effect of members' cognitive styles on team processes and performance", *Organizational Behavior and Human Decision Processes*, Vol. 122, pp. 92-99, https://doi.org/10.1016/j.obhdp.2013.04.003. [20]

Aggarwal, I. et al. (2019), "The impact of cognitive style diversity on implicit learning in teams", *Frontiers in Psychology*, Vol. 10/112, http://dx.doi.org/10.3389/fpsyg.2019.00112. [7]

Anderson, S., J. Goeree and C. Holt (2001), "Minimum-effort coordination games: Stochastic potential and logit equilibrium", *Games and Economic Behavior*, Vol. 34/2, pp. 177-199, http://people.virginia.edu/~cah2k/mineff.pdf. [37]

Ausburn, L. and F. Ausburn (1978), "Cognitive styles: Some information and implications for instructional design", *Educational Communications & Technology Journal*, Vol. 26, pp. 337-354. [13]

Barlow, J. and A. Dennis (2016), "Not as smart as we think: A study of collective intelligence in virtual groups", *Journal of Management Information Systems*, Vol. 33/3, pp. 684-712, http://dx.doi.org/10.1080/07421222.2016.1243944. [39]

Bates, T. and S. Gupta (2017), "Smart groups of smart people: Evidence for IQ as the origin of collective intelligence in the performance of human groups", *Intelligence*, Vol. 60, pp. 46-56, https://doi.org/10.1016/j.intell.2016.11.004. [41]

Blazenkova, O., M. Kozhevnikov and M. Motes (2006), "Object-spatial imagery: A new self-report imagery questionnaire", *Applied Cognitive Psychology*, Vol. 20/2, pp. 239-263, http://dx.doi.org/10.1002/acp.1182. [10]

Blazhenkova, O., M. Becker and M. Kozhevnikov (2011), "Object–spatial imagery and verbal cognitive styles in children and adolescents: Developmental trajectories in relation to ability", *Learning and Individual Differences*, Vol. 21/2, pp. 281-287, https://doi.org/10.1016/j.lindif.2010.11.012. [17]

Blazhenkova, O. and M. Kozhevnikov (2020), "Creative processes during a collaborative drawing task in teams of different specializations", *Creative Education*, Vol. 11/9, p. 1751, http://dx.doi.org/10.4236/ce.2020.119128. [12]

Blazhenkova, O. and M. Kozhevnikov (2008), "The new object-spatial-verbal cognitive style model: Theory and measurement", *Applied Cognitive Psychology*, Vol. 23/5, pp. 638-6563, http://dx.doi.org/10.1002/acp.1473. [18]

Chikersal, P. et al. (2017), "Deep structures of collaboration: Physiological correlates of collective intelligence and group satisfaction", *Proceedings of the 20th ACM Conference on Computer-Supported Cooperative Work and Social Computing*, pp. 873-888, https://doi.org/10.1145/2998181.2998250. [22]

Credé, M. and G. Howardson (2017), "The structure of group task performance – a second look at 'collective intelligence': Comment on Woolley et al. (2010)", *Journal of Applied Psychology*, Vol. 102/10, pp. 1483-1492, https://doi.org/10.1037/apl0000176. [40]

DeChurch, L. and J. Mesmer-Magnus (2010), "The cognitive underpinnings of effective teamwork: A meta-analysis", *Journal of Applied Psychology*, Vol. 95/1, pp. 32-53, http://dx.doi.org/10.1037/a0017328. [25]

Deming, D. (2017), "The growing importance of social skills in the labor market", *The Quarterly Journal of Economics*, Vol. 132/4, pp. 1593-1640, http://dx.doi.org/10.1093/qje/qjx022. [34]

Engel, D. et al. (2014), "Reading the mind in the eyes or reading between the lines? Theory of mind predicts collective intelligence equally well online and face-to-face", *PLoS ONE*, Vol. 9/12, p. e115212, http://dx.doi.org/10.1371/journal.pone.0115212. [5]

Glikson, E. et al. (2019), "Visualized automatic feedback in virtual teams", *Frontiers in Psychology*, Vol. 16 April, https://doi.org/10.3389/fpsyg.2019.00814. [8]

Gupta, P. et al. (2019), "Digital nudging team processes to enhance collective intelligence", Proceedings of Collective Intelligence 2019. [42]

Gupta, P. and A. Woolley (2021), "Articulating the role of artificial intelligence in collective intelligence: A transactive systems framework.", No. 65, Proceedings of the Human Factors and Ergonomics Society. [26]

Hackman, J. (1987), "The design of work teams", in Lorsch, J. (ed.), *Handbook of Organizational Behavior*, Prentice Hall, Englewood Cliffs, NJ. [27]

Hodzic, S. et al. (2018), "How efficient are emotional intelligence trainings: A meta-analysis", *Emotion Review*, Vol. 10/2, pp. 138-148, http://dx.doi.org/10.1177/1754073917708613. [45]

Hughes, A. et al. (2016), "Saving lives: A meta-analysis of team training in healthcare", *Journal of Applied Psychology*, Vol. 10/9, pp. 1266-1304, http://dx.doi.org/10.1037/apl0000120. [1]

Kidd, D. and E. Castano (2017), "Panero et al. (2016): Failure to replicate methods caused the failure to replicate results", *Journal of Personality and Social Psychology*, Vol. 11/3, pp. e1–e4, http://dx.doi.org/10.1037/pspa0000072. [49]

Kidd, D. and E. Castano (2013), "Reading literary fiction improves theory of mind", *Science*, Vol. 342/6156, pp. 377-380, http://dx.doi.org/10.1037/apl0000120. [44]

Kim, Y. et al. (2017), "What makes a strong team? Using collective intelligence to predict performance of teams in League of Legends", *Proceedings of the 20th ACM Conference on Computer-Supported Cooperative Work and Social Computing*, Vol. 30 November, http://dx.doi.org/doi.org/10.5465/ambpp.2016.13564abstract. [6]

Kotsou, I. et al. (2019), "Improving emotional intelligence: A systematic review of existing work and future challenges", *Emotion Review*, Vol. 11/2, pp. 151-165, http://dx.doi.org/10.1177/1754073917735902. [46]

McGrath, J. (1984), *Groups: Interaction and Performance*, Prentice-Hall, Englewood Cliffs, NJ. [28]

Meslec, N., I. Aggarwal and P. Curşeu (2016), "The insensitive ruins it all: Compositional and compilational influences of social sensitivity on collective intelligence in groups", *Frontiers in Psychology*, Vol. 9 May, http://dx.doi.org/doi.org/10.3389/fpsyg.2016.00676. [38]

Murphy, K. (1989), "Is the relationship between cognitive ability and job performance stable over time?", *Human Performance*, Vol. 2/3, pp. 183-200, http://dx.doi.org/10.1207/s15327043hup0203_3. [33]

Paivio, A. (1979), *Imagery and Verbal Processes*, Lawrence Erlbaum Assoc Inc., Hillsdale, NJ. [15]

Panero, M. et al. (2017), "No support for the claim that literary fiction uniquely and immediately improves theory of mind: A reply to Kidd and Castano's commentary on Panero et al. (2016)", *Journal of Personality and Social Psychology*, Vol. 112/3, http://dx.doi.org/10.1037/pspa0000079. [48]

Panero, M. et al. (2016), "Does reading a single passage of literary fiction really improve theory of mind? An attempt at replication", *Journal of Personality and Social Psychology*, Vol. 111/5, pp. e46–e54, http://dx.doi.org/10.1037/pspa0000064. [47]

Paulus, P. and H. Yang (2000), "Idea generation in groups: A basis for creativity in organizations", *Organizational Behavior and Human Decision Processes*, Vol. 82/1, pp. 76-87, http://dx.doi.org/doi.org/10.1006/obhd.2000.2888. [36]

Pearce, C. (2004), "The future of leadership: Combining vertical and shared leadership to transform knowledge work", *Academy of Management Executive*, Vol. 18/1, pp. 47-57, http://dx.doi.org/10.5465/ame.2004.12690298. [2]

Phillips, K. and D. Loyd (2006), "When surface and deep-level diversity collide: The effects on dissenting group members", *Organizational Behavior and Human Decision Processes*, Vol. 99/2, pp. 143-160, http://dx.doi.org/10.1016/j.obhdp.2005.12.001. [23]

Premack, D. and G. Woodruff (1978), "Does the chimpanzee have a theory of mind?", *Behavioral and Brain Sciences*, Vol. 1/4, p. 515, http://dx.doi.org/10.1017/S0140525X00076512. [9]

Richardson, A. (1977), "Verbalizer-visualizer: A cognitive style dimension", *Journal of Mental Imagery*, Vol. 1/1, pp. 109-126. [14]

Riedl, C. et al. (2021), "Quantifying collective intelligence in human groups.", *Proceedings of the National Academy of Sciences of the United States of America*, Vol. 118/21, http://dx.doi.org/10.1073/pnas.2005737118. [35]

Schmidt, F. and J. Hunter (1998), "The validity and utility of selection methods in personnel psychology: Practical and theoretical implications of 85 years of research findings", *Psychological Bulletin*, Vol. 124/2, p. 262, http://dx.doi.org/10.1037/0033-2909.124.2.262. [32]

Sommers, S. (2006), "On racial diversity and group decision making: Identifying multiple effects of racial composition on jury deliberations", *Journal of Personality and Social Psychology*, Vol. 90/4, pp. 597-612, http://dx.doi.org/doi.org/10.1037/0022-3514.90.4.597. [24]

Steiner, I. (1972), *Group Process and Productivity*, Academic Press, New York. [30]

Thaler, R. and C. Sunstein (2009), *Nudge: Improving decisions about health, wealth, and happiness*, HeinOnline. [43]

Thompson, J. (1967), "The structure of complex organizations", in *Organization Theory*, Penguin Books, Harmondsworth. [29]

Ungerleider, L. and M. Mishkin (1982), "Two cortical visual systems", in Ingle, D., M. Goodale and R. Mansfield (eds.), *Analysis of Visual Behavior*, MIT Press, Cambridge, MA. [16]

Voss, C. et al. (2016), "Superpower glass: Delivering unobtrusive real-time social cues in wearable systems", *Proceedings of the 2016 ACM International Joint Conference on Pervasive and Ubiquitous Computing: Adjunct. New York*, pp. 1218-1226, https://arxiv.org/abs/2002.06581v1. [50]

Wageman, R. (1995), "Interdependence and group effectiveness", *Administrative Science Quarterly*, Vol. 40, pp. 145-180, http://dx.doi.org/doi.org/10.2307/2393703. [31]

Woolley, A. et al. (2010), "Evidence for a collective intelligence factor in the performance of human groups", *Science*, Vol. 330/6004, pp. 686-688, http://dx.doi.org/10.1126/science.1193147. [4]

Woolley, A. et al. (2008), "Bringing in the experts: How team composition and work strategy jointly shape analytic effectiveness", *Small Group Research*, Vol. 39/3, pp. 352-371, http://dx.doi.org/10.1177/1046496408317792. [19]

Woolley, A. et al. (2007), "Using brain-based measures to compose teams: How individual capabilities and team collaboration strategies jointly shape performance", *Social Neuroscience*, Vol. 2, pp. 96-105, http://dx.doi.org/10.1080/17470910701363041. [21]

Wuchty, S., B. Jones and B. Uzzi (2007), "The increasing dominance of teams in production of knowledge", *Science*, Vol. 316/5827, pp. 1036-1039, http://dx.doi.org/10.1126/science.1136099. [3]

7. Skills assessments in education

Samuel Greiff, University of Luxembourg

Jan Dörendahl, University of Luxembourg

This chapter describes skills typically included in large-scale educational assessments and discusses how such assessments can be used for measuring the capabilities of artificial intelligence (AI). It focuses on two major skill domains covered in educational assessments: core domain skills such as mathematics, reading and science literacy, and transversal skills such as problem solving, collaboration and creativity. The chapter provides an overview of the skills in each domain, as well as their theoretical underpinning and measurement. In addition, it examines the role of these skill domains in occupational settings by drawing links to taxonomies of skill requirements in the workplace. The chapter concludes with recommendations regarding the use of education tests for scaling AI capabilities.

Introduction

Educational systems must provide environments that foster and facilitate the highly diverse skill sets needed for the digital world. These skill sets can generally be represented by three different skill domains. First, students require literacy in core domains such as mathematics, reading and science (OECD, 2017[1]). Second, they need skills in transversal domains that span situations and contexts such as problem solving, collaboration and creativity (OECD, 2013[2]; 2017[3]; 2019[4]). Finally, they need basic cognitive skills such as general mental ability, fluid reasoning or working memory (McGrew, 2009[5]); these are often considered fundamental for acquiring more complex skill sets in the other two domains.

In response to this three-part challenge, global student assessment initiatives, such as the Programme for International Student Assessment (PISA), have included skills from the transversal domain in addition to domain-specific knowledge (OECD, 2019[6]). Transversal skills are now part of many educational large-scale assessments and considered important markers of educational achievement (OECD, 2013[2]; 2017[1]; 2019[6]). However, basic cognitive skills have been included to a lesser extent. On average, evidence suggests that countries improve with respect to the core domain skills, but basic cognitive skills are malleable to a lesser extent within education. This, in turn, has lessened the interest of practitioners and policy makers in basic cognitive skills.

For all the importance of developing skills in the core domains, transversal skills and basic cognitive skills, educational systems face another challenge. Given the emergence of artificial intelligence (AI) and robotics, which skills will become obsolete for humans, both as a requirement of the workforce and as an educational goal?

To approach this question, a taxonomy of skills is needed to assess and scale AI-related capabilities. Ideally, this taxonomy will relate to skill frameworks and the tasks associated with them, for instance, from international large-scale assessments. It also needs to distinguish skills from core domains (OECD, 2017[1]), transversal skills (OECD, 2013[2]; 2017[3]; 2019[4]) and basic cognitive skills (McGrew, 2009[5]).

This chapter focuses on core domains and transversal skills, leaving basic cognitive skills for Chapter 3. First, it provides a brief overview of skills from core domains and the transversal domain. It also looks at specific skills typically assessed in education and provides a brief background on the underlying theories. Second, it presents assessment instruments to measure these skills with a focus on innovative and technology-based instruments. Drawing on these two points, it then assesses the extent to which such tests could assess the capabilities of AI.

Educationally relevant skills

Core domains and transversal skills have different theoretical backgrounds, partly due to disparate research traditions. Research on reading literacy (skill set: core domains), for example, originates in the educational and learning sciences. Conversely, research on collaborative problem solving (skill set: transversal domain) is largely rooted in educational large-scale assessments and social psychology.

Although the two skill domains and the nature of the associated skills vary in complexity, they might be equally important for success in life. Both domains relate to recognising, interacting with and solving real-world situations. While core domains and transversal skills are interdependent (OECD, 2014[7]), this chapter considers them separately for ease of interpretation and readability.

Assessment of the two skill sets: General concepts

There is broad consensus that the two skill sets – core domain and transversal skills – are important across several outcomes. Thus, there is a strong need for measurement and assessment to keep track of them and their development. This could include, for instance, international comparisons of educational systems or tracking individual student progress across grade levels.

Several skill sets continue to be the focus of international large-scale assessments, as well as of comprehensive research efforts. Some examples of specific measurements appear below. However, there are many ways of assessing the two skill sets, including classical paper-pencil assessments and highly innovative computer-simulations.

This section focuses on innovative item types and the potential of such items for assessing AI skills. Innovative item types often provide additional information such as behavioural patterns of students when working with dynamically changing problem-solving items such as MicroDYN (see Greiff and Funke (2009[8]); skill set of transversal skills). Similarly, navigating complex texts in an online environment such as digital reading items (skill set of core domain skills) can also reveal behavioural patterns.

Specifically, the chapter considers the field of international large-scale assessments:

- PISA (OECD, 2013[2]; 2017[1]; 2019[9])
- Programme for the International Assessment of Adult Competencies (PIAAC); (PIAAC Expert Group in Problem Solving in Technology-Rich Environments, 2009[10]); (PIAAC Literacy Expert Group, 2009[11])
- National Assessment of Educational Progress (NAEP) (National Assessment and Governing Board, 2019[12]; 2019[13]; 2019[14])
- Trends in International Mathematics and Science Study (TIMSS) (Mullis and Martin, 2017[15])
- Graduate Record Examination (GRE) and the SAT (formerly known as Scholastic Assessment Test).[1]

For each of the two skill domains, this chapter describes typical skills for the respective overarching set of skills; provides an overview and examples of assessments that include these skills; and summarises the theoretical backgrounds and sub-processes for the respective skills. After these subsections, the chapter presents possible dimensions of a skill taxonomy in relation to scaling AI capabilities.

Core domain skills

The core domain skills [i.e. mathematic literacy, reading literacy, science literacy; OECD (2017[1])] focus on knowledge and processes closely related to scholastic domains. Although the labels for these skills are not consistent across assessments, they are similar and show strong overlap (Table 7.1). While definitions and sub-processes of each skill might differ slightly across assessments (and even across different cycles within one assessment), their overlap is substantial. Table 7.1 summarises the definitions of the three skills and their sub-processes. Additionally, it provides an overview of several other large-scale assessments where these skills have been assessed.

Several frameworks exist for each of the skills, including those developed by scientific expert groups within PISA through expert opinions and from the scientific literature. The framework documents are constantly refined as the assessment cycles progress. This provides a theoretical foundation in defining the skills and fans out sub-processes. The frameworks also make suggestions and provide specific guidance on how the theoretical background can be translated into specific and actionable assessments that are ultimately run in the respective assessments such as PISA or PIAAC.

Table 7.1. A comparison of large-scale assessment frameworks

Skill	Definition	Sub-process	Examples of large-scale assessments assessing these skills (and labels used in the assessment)
Mathematical literacy	An individual's capacity to formulate, employ and interpret mathematics in a variety of contexts.	• Formulate mathematics • Employ mathematics • Interpret mathematical results	PISA (Mathematical literacy) PIAAC (Numeracy) NAEP (Mathematics) TIMSS (Mathematics) GRE (Quantitative fluid reasoning) SAT (Mathematics)
Reading literacy	The ability to make use of written texts, to achieve one's goals, to increase one's knowledge and potential, and to participate in society.	• Access and retrieve • Integrate and interpret • Reflect and evaluate	PISA (Reading literacy) PIAAC (Literacy) NAEP (Reading) GRE (Verbal fluid reasoning; analytical writing) SAT (English; Languages)
Science literacy	The ability to engage with science-related issues and with the ideas of science.	• Explain phenomena scientifically • Evaluate and design scientific enquiry • Interpret data and evidence scientifically	PISA (Science literacy) NAEP (Science) TIMSS (Science) SAT (Science)

Note: Definitions are partly direct quotes.
Source: OECD (2017[1]).

Figure 7.1 provides an example of a set of items (labelled "unit" in the PISA context) that assesses mathematical literacy, as used in the PISA 2012 cycle. The unit consists of three items in a real-world context. The students need to solve them by interpreting and comparing the numbers in the table and performing calculations themselves. More technically, they use sub-processes indicated in Table 7.1 to interpret mathematical results for items 1 and 2, and then to employ them for item 3.

Figure 7.2 presents an example item assessing reading literacy in PISA 2018 (OECD, 2017[3]). Test takers need to reflect on and evaluate three texts on space exploration. They then write a comment on its benefits afterwards (i.e. employing the "reflect and evaluate" sub-process; see Table 7.1).

Figure 7.3 displays an example item assessing science literacy in PISA 2012 (OECD, 2014[16]). Test takers are asked to interpret scientific information and explain why they do not support the conclusion of a fellow student (i.e. employing the "explain phenomena scientifically" sub-process; see Table 7.1).

Figure 7.1. Sample item assessing mathematical literacy

WHICH CAR? – a unit from the PISA 2012 main survey

WHICH CAR?

Chris has just received her car driving licence and wants to buy her first car.
This table below shows the details of four cars she finds at a local car dealer.

Model:	**Alpha**	**Bolte**	**Castel**	**Dezal**
Year	2003	2000	2001	1999
Advertised price (zeds)	4 800	4 450	4 250	3 990
Distance travelled (kilometres)	105 000	115 000	128 000	109 000
Engine capacity (litres)	1.79	1.796	1.82	1.783

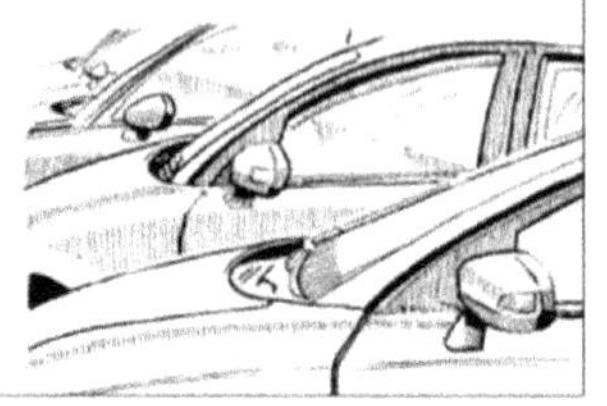

WHICH CAR? – *QUESTION 1*

Chris wants a car that meets **all** of these conditions:

- The distance travelled is **not** higher than 120 000 kilometres.
- It was made in the year 2000 or a later year.
- The advertised price is **not** higher than 4 500 zeds.
- Which car meets Chris's conditions?

A. Alpha
B. Bolte
C. Castel
D. Dezal

WHICH CAR? – *QUESTION 2*

Which car's engine capacity is the smallest?

A. Alpha
B. Bolte
C. Castel
D. Dezal

WHICH CAR? – *QUESTION 3*

Chris will have to pay an extra 2.5% of the advertised cost of the car as taxes.

How much are the extra taxes for the Alpha?

Extra taxes in zeds: ....

Source: OECD (2014[16]).

Figure 7.2. Sample item assessing reading literacy in PISA 2018

PISA 2018

Unit Title: Space Exploration
Question 5/5

Refer to the articles on the right. Type your answer to the questions in the space provided.

Think about how Scott Huffington wrote his article and the commenters responded. Based on this information, write a comment that explains two primary benefits of space exploration? Support your answer with details from the articles.

Text 1 | **Text 2** | **Text 3**

Is the Golden Era of Space Exploration Over?
by Scott Huffington • May 16, 201

Beginning with the launch of Sputnik in 1957 the focus of space exploration had one aim: be the first to go where no human had gone before. In 1961 Yuri Gagarin became the first man in space sparking an intense competition where astronauts and cosmonauts battled to break records, expand frontiers, and bring notoriety to their countries of origin. However, since July 22nd 1969 and Neil Armstrong's historic leap for mankind, space exploration has slowed.

Since then, space programs have focused on creating a sustainable presence in low-Earth Orbit through the development and maintenance of space craft, space stations, and satellites. The Russian space station Mir and the US Skylab were the first space stations but proved too expensive to operate independently. We now have the International Space Station (ISS), an impressive international collaborative effort led by the United States, Russia, Canada, and Japan. Yet, the station was meant to be a stepping stone to bolder projects including a manned mission to Mars. Thirty years later, we are still maintaining the space station but we are no closer to achieving a manned mission to Mars.

For decades, the idea of human space exploration has widely been seen as the exclusive domain of government agencies like the Russian Federal Space Agency (RKA), the National Aeronautics and Space Administration (NASA) in the United States, and the European Space Agency (ESA) with 22 member countries. However, the rise of private companies making serious steps toward successful commercial space flights has many people questioning the relevance and necessity of government run and publicly funded space exploration programs. Add the highly publicized U.S. space shuttle disasters in 1986 and 2003 and the enthusiasm and commitment for space exploration has further eroded.

All of this leads me to conclude that the world has lost the focus and drive to explore new frontiers. I fear that the golden age of space exploration has passed, and we are rapidly progressing toward a decidedly

Source: OECD (2019[9]).

Figure 7.3. Sample item assessing science literacy in PISA 2012

Read the texts and answer the questions that follow.

THE GREENHOUSE EFFECT: FACT OR FICTION?

Living things need energy to survive. The energy that sustains life on the Earth comes from the Sun, which radiates energy into space because it is so hot. A tiny proportion of this energy reaches the Earth.

The Earth's atmosphere acts like a protective blanket over the surface of our planet, preventing the variations in temperature that would exist in an airless world.

Most of the radiated energy coming from the Sun passes through the Earth's atmosphere. The Earth absorbs some of this energy, and some is reflected back from the Earth's surface. Part of this reflected energy is absorbed by the atmosphere.

As a result of this the average temperature above the Earth's surface is higher than it would be if there were no atmosphere. The Earth's atmosphere has the same effect as a greenhouse, hence the term greenhouse effect.

The greenhouse effect is said to have become more pronounced during the twentieth century.

It is a fact that the average temperature of the Earth's atmosphere has increased. In newspapers and periodicals the increased carbon dioxide emission is often stated as the main source of the temperature rise in the twentieth century.

A student named André becomes interested in the possible relationship between the average temperature of the Earth's atmosphere and the carbon dioxide emission on the Earth.

In a library he comes across the following two graphs.

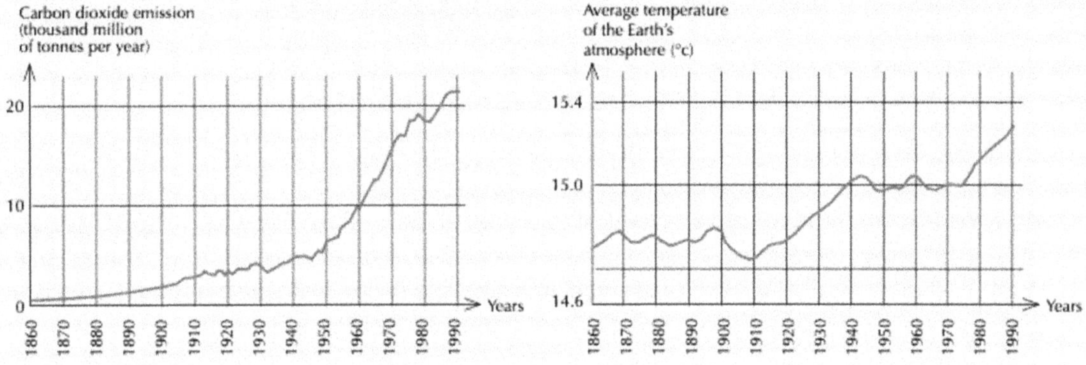

André concludes from these two graphs that it is certain that the increase in the average temperature of the Earth's atmosphere is due to the increase in the carbon dioxide emission.

Another student, Jeanne, disagrees with André's conclusion. She compares the two graphs and says that some parts of the graphs do not support his conclusion.

Give an example of a part of the graphs that does not support André's conclusion. Explain your answer.

....

....

....

Source: OECD (2014[16]).

Transversal skills

Transversal skills such as problem solving, collaboration, creative thinking, learning in a digital world and global competence are centred around two capacities (OECD, 2013[2]; 2017[3]; 2019[4]). First, they allow people to navigate successfully in dynamically changing environments. Second, they allow them to act competently both on a cognitive and a non-cognitive level in today's world.

By definition, these skills are important in a variety of situations (i.e. domain-general) and involve adaptability and flexibility as defining parts (Griffin and Care, 2015[17]). For instance, adaptability and flexibility are inherent parts of problem-solving activities that require dealing with unknown situations and successfully choosing the right actions to solve the problem. Indeed, adaptability and flexibility are at the core of adaptive problem solving that is planned for the PIAAC 2022 assessment (Greiff et al., 2017[18]).

Large-scale assessments have repeatedly focused on transversal skills. PISA 2012, for example, focused on creative problem solving. Meanwhile, PIAAC 2012 looked at problem solving in technology-rich environments, and PIAAC 2022 will look at "adaptive problem solving" (PIAAC Expert Group in Problem Solving in Technology-Rich Environments, 2009[10]; OECD, 2013[2]; Greiff et al., 2017[18]). PISA 2015 assessed collaborative problem solving (OECD, 2017[3]) and PISA 2018 looked at global competence (OECD, 2019[9]). PISA 2022 envisages assessment of other and equally diverse transversal skills such as creative thinking (OECD, 2019[19]), while PISA 2025 plans to assess learning in a digital world.

As with core domain skills, conceptual frameworks exist or are in development for each of the transversal skills. Again, these have been derived through expert opinions and from the scientific literature. They provide a theoretical foundation both in defining the skills and their sub-processes, and in guiding the developments of appropriate assessment instruments.

Table 7.2 displays how frameworks from the completed cycles and drafts for the planned cycles define the above-mentioned transversal skills. It also displays definitions and sub-processes for the transversal skills of problem solving, collaboration, creative thinking, learning in a digital world and global competence on the basis of PISA 2012, 2015 and 2018 assessment frameworks, as well as for the draft of PISA 2022. Additionally, it provides an overview of the large-scale assessments where similar skills have been assessed or will be assessed in future cycles (comparable to Table 7.1 above).

Again, as with the core domain skills, the sub-processes of the transversal skills displayed in Table 7.2 largely define what items ultimately need to assess. However, unlike for core domain skills, innovative item formats are generally required for assessment of transversal skills. Both PISA and PIAAC typically use dynamic, scenario-based approaches where the assessment situation changes through actions of the test taker. In these scenarios, test takers are presented with a problem in a real-world setting, such as working out how to use a new air conditioner (see sample item presented in Figure 7.4).

Innovative item formats provide a more realistic simulation of real-world situations. They are particularly relevant for scaling AI as skills in general are measured against their real-world (and less their theoretical) relevance. In addition, the availability of such item types has implications both for face validity and for how well an assessment can represent the underlying theoretical concept.

To this end, innovative item formats come with several advantages. First, they provide dynamically changing and interactive environments that cannot be simulated using traditional paper-pencil formats. In this way, they allow for assessment of new constructs such as transversal skills [i.e. problem solving, collaboration, creative thinking, learning in a digital world and global competence; OECD (2013[2])]. Second, these innovative items can be easily constructed in divergent ways and can include direct simulations of complex real-world situations. As such, they allow for an increasing construct coverage (Greiff and Funke, 2009[8]). Third, they record the test takers' behaviour, saving it into so-called log files (these come in addition to the final test score). This allows gathering of insights about the processes operating in solving the items (Greiff et al., 2016[20]). Finally, as innovative item formats allow to simulate everyday situations, they offer increasing face validity and engagement, and increased ecological validity (Greiff and Funke, 2009[8]).

Table 7.2. Definitions and sub-processes for five key skills in large-scale assessments

Skill	Definition	Sub-process	Examples of large-scale assessments assessing/planning to assess these skills (and labels used in the assessment)
roblem solving	The ability to engage in cognitive processing to understand and resolve problem situations where a solution is not immediately obvious.	• Explore and understand the problem. • Represent and formulate a mental model of the problem. • Plan and execute strategies to solve the problem. • Monitor and reflect the progress made in solving the problem.	PISA (Creative problem solving, 2012 cycle) PIAAC (Cycle 1 in 2012: Problem solving in technology-rich environments; Cycle 2 in 2022: Adaptive problem solving)
ollaboration	The ability of an individual to engage effectively in a process whereby two or more agents attempt to solve a problem.	• Establish and maintain a shared understanding of the problem. • Take appropriate actions to jointly solve the problem. • Establish and maintain team organisation.	PISA (2015 cycle: Collaborative problem solving)
reative thinking	The ability to generate, evaluate and improve ideas directed towards original and effective solutions, advances in knowledge and impactful expressions of imagination.	• Creative expression (written and visual). • Problem solving (scientific and social).	PISA (2022 cycle: Creative thinking)
earning in a digital world	Forthcoming.	Forthcoming	PISA (2025 cycle: Learning in a digital world)
obal competence	A combination of skills, knowledge, values and attitudes facilitating individuals to act and interact respectfully, successfully and in sustainable manner on a local, global and intercultural level.	• Examine local, global and intercultural issues. • Understand and appreciate different perspectives and worldviews. • Interact successfully and respectfully with others. • Take responsible action towards sustainability and collective well-being.	PISA (2018 cycle: Global competence)

Note: Definitions are partly direct quotes.
Source: OECD (2013[2]; 2014[7]; 2019[9]; 2019[19]); PIAAC Expert Group in Problem Solving in Technology-Rich Environments (2009[10]).

Figure 7.4 provides an example item from PISA 2012 assessment of creative problem solving (OECD, 2014[7]). In a simulated microworld, test takers need to figure out how to use a new air conditioner without further instructions. To this end, they must first explore how the three input variables (i.e. top control, central control and bottom control) influence the outcome variables of temperature and humidity (i.e. sub-process "explore and understand the problem"; see Table 7.2).

In a mental model (see bottom part of Figure 7.4), the test takers then represent how input and output variables are connected based on their explorations. This engages the "represent and formulate a mental model of the problem" sub-process; see Table 7.2). Based on this mental model, test takers derive strategies for solving the problem (i.e. reaching certain target values for temperature and humidity) and subsequently execute them (i.e. plan and execute strategies). At the same time, they monitor their progress (i.e. "monitor and reflect the progress" sub-process; see Table 7.2).

Figure 7.4. Sample item assessing problem solving in PISA 2012

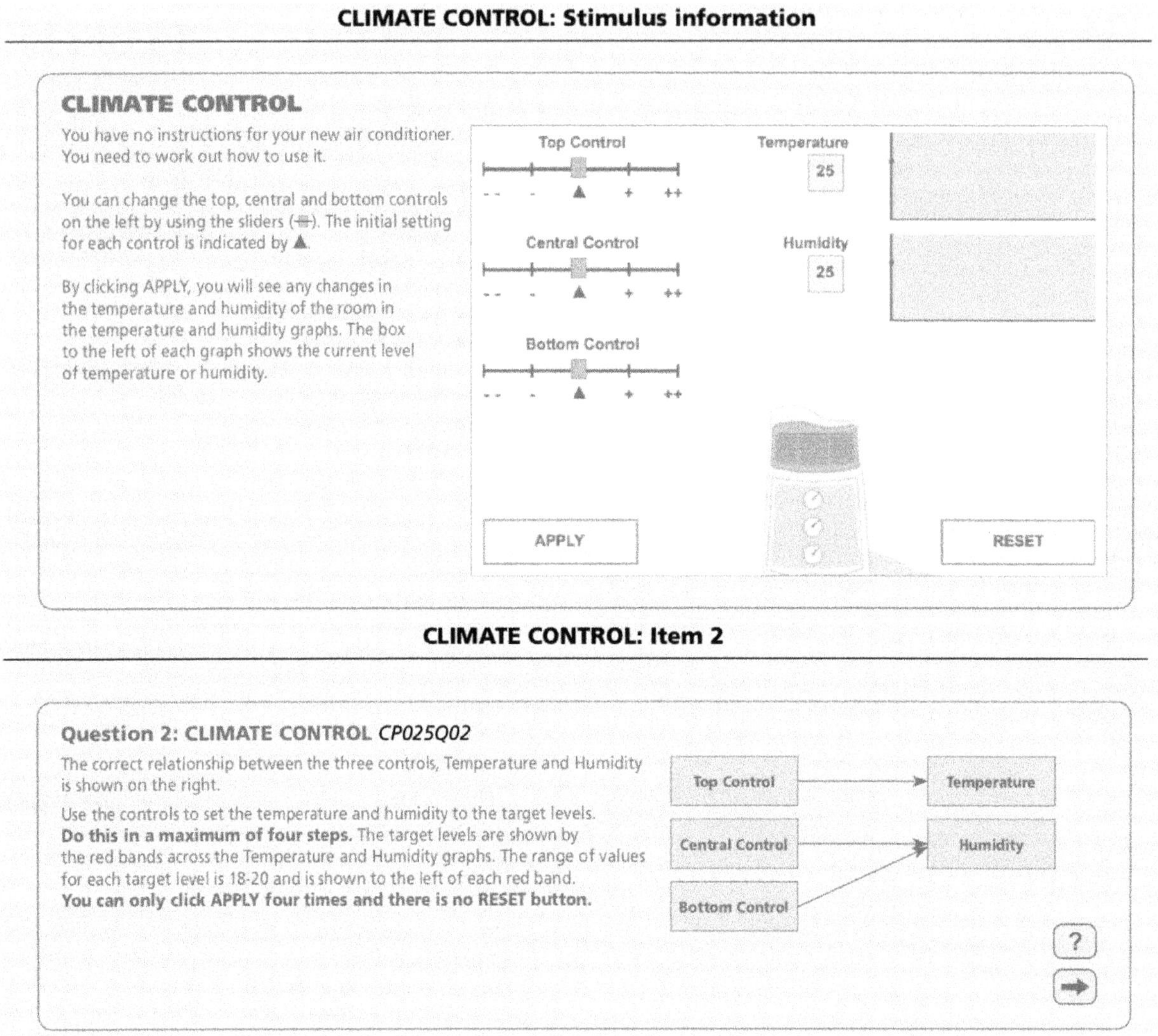

Source: OECD (2014[7]).

Figure 7.5 displays an example item assessing collaboration in PISA 2015 (OECD, 2017[3]). The item requires test takers to work in teams to gather information about a fictional country named Xandar. To this end, the test takers need to interact with computer agents in a chat to establish and maintain a shared understanding of the problem. They subsequently plan and execute strategies to solve the problem together with team members (see sub-processes in Table 7.2).

Figure 7.5. Sample item assessing collaboration in PISA 2015

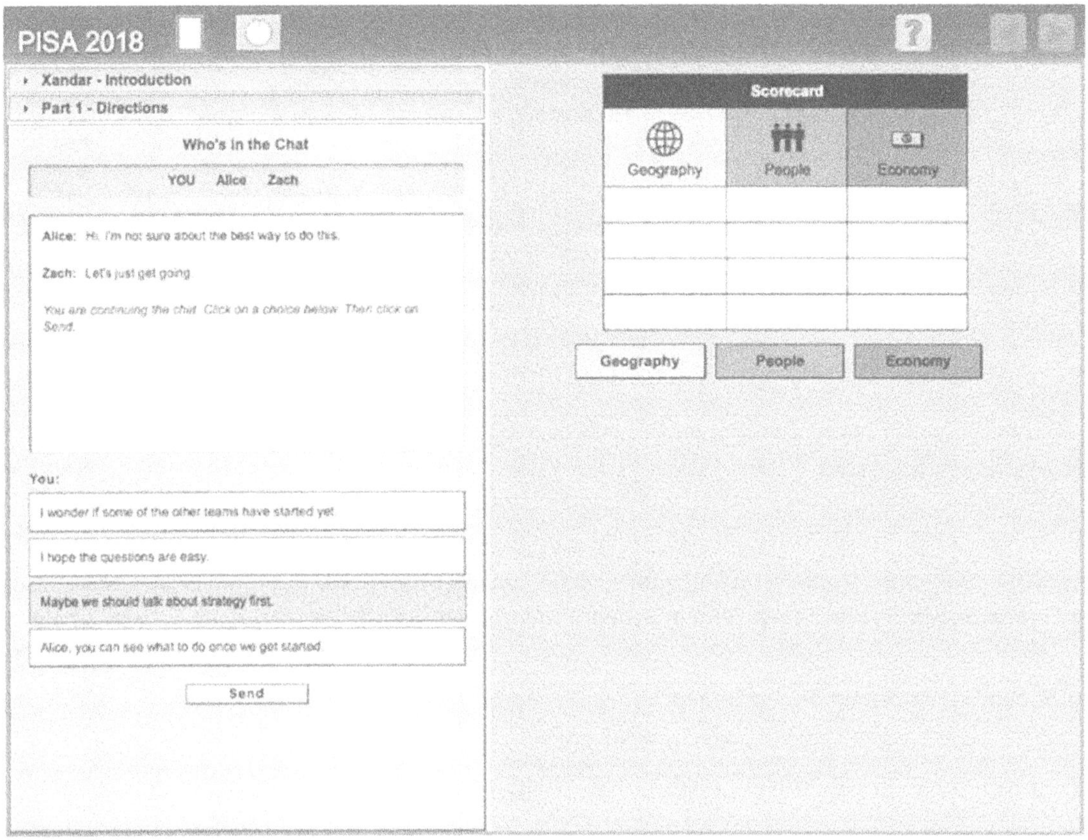

Source: (OECD, 2017[3]).

Role of the two skill sets in occupational settings

This section analyses how core domain and transversal skills relate to explicit skill models in the workforce. It focuses on commonalities and overlap between skills in education and in the workforce. To this end, it introduces two skill models from the workforce (ISCO-08 and O*NET). It also connects the two skill domains with the two skill models to provide an overview of which skills will be important for which types of jobs.

Given the focus of this chapter on scaling AI capabilities, it focuses only on skills found in education and relevant for the workforce. For skill requirements on the job, comprehensive skill models from the workforce, such as ISCO-08 (ILO, 2012[21]) and O*NET (National Center for O*NET Development, 2020[22]), have been derived. In ISCO-08, four skill levels with increasing complexity are distinguished (see Table 7.3).

Table 7.3. Skill levels and their description in the workforce skill model ISCO-08

Skill level	Description	Example tasks requiring the skills
1	Skills for performing simple and routine physical or manual tasks.	Cleaning; carrying materials; performing earthworks.
2	Skills for interacting with machines and information.	Operating, maintaining and repairing machines and electronic devices; manipulating, ordering and storing information.
3	Skills for performing complex technical and practical tasks that require extensive factual, technical and specialised knowledge.	Resource calculations for projects; technical support for professionals; ensuring compliance with regulations and schedules.
4	Skills involving complex problem solving, decision making and creativity.	Understanding and communication of complex information; research and diagnose; designing buildings and machines.

Note: Cell content is partly direct quotes.
Source: ILO (2012[21]).

In the O*NET taxonomy (National Center for O*NET Development, 2020[22]), skills are not arranged in levels of complexity but rather combined in six different groups (see Table 7.4).

Table 7.4. Skill groups and their description in the workforce skill model O*NET

Skill group	Description	Example skills
1	Basic skills	Mathematics; reading comprehension; writing; science.
2	Complex problem solving	Complex problem solving.
3	Resource management	Management of financial, material, human and time resources.
4	Social skills	Co-ordination; negotiation; persuasion.
5	System skills	Judgement and decision making; systems analysis; systems evaluation.
6	Technical skills	Equipment maintenance; operation and control; repairing.

Note: Cell content is partly direct quotes.
Source: National Center for O*NET Development (2020[22]).

Table 7.5 summarises the connection of the two skill sets relevant for education (and their components) with the ISCO-08 levels and O*NET groups. In sum, core domain skills are relevant in almost any type of job according to ISCO-08 and O*NET (i.e. no isomorphic mapping). In contrast, transversal skills are mainly required in non-routine, cognitive occupations that are associated with higher ISCO-08 skill levels and the O*NET groups of complex problem solving, social skills and resource management skills.

For the core domain skills, mathematical and reading literacy are important skills in almost any occupation and can therefore be linked to all ISCO-08 skill levels. However, their involvement may vary across skill levels. For instance, ISCO-08 Skill Level 1 requires only minimal mathematical literacy and reading literacy whereas Skill Level 4 requires extensive proficiency in these skills. In contrast, science literacy is only required at the higher ISCO-08 Skill Levels 3 and 4.

With respect to the O*NET taxonomy, some skills are directly allocated (labelled as mathematics, reading comprehension, writing and science) in the group of basic skills. However, mathematical and reading literacy are required for all other skills groups except for group social skills.

Transversal skills are essential for performing non-routine cognitive tasks, and thus not required on the lowest ISCO-08 skill level. In general, these skills are required in ISCO-08 Skill Levels 3 and 4. However, collaboration and global competence might already be useful on Skill Level 3. This level includes jobs such as bus driver or police officer that involve social interactions and require individuals to understand their role and responsibilities in greater groups and society as a whole. In contrast, creative thinking is only connected to ISCO-08 Skill Level 4.

Table 7.5. Integration of core domain skills and transversal skills into the workforce skill models ISCO-08 and O*NET

Skill domain	Skill	ISCO-08 Skill level	O*NET Skill group
Core domain skills	Mathematical literacy	1-4	Primarily basic skills, but also complex problem-solving skills, resource management skills, systems skills and technical skills
	Reading literacy	1-4	Primarily basic skills, but also complex problem-solving skills, resource management skills, systems skills and technical skills
	Science literacy	3-4	Basic skills
Transversal skills	Problem solving	3-4	Complex problem-solving skills
	Collaboration	2-4	Social skills; resource management skills
	Creative thinking	4	Basic skills
	Global competence	2-4	Social skills; resource management skills

How to scale the capabilities of artificial intelligence? Some recommendations

This chapter identified two sets of skills relevant for concurrent educational efforts: skills from the core domains and transversal skills. It has provided definitions for each of these skill sets, referenced theoretical frameworks and provided examples of items. It has also drawn connections to ISCO-08 and O*NET, two commonly used frameworks to describe demands on the job. Through these frameworks, it has shown that many skills relevant in education can be found, in one way or another, in taxonomies of work requirements.

From a content perspective, both skill sets (and the specific skills therein) are relevant for educational success and beyond. The same holds for the set of basic cognitive skills covered in Chapter 3. Thus, in principle, all of them can be used to assess and scale the capability of AI. This is comparable to the assessment of the capacity of a student or educational system on different levels and skills. In fact, it is for good reasons that international large-scale assessments measure different dimensions to gain an adequate overview. The same approach should be considered when assessing AI capabilities (i.e. looking at different dimensions from all available skill sets).

In addition to a broad content coverage of different skill sets from the set of domain-specific, transversal and basic skills, educational tests to scale AI capabilities should consider these five recommendations.

Recommendations

- **Select skills based on established relevance**

There is a limited number of skills towards which the capability of AI to reproduce human capabilities can be assessed and, subsequently, scaled. With this in mind, selection of skills should be driven by the available body of existing theory-driven empirical research. Moreover, it should include only skills for which there is a minimal level of agreement across researchers, practitioners and policy makers about their theoretical, empirical and curricular relevance. For instance, mathematics and science are clearly relevant skills that have been part of curricular teaching across the globe for decades. Conversely, transversal skills have accumulated less research and are not (yet) consistently included in school curricula. A hierarchy of skills that considers consensus on theoretical, empirical and practical levels, and curricular relevance in education and the workforce, will be beneficial when assessing and scaling AI-related capabilities.

- **Ensure enough high-quality items are available to measure the skill**

AI capabilities, as all other skills, should not be judged on specific items and their content but rather on underlying psychological constructs (so-called latent traits). To this end, assessments need sufficiently high numbers of items that all tap into the same construct in a reliable and valid manner. More specifically, this implies that empirical research has identified a set of items as reliable and valid indicators of a measurement. Moreover, many items are needed to allow for testing across different item contexts. This becomes particularly challenging when looking at scenario-based, computer-administered item types. Such item types are laborious to develop and so usually fewer are available, even though they are better representations of a real-world scenario. Thus, when choosing assessments to scale AI capabilities, constructs with large and empirically tested item banks should be preferred.

- **Use only skills linked to a measurement theory to scale AI-related capabilities**

Only skills for which items as the direct observable entities and the construct as the latent entity are explicitly linked through a measurement theory should be used to scale AI-related capabilities. Different measurement theories are available but most large-scale assessments use Item-Response-Theory, which links specific item responses to the underlying traits in a probabilistic way. The recommendation primarily relates to measurement theory but extends to the need for scoring rules and linking procedures that are clearly spelled out and theoretically justified. Linking of item banks from different studies is difficult and requires specific statistical procedures. It is, for instance, not possible to link items of different assessments empirically such as PISA or GRE, even if they theoretically claim to measure the same concept. Thus, skills for which measurement is rooted in an established psychological measurement theory and for which enough items are empirically linked to each other are preferred.

- **Ensure the underlying process towards the correct solution of item can be described**

Scaling AI capabilities is unlikely to stop at the evaluation of whether an AI algorithm can solve a particular item or where it stands on a dimensionally measured construct. It will be more informative to measure and describe at what point an AI algorithm fails in solving an item and its distance from a pre-specified goal (i.e. the solution). Scenario-based items usually provide information that goes beyond a correct/incorrect judgement of the response and bear the potential to explicate more fine-grained information. Of note, it would also be interesting to see whether AI can improve after receiving feedback (related to the concept of formative assessment). To this end, preference should be given to assessments that provide information beyond the mere correctness of a response and provides data on the underlying solution process.

- **Use items to assess skills that AI can understand and perform**

Only tests that involve tasks AI can understand and that contain operators that, in principle, AI can perform should be used. For instance, it is not meaningful to confront an AI algorithm with a task that requires some physical intervention.

The five recommendations should be carefully weighed against each other when scaling AI capabilities for educationally relevant skills. This exercise allows choosing a taxonomy that predicts which skills AI can replace and make human input – at least to some extent – superfluous.

This is an interdisciplinary undertaking that requires substantial evaluation and great expertise. It should involve experts from relevant fields including education, cognitive science, computer science, AI and machine learning, and economy with both scientific and policy perspective. Together, they can fill such a taxonomy with the aim of making informed judgements on how the state of the art allows to predict the role of AI in education and the workforce within the next decades.

References

Autor, D., F. Levy and R. Murnane (2003), "The skill content of recent technological change: An empirical exploration", *The Quarterly Journal of Economics*, Vol. 118/4, pp. 1279-1333, https://doi.org/10.1162/003355303322552801. [23]

Greiff, S. and J. Funke (2009), "Measuring complex problem solving: The MicroDYN approach", in Scheuermann, F. and J. Björnsson (eds.), *The Transition to Computer-based Assessment: New Approaches to Skills Assessment and Implications for Large-scale Testing*, OPOCE, Luxembourg. [8]

Greiff, S. et al. (2016), "Understanding students' performance in a computer-based assessment of complex problem solving: An analysis of behavioral data from computer-generated log files", *Computers in Human Behavior*, Vol. 61, pp. 36-46, https://doi.org/10.1016/j.chb.2016.02.095. [20]

Greiff, S. et al. (2017), "Adaptive problem solving: Moving towards a new assessment domain in the second cycle of PIAAC", *OECD Education Working Papers*, No. 156, OECD Publishing, Paris, https://dx.doi.org/10.1787/90fde2f4-en. [18]

Griffin, P. and E. Care (2015), *Assessment and Teaching of 21st century Skills: Methods and Approach (1st ed.)*, Springer, https://doi.org/10.1007/978-94-017-9395-7. [17]

ILO (2012), *International Standard Classification of Occupations: ISCO-08*, International Labour Organization, Geneva. [21]

International Baccalaureate Organization (n.d.), "Curriculum", webpage, https://www.ibo.org/programmes/diploma-programme/curriculum/ (accessed on 20 October 2021). [27]

International Baccalaureate Organization (n.d.), "Why the IB is Different", webpage, https://www.ibo.org/benefits/why-the-ib-is-different/ (accessed on 20 October 2021). [28]

Mainert, J. et al. (2019), "The incremental contribution of complex problem-solving skills to the prediction of job level, job complexity and salary", *Journal of Business Psychology*, Vol. 34, pp. 825-845, https://doi.org/10.1007/s1. [24]

McGrew, K. (2009), "CHC theory and the human cognitive abilities project: Standing on the shoulders of the giants of psychometric intelligence research", *Intelligence*, Vol. 37/1, pp. 1-10, https://doi.org/10.1016/j.intell.2008.08.004. [5]

Mullis, I. and M. Martin (eds.) (2017), *TIMSS 2019 Assessment Frameworks*, TIMSS & PIRLS International Study Center, Lynch School of Education, Boston College and International Association for the Evaluation of Educational Achievement, Amsterdam. [15]

National Assessment and Governing Board (2019), *Mathematics Framework for the 2019 National Assessment of Educational Progress*, National Assessment Governing Board, Washington, DC. [12]

National Assessment and Governing Board (2019), *Reading Framework for the 2019 National Assessment of Educational Progress*, National Assessment Governing Board, Washington, DC. [13]

National Assessment and Governing Board (2019), *Science Framework for the 2019 National Assessment of Educational Progress*, National Assessment Governing Board, Washington, DC. [14]

National Center for O*NET Development (2020), "O*NET OnLine", webpage, https://www.onetonline.org/ (accessed on 20 October 2021). [22]

OECD (2019), *PISA 2018 Assessment and Analytical Framework*, PISA, OECD Publishing, Paris, https://dx.doi.org/10.1787/b25efab8-en. [9]

OECD (2019), *PISA 2018 Results (Volume I): What Students Know and Can Do*, PISA, OECD Publishing, Paris, https://dx.doi.org/10.1787/5f07c754-en. [6]

OECD (2019), *PISA 2021 Creative Thinking Framework (Third Draft)*, PISA, OECD Publishing, Paris. [19]

OECD (2019), *Skills Matter: Additional Results from the Survey of Adult Skills*, OECD Skills Studies, OECD Publishing, Paris, https://dx.doi.org/10.1787/1f029d8f-en. [4]

OECD (2017), *PISA 2015 Assessment and Analytical Framework: Science, Reading, Mathematic, Financial Literacy and Collaborative Problem Solving*, PISA, OECD Publishing, Paris, https://dx.doi.org/10.1787/9789264281820-en. [1]

OECD (2017), *PISA 2015 Results (Volume V): Collaborative Problem Solving*, PISA, OECD Publishing, Paris, https://dx.doi.org/10.1787/9789264285521-en. [3]

OECD (2014), *PISA 2012 Results: Creative Problem Solving (Volume V): Students' Skills in Tackling Real-Life Problems*, PISA, OECD Publishing, Paris, https://dx.doi.org/10.1787/9789264208070-en. [7]

OECD (2014), *PISA 2012 Results: What Students Know and Can Do (Volume I, Revised edition, February 2014): Student Performance in Mathematics, Reading and Science*, PISA, OECD Publishing, Paris, https://dx.doi.org/10.1787/9789264208780-en. [16]

OECD (2013), *PISA 2012 Assessment and Analytical Framework: Mathematics, Reading, Science, Problem Solving and Financial Literacy*, PISA, OECD Publishing, Paris, https://dx.doi.org/10.1787/9789264190511-en. [2]

Pellegrino, J. and M. Hilton (2013), *Education for Life and Work: Developing Transferable Knowledge and Skills in the 21st Century*, National Academies Press, Washington, D.C. [25]

PIAAC Expert Group in Problem Solving in Technology-Rich Environments (2009), "PIAAC Problem Solving in Technology-Rich Environments: A Conceptual Framework", *OECD Education Working Papers*, No. 36, OECD Publishing, Paris, https://dx.doi.org/10.1787/220262483674. [10]

PIAAC Literacy Expert Group (2009), "PIAAC Literacy: A Conceptual Framework", *OECD Education Working Papers*, No. 34, OECD Publishing, Paris, https://dx.doi.org/10.1787/220348414075. [11]

PIAAC Numeracy Expert Group (2009), "PIAAC Numeracy: A Conceptual Framework", *OECD Education Working Papers*, No. 35, OECD Publishing, Paris, https://dx.doi.org/10.1787/220337421165. [26]

Notes

[1] Another assessment program, the *International Baccalaureate* (International Baccalaureate Organization, n.d.[28]), is not strictly a large-scale assessment. The IB aims at equipping students with subject-specific as well as subject-general skills through different programs. In the Diploma Programme, for example, students complete three core disciplines: theory of knowledge (i.e. reflecting on the concept of knowledge), extended essay (i.e. conducting and reporting a piece of research), creativity, activity, and service (International Baccalaureate Organization, n.d.[27]). In addition, students choose one subject out of each of the following six subject groups: Studies in language and literature, Language acquisition, Individuals and societies, Sciences, Mathematics and the arts (International Baccalaureate Organization, n.d.[27]).

8. Abilities and skills: Assessing humans and artificial intelligence/robotics systems

Phillip L. Ackerman, Georgia Institute of Technology

This chapter makes recommendations towards an approach for comparing human and artificial intelligence (AI) capabilities. It stresses the need to compare domain knowledge and skills rather than broad, higher-order abilities such as intelligence. The chapter provides the theoretical and empirical foundations of domain knowledge and skills assessments. It discusses methodological challenges arising from humans' use of tools, differences in learning between humans and AI, and the inaccuracy of skills assessments at high performance levels. Finally, the chapter proposes an assessment strategy that draws on tests developed for jobs subject to licensing examination.

Introduction

This chapter argues the pedagogical rather than the psychological method of assessment holds more promise for assessing the respective capabilities of human and artificial intelligence (AI)/robotics systems. A psychological approach assesses underlying abilities identified from research on human intelligence. Conversely, a pedagogical approach seeks to understand and assess specific knowledge and skills of an individual that allow for the successful (or unsuccessful) accomplishment of real-world tasks. The latter approach will usefully allow a contrast between the relevant respective capabilities of both human and AI/robotics systems. These knowledge and skills repertoires typically have both declarative knowledge and procedural knowledge components. In many cases, they also have tacit knowledge involvement [see (Polanyi, 1966/1983[1])].

The chapter provides the theoretical and empirical foundations of such assessments, especially in the domain of certification tests. Different types of knowledge, along with issues of tool use, learning and differentiating between competence and expertise are discussed. Finally, it proposes a strategy for developing a sampling of tests and tasks for comparative assessments.

Intelligence assessment from psychological and pedagogical methods

The mainstream theory and application of human intelligence/abilities assessment are fundamentally mismatched with assessing the adequacy of AI/robotics systems to replace human systems. This mismatch derives from how psychologists have explored the taxonomy of human abilities and used ability assessments to predict individual differences in learning and task/job performance.

Starting with Binet and Simon (1961[2]) in 1905 and 1908, and even with Spearman (1904[3]), abilities researchers and practitioners have primarily sought to determine the factors underlying human capabilities and limitations to help develop respective tests and measurements. The results could then be used to predict individual differences in success or failure in *future* academic and occupational situations. Binet was interested in predicting failures of children in overall academic performance. Spearman, in contrast, wanted to determine the fundamental characteristics of individual differences in a general intellectual ability – sometimes called a "mental engine".

Tests and measurements derived from both approaches' attempt to assess the underlying abilities that give rise to individual differences in learning and performance. However, they say little about whether an individual can perform any particular task beyond the tests themselves. For example, Spearman's advocates commonly use the Raven's Progressive Matrices test. This measure requires inductive reasoning with non-verbal test content [e.g. Burke (1958[4])]. The test has obvious limitations for predicting academic or job performance [e.g. Vernon and Parry (1949[5])]. Moreover, the test scores provide no useful information about the individual's knowledge and skills for any other task.

Broad intelligence tests, such as the Stanford-Binet or one of the Wechsler tests, have some advantages over the Raven's test. Stanford-Binet and Wechsler provide more detailed information about a range of different abilities than the Raven's test. For example, they assess spatial, verbal and math abilities, as well as some general/cultural knowledge.

Yet, like the Raven's test, the broad intelligence tests fall short in key areas. For example, they fail to indicate whether the examinee is good at physics, carpentry, medicine, plumbing or just about any other occupation. Furthermore, the design of these tests actually precludes assessment of basic literacy skills.

Information processing tests, such as assessments of simple and choice reaction time, perception, attention and working memory, may attempt to determine individual differences in the "building blocks" for human intelligence. However, they too provide little insight into the knowledge and skills of the individual examinees.

At best, all these tests may simply predict an individual's likelihood to succeed in an academic learning or occupational training programme. The tests resemble Aristotle's depiction of a block of bronze in terms of "potentiality" and "actuality" (Ackerman, 2008[6]). In other words, intelligence tests only assess current performance (and not actual "potential").

However, the use of intellectual ability tests is fundamentally a matter of prediction of some criterion, such as future academic or occupational performance. These tests do not directly measure the knowledge, skills and abilities needed for any specific task, except in highly limited and artificial circumstances (e.g. mental arithmetic).

The idea of "potentiality" and "actuality" can be transposed to the AI/robotics domain. A central processing unit and a six-axis robot arm have enormous "potential" to perform basic or advanced human tasks/jobs. However, their "actual" capability is limited by lack of software to guide them in such tasks. Thus, the discussion about potential is practically useless in determining what tasks the systems can accomplish today, and likely limited in predicting tasks they will accomplish tomorrow.

The pedagogical method of intelligence assessment

Binet referred to his method of intelligence as the "psychological method" of assessing intelligence. Such an approach assesses higher-order mental abilities (reasoning, memory, comprehension). He contrasted this with the "pedagogical method" [see Ackerman (1996[7])], which assesses "intelligence" by *examining what the individual knows*.

The shortcomings of psychological tests have been known for more than a century. As early as the 1910s, when psychologists developed the first "trade" tests, they recognised the limitations of the psychological method for assessing *current* capabilities [see Chapman (1921[8])]. In the First World War, trade tests aimed to determine which individuals had expertise in trades such as electronics, automotive mechanics, butcher, cook, barber, etc. The pedagogical approach allows individuals to be assigned to a specialty and perform the job without additional training. These tests took a variety of approaches, including self-report background, open-ended tests, multiple-choice questions and hands-on performance. The key theme is that the tests were designed to sample the knowledge and skills of individual examinees. The most successful versions were those that required actual hands-on demonstration of expertise and thus a direct assessment of skills.

Such techniques are still used in competence assessments (see Chapter 9). Similar assessments might be used across a variety of jobs/tasks[1] to compare human and AI/robotics systems for at least two of the three major types of knowledge: declarative, procedural and tacit. These are discussed below.

Declarative knowledge (knowing that)

Declarative knowledge is essentially factual knowledge. In earlier eras, such information could be found in an encyclopaedia. Today, the search for facts on a country, historic figures, literature and so on is often easily accessible through an Internet search.

Declarative knowledge may be isolated facts but can also consist of principled and organised information. Examples of organised information include the Periodic Table of chemical elements and classification systems in botany. Adler (1974[9]), for example, outlined five branches of knowledge that mainly represent declarative knowledge (Logic, Mathematics, Science, History and the Humanities, and Philosophy).

A wide range of paper-and-pencil occupational certification tests is designed to assess job domain-specific declarative knowledge. Although multiple-choice tests are common, Carroll (1982[10]) and others have discussed their limitations. Apart from their dependence on literacy/reading comprehension, these tests depend mainly on "recognition" of correct answers.

Conversely, a reliance on "recall" requires deeper knowledge of human examinees and may place similar demands on AI/robotics systems. Other techniques have also been used, such as oral examinations and extensive use of photographs or physical objects (Figure 8.6).

Figure 8.6. Selected item from Carpenter trade test

THE PICTURE TRADE TEST METHOD 201

PICTURE 18

22. Q. Where is the sill?
A. I.
23. Q. Where are the studs?
A. A.
24. Q. Where is the sheathing?
A. C.
25. Q. Where is the water table?
A. H.

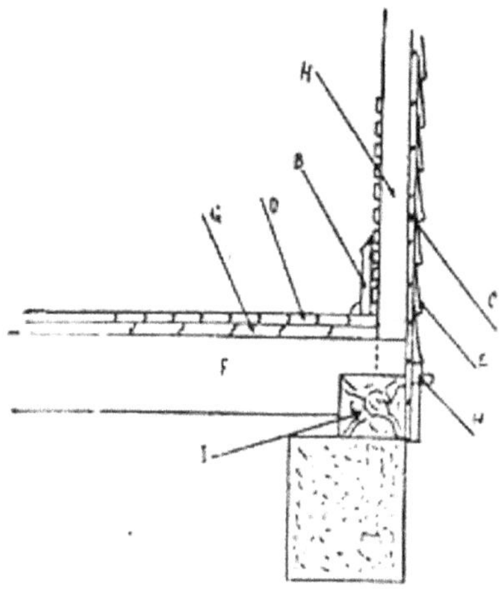

IDENTIFICATION OF DIFFERENT KINDS OF WOOD

Say to the candidate: "Here are a number of different kinds of wood. Tell me the name of each kind. Begin with number 1 and go right through."

26. (1) A. White Pine.
27. (2) A. Hackmatack (Larch) (Tamarack).
28. (3) A. Cypress.
29. (4) A. Basswood.
30. (5) A. Maple.
31. (6) A. Cherry.
32. (7) A. Elm (Butternut).
33. (8) A. Ash.
34. (9) A. Chestnut.
35. (10) A. Oak.
36. (11) A. Red Oak.
37. (12) A. Gum (Hazel) wood.
38. (13) A. Walnut.
39. (14) A. Mahogany (Baywood).
40. (15) A. Teak (Pasanda).

Source: Chapman (1921[8]).

Procedural knowledge (knowing how)

Procedural knowledge consists of sequences of actions (e.g. baking a cake, operating a table saw). In some cases, procedural knowledge of a sequence can be represented as declarative knowledge (e.g. a musical score or a recipe for a meal). However, for procedural knowledge, the action sequence must be performed to assess the individual's competence.

Thus, someone who can write down from memory the score from a Beethoven sonata could only be said to have declarative knowledge of the sonata. Someone who can play it competently in real time on a piano could be said to have procedural knowledge of the sonata (regardless of whether that person could write down the score).

The task list to acquire a barber's licence in the state of New York (United States) offers a suitable example of procedural knowledge assessment. Among 14 areas of job tasks, for example, the examination tests "haircutting techniques". This, in turn, has the following subcomponents: "comb the head, taper the nape area, sideburns, top, back & sides, arching, re-drape, and prepare for finish [the hairstyle]" (Government of New York, 2020[11]).

Tacit knowledge (knowing with)

Human adults are said to have acquired various degrees of knowledge that cannot be easily subsumed under either declarative or procedural knowledge categories. Polanyi (1966/1983[1]) described this as "tacit knowing". This kind of knowledge is not typically articulated. Frequently, it is not even accessible through personal introspection.

Broudy (1977[12]) called this type of knowledge as "knowing with". He proposed that, for educated people, this knowledge would be what the individual "thinks, perceives and judges with everything that he has studied in school, even though he cannot recall these learnings on demand" (Broudy, 1977, p. 12[12]).

These concepts of tacit knowledge share similarities with Gestalt principles of organisation [e.g. Köhler (1947[13])]. However, Bransford and Schwartz (2000[14]) offer more explicit examples of tacit knowledge in the context of transfer of knowledge/training from one domain to another. For a discussion, see Ackerman (2008[6]).

Job/task-relevant tacit knowledge may contribute to success in some occupations. However, it is not yet possible to develop assessments that reveal individual differences in tacit knowledge. This is mainly because it is difficult to articulate this type of knowledge to begin with.

Task-independent components of tacit knowledge may exist, such as determining how to manage or motivate particular employees or interact with customers. However, there may also be task/job-specific elements, too. Ultimately, attention to this type of knowledge may be necessary to compare human and AI/robotic systems for effectiveness.

Tool use

One of the most salient elements in the human history of work has been the development and application of tools. The first tools augmented human muscles (e.g. the lever, pulley). The next tools augmented sensory and perceptual limitations (e.g. telescopes, radar). Finally, tools were developed to augment cognitive limitations (calculators and computers).

Most recently, of course, are the tools made available by the Internet. These tools provide both declarative knowledge of the sort available in news websites or Wikipedia. However, they also entail information relevant to the acquisition of procedural knowledge (such as explanatory YouTube videos).

Historically, ability assessments have limited the availability of most tools for completion of problems. Notable exceptions include paper and pencil for group tests, and hand-held calculators for tests like the Scholastic Assessment Test (SAT) [e.g. Ackerman (2018[15])].

Tools in occupations serve different uses. They may be required for day-to-day task completion (such as for a carpenter, electrician or surgeon). Tools may also serve mainly to augment individual capabilities (such as computerised spell checkers and grammar advisers for writing tasks).

For many jobs, removing access to such tools might make required tasks difficult or impossible to complete. In these cases, individuals might need an entirely new set of skills or to relearn an old set of skills. This could happen, for example, if a dentist desired to diagnose the presence or absence of a cavity without being able to employ X-ray technology.

On one hand, AI/robotics systems could be considered to be tools rather than agents. On the other hand, if AI/robotics systems are considered to be independent agents compared to humans, what would be a "fair" comparison? Would AI/robotics systems be assessed solely by what can be accomplished with a common set of "tools"? Would they be assessed without reference to other computerised sources (e.g. natural language processors or external databases)? It will be difficult to establish the boundary conditions for allowing either system to make use of tools during the assessment.

Competence vs. expertise

There are numerous examples of thorough assessment procedures for assessing "competence" of human jobs/tasks across many occupations. However, there is a significant limitation associated with using such assessments for comparing AI/robotics and human capabilities. This is because most competence assessment instruments and procedures are designed to determine whether the examinee has a "minimum competence" for certification.

Such certifications range from law and health care (medical doctors, nurses, psychologists) to plumbing, electricians, taxi-cab drivers and practice in a variety of other occupations. Depending on the country, they could even include obtaining a high-school diploma. This is a reasonable threshold for determining whether the individual can perform a task at an acceptable level. However, when two individuals have obtained a passing grade, such assessments do not ordinarily determine whether one is more "expert" than the other.

Some assessments set a higher threshold than "acceptable" performance (e.g. Board Certification for medical doctors in the United States). However, rather than assigning differential scores to individuals, these assessments mainly just raise the passing threshold to a higher standard.

Ideally, assessments for comparing AI/robotics systems against human operators should distinguish between the two on continuous-graded scales. If not, they should at least be capable of reliable and valid assignment of categorical ratings, such as "novice, apprentice, journeyman and expert" Chapman (1921[8]).

It may be difficult to use such assessments in a manner that allows for rank-ordering of expertise among human examinees. Various stakeholders (e.g. unions, individual employees) may resist disclosure of information beyond the certification. (For example, consider the doctor who received only a "barely passing" score on a Board-certification exam.) The employees may also fear the information might be used in ways that are deleterious or cannot be anticipated, such as when clients or customers do not know how to evaluate varying levels of performance on the certification exams. For example, in the United States, airline pilots are re-certified periodically to maintain their licences. However, in accordance with union agreements, all other data beyond whether they pass or fail are scrubbed after re-certification.

One could adapt many certification or competence assessments to assess the relative strengths and weaknesses of individual humans or AI/robotics systems. However, the traditional psychometric approach provides the maximal discrimination at the competent/not competent cut-off levels in these certification assessments. (In this sense, it is similar to how the original IQ tests were designed to differentiate more precisely at lower levels of performance. They give little attention to distinguishing among higher-ability individuals). Thus, a redesign of such assessments might be needed prior to use. This would remove ceiling-effect limitations and make more precise evaluations at the higher-end of performance.

The "critical incident technique" (Flanagan, 1954[16]) develops/adapts assessments with greater focus on measuring varying levels of expertise rather than competence. This allows the measurement professionals to generate assessment scenarios that have historically been associated with exceptionally high/low levels of effectiveness for specific tasks/jobs. These scenarios can then be adapted for future assessments.

The technique draws on descriptions of situations or scenarios with an outsized effect on the success or failure of a job task. From the collection of critical incidents, assessments can be designed to go beyond the "basic" requirements of the tasks/jobs, which typically focus on day-to-day requirements. Rather, they can focus on a more holistic assessment. These would look at situations that are likely to have greater impacts on performance than are identified through a more traditional job analysis.

Learning

The speed of acquisition of knowledge and skills (e.g. learning) of the respective systems is another consideration. The respective strengths and weaknesses of human and AI/robotics systems likely diverge in terms of the underlying characteristics for tasks of different learning demands.

For some high-knowledge tasks, such as radiological diagnoses, humans can only attain successful performance through long and varied training and practice. Conversely, AI/robotics systems only need access to a substantial database to find patterns and draw accurate conclusions. As a result, humans may be highly limited in contrast to AI/robotics systems.

Jobs/tasks that require relatively little training/learning for accomplishment pose an interesting contrast between human and AI/robotics systems. Human operators bring a basic repertoire of knowledge and skills to such tasks, such as sorting mail or "bussing" a restaurant table. Yet AI/robotics systems may be especially challenged for these same tasks.

There is no suitable taxonomy of knowledge and skills that has a high prevalence or is nearly universal among human adults. It clearly runs the gamut – from gross body movements (e.g. walking, running, picking up widely different objects) and a variety of manual dexterity tasks (e.g. eating, bathing, cooking), to basic personal tool use (e.g. toothbrush, knife, fork, spoon or chopsticks) and occupational tool use (e.g. screwdriver, hammer).

Similarly, although basic literacy skills are not nearly universal in many areas, humans have a wide prevalence of skills for recognising letters and numbers. This means many individuals can learn mail-sorting skills without substantial investment in training. Conversely, machine learning of mail-sorting was a significant challenge to the early AI/robotics community. This was especially the case with hand-addressed mail rather than mail with bar-coded address labels. In addition, most jobs/tasks also require understanding of natural language and other near-universal human skills. Such differences in competence must be considered in assessing the overall effectiveness of AI/robotics systems in comparison to human systems.

Recommendations

The OECD wants to use taxonomies of human abilities as a starting point, presumably as a tool to provide the foundation for sampling of particular assessments to compare human and AI/robotics systems. However, the disconnection between ability taxonomies [e.g. Carroll (1993[17])] and real-world job knowledge/skills is a significant impediment for such an approach.

The problem does not seem insurmountable. The following approach, followed in sequence, might be suitable:

- **Survey OECD countries for jobs that require explicit testing**

Survey OECD countries to find which jobs require explicit testing. Of particular interest are tests that combine domain-specific declarative knowledge assessments (e.g. written or oral tests) with hands-on procedural knowledge assessments (i.e. that require more than completion of course work or apprenticeship/supervised clinical hours).

- **Determine nature of analysis (jobs vs. tasks)**

Determine whether the analysis will take place at the broader level of "jobs" or the narrower level of "tasks" within jobs.

- **List jobs/tasks and sort by ability/skill**

Obtain a list of these jobs/tasks, and then have subject matter experts sort the jobs/tasks by similarity of underlying ability/skill demands.[2]

- **Conduct a cluster analysis**

Subject the aggregated sorting data to cluster analysis (Hartigan, 1975[18]; Arabie and Carroll, 1980[19]) to determine groups of similarly situated jobs/tasks.

- **Create representative set of jobs/tasks for use/adaptation**

Sample from the obtained clusters and then create a representative set of jobs/tasks from which certification/competence assessments can be used/adapted for comparing human and AI/robotics systems.

It will not be easy to compare humans and AI/robotics head-to-head on a job level. This is because many jobs involve a multitude of tasks, some of which may be idiosyncratic to a particular individual and/or a particular day. Comparing humans and AI/robotics head-to-head will therefore likely need to start at the task level. At a later stage, accounting for an increasing number of tasks may approximate the majority of particular job requirements.

References

Ackerman, P. (2018), "Intelligence as potentiality and actuality", in Sternberg, R. (ed.), *The Nature of Human Intelligence*, Cambridge University Press, Cambridge, UK. [15]

Ackerman, P. (2008), "Knowledge and cognitive aging", in Craik, F. and T. Salthouse (eds.), *The Handbook of Aging and Cognition: Third Edition*, Psychology Press, New York. [6]

Ackerman, P. (1996), "A theory of adult intellectual development: Process, personality, interests, and knowledge", *Intelligence*, Vol. 22/2, pp. 227-257, https://doi.org/10.1016/S0160-2896(96)90016-1. [7]

Adler, M. (1974), "The circle of learning", in *Encyclopædia Britannica*, Encyclopædia Britannica, Inc., Chicago, IL. [9]

Arabie, P. and J. Carroll (1980), "MAPCLUS: A mathematical programming approach to fitting the ADCLUS model", *Psychometrika*, Vol. 45/2, pp. 211-235, https://doi.org/10.1007/BF02294077. [19]

Binet, A. and T. Simon (1961), "Méthodes nouvelles pour le diagnostic du niveau intellectuel des anormaux (Année Psychologique, 11, 191-336); and Le développement de l'intelligence chez les enfants (Année Psychologique, 14, 1-94)", in Jenkins, J. and D. Paterson (eds.), *Studies in Individual Differences: The Search for Intelligence*, Appleton-Century-Crofts, New York, https://doi.org/10.1037/11491-000. [2]

Bransford, J. and D. Schwartz (2000), "Rethinking transfer: A simple proposal with multiple implications", *Review of Research in Education*, Vol. 24, pp. 61-100, https://doi.org/10.3102/0091732X024001061. [14]

Broudy, H. (1977), "Types of knowledge and purposes of education", in Anderson, R., R. Spiro and W. Montague (eds.), *Schooling and the Acquisition of Knowledge*, Erlbaum, Hillsdale, NJ. [12]

Burke, H. (1958), "Raven's progressive matrices: A review and critical evaluation", *Journal of Genetic Psychology*, Vol. 93/2, pp. 199-228, https://doi.org/10.1080/00221325.1958.10532420. [4]

Carroll, J. (1993), *Human Cognitive Abilities: A Survey of Factor-analytic Studies*, Cambridge University Press, New York. [17]

Carroll, J. (1982), "The measurement of intelligence", in Sternberg, R. (ed.), *Handbook on Human Intelligence*, Cambridge University Press, New York. [10]

Chapman, J. (1921), *Trade Tests: The Scientific Measurement of Trade Proficiency*, Henry Holt and Company, New York. [8]

Flanagan, J. (1954), "The critical incident technique", *Psychological Bulletin*, Vol. 51/4, pp. 327-358, https://doi.org/10.1037/h0061470. [16]

Government of New York (2020), *Barber License Examination*. [11]

Hartigan, J. (1975), *Clustering Algorithms*, John Wiley and Sons, New York. [18]

Köhler, W. (1947), *Gestalt Psychology*, Liveright Publishing, New York. [13]

Polanyi, M. (1966/1983), *The Tacit Dimension*, Peter Smith, Gloucester, MA. [1]

Spearman, C. (1904), "'General intelligence', objectively determined and measured", *American Journal of Psychology*, Vol. 15, pp. 201-293, https://doi.org/10.2307/1412107. [3]

Vernon, P. and J. Parry (1949), *Personnel Selection in the British Forces*, University of London Press, London. [5]

Notes

[1] "Jobs/tasks" is meant to capture the range of cognitively demanding activities associated with a particular occupation. One can consider "tasks" as the lowest-meaningful component of a job, and a "job" as typically subsuming a number of different tasks required of the employee. That is, in a particular job, the employee would be expected to perform a range of tasks. Job analysis and task analysis procedures have typically been used to decompose the requirements of a job into identifiable elements. These individual elements can be described at several levels of analysis. For example, psychologists have used many job and task analyses to hypothesise the underlying ability, knowledge and skill requirements for a particular occupation.

[2] Traditionally, with a set of 50 or so items to be sorted, each subject matter expert will create 8-12 sets of items that are conceptualised to be similar (within sets) and different (between sets). The resulting "data" for each subject matter expert will be a co-occurrence matrix (0's indicate a pair of items is in different sets, and 1's indicate the pair of items is in the same set). Data are then aggregated across the subject matter experts. In this way, each entry in the co-occurrence matrix ranges from 0 (all experts agree the items are different) to n, where n represents the total number of experts (all experts agree the items are similar).

9. Competence assessment in German vocational education and training

Britta Rüschoff, FOM University of Applied Sciences for Economics and Management

This chapter reviews the methods of competence assessment in German vocational education and training (VET) and discusses their suitability for assessing the capabilities of artificial intelligence (AI) and robotics. It describes how vocational competences are defined, the instruments used to assess them and how VET examinations are developed and administered. In addition, it discusses the validity and reliability of assessment instruments in VET, providing concrete examples of assessments. Finally, the chapter indicates the advantages of using VET tests for assessing AI capabilities and concludes with several considerations for applying VET examinations on machines.

Introduction

Developments in artificial intelligence (AI) and robotics will have a substantial impact on future working environments. Technologies will partially or completely take over tasks previously performed by humans. Such developments will change the professional skills and competences required of future generations of workers.

Which professional tasks are likely to be partially or entirely performed by AI or robotics? How can AI and robotics be trained to perform these tasks? What methods and tools can be used to assess their performance, particularly in comparison with humans?

One approach to these questions is to review methods commonly used to train and assess competences in humans. These methods could then inform the training and assessment of AI and robotics. This chapter provides an overview of how competences are assessed in vocational education and training (VET) in Germany. It looks at how skills and competences are defined; methods and tools in use; and examples of tasks used in VET. Finally, it reflects on the suitability of these methods and tools to assess the capabilities of AI and robotics and offers recommendations.

Vocational education and training in Germany

Germany has 325 occupations that are state-recognised or deemed state-recognised under the Vocational Training Act (BBiG) or the Crafts Code (HwO) (BIBB, 2020[1]). Some occupations, such as in the medical field, are regulated in special laws (e.g. Nursing Act, Geriatric Care Act). Most vocational training programmes completed in Germany are based on the dual model (Baumgarten, 2020[2]). In these cases, the apprentices alternate between the vocational school and their respective training company with which they have made a contract for their training. In dual VET, framework curricula (*Rahmenlehrpläne*) regulate school-based vocational education. They provide a detailed overview over the learning fields and learning outcomes for each VET-occupation. In-company training is regulated by the Vocational Training Act (BMBF, 2019[3]) and the respective training regulations (*Ausbildungsordnungen*) for an occupation. Both the framework curricula and the training regulations are available for public inspection through the Standing Conference of the Ministers of Education and Cultural Affairs (KMK).

This chapter refers primarily to dual VET since it is the most common training model in Germany.[1]

Definitions of competence in German vocational education and training

Empirical and normative definitions of competence

The objective of VET is the acquisition of vocational competences. However, these vary on areas of application, and there is no uniform definition in VET (Koeppen et al., 2008[4]; Zlatkin-Troitschanskaia and Seidel, 2011[5]; Rüschoff, 2019[6]).This section briefly discusses definitions of vocational competences in different contexts to provide a framework for the illustrations of VET in Germany. A rough distinction can be made between empirical and normative definitions of competence.

Empirical definitions are applied particularly when competences are to be measured. In empirical contexts, vocational competences are frequently described as "context-specific cognitive performance dispositions, which in functional terms relate to situations and requirements in certain domains" (Klieme and Leutner, 2006[7]). Hence, they are context-specific dispositions for action (Weinert, 2001[8]; Klieme and Leutner, 2006[7]; Rüschoff, 2019[6]). Such approaches usually aim at developing standardised assessment instruments. These would capture individual facets or sub-dimensions of vocational competence. In so doing, they draw quantifying conclusions about the respective (sub-) competences of an individual, e.g. about specific knowledge or skills (Nickolaus, 2018[9]).

Normative definitions of competence are used mostly in political and legislative debates when it comes to general educational objectives. Normative definitions may go well beyond the immediate work context. Thus, they might also consider non-occupational (e.g. societal) contexts as possible areas to apply vocational competences (KMK, 1996, 2007, 2011, 2017[10]; Roth, 1971[11]). Such comprehensive definitions are almost impossible to capture empirically due to their extensive scope. However, they are not generally designed with the aim of empirical measurement. Rather, they serve as an orientation framework for policy making and the description of educational goals such as by the Standing Conference of the Ministers of Education and Cultural Affairs of the Länder in the Federal Republic of Germany (KMK, 1996, 2007, 2011, 2017[10]), the German Vocational Training Act (BMBF, 2019[3]), and the German Qualifications Framework for Lifelong Learning (DQR, 2011[12]; BMBF, 2013[13]).

This difference between normative and empirical definitions has an important implication. While normative approaches often describe rather broad competence dimensions, empirical approaches will break these dimensions down into sub-dimensions to make them accessible for measurement. This distinction becomes relevant in the sections that follow that present both the normative frameworks of vocational competence and examples of different instruments for assessing competence.

Dimensions of competence in vocational education and training

At the policy level, the dimensions and levels of competence in German VET are largely based on the German Qualifications Framework (DQR, 2011[12]; BMBF, 2013[13]). This, in turn, is based on the European Qualifications Framework (EQF) (Cedefop, 2020[14]).

The EQF offers a shared European reference framework to facilitate the recognition of qualifications across different countries and educational systems. It covers qualifications at all educational levels and in all education and training sub-systems. In so doing, it provides a comprehensive overview of the qualifications in the countries involved in its implementation.

When applied in individual countries, the EQF is implemented in National Qualifications Frameworks that consider country-specific contexts and characteristics (BMBF, 2013[13]). The German Qualifications Framework, for example, presents a general taxonomy of competence dimensions that applies at all educational levels. It distinguishes between Professional Competence (Knowledge and Skills) and Personal Competence (Social Competence and Autonomy). It further distinguishes between eight educational levels – from pre-vocational training at Level 1 to a doctoral degree at Level 8 (BMBF, 2013[13]; DQR, 2011[12]).

The general competence dimensions remain the same at all educational levels. However, the required learning outcomes increase. Thus, the complexity with which graduates at each level need to master the competences also increases. German VET is located either at Level 3 for occupations that require two years of training or at Level 4 for occupations that require 3-3.5 years of training (DQR, 2011[12]); (BMBF, 2013[13]).

Table 9.1 depicts the overarching competence dimensions along with the required mastery of these competences at Levels 3 and 4. Table 9.2 depicts an example of how these competence dimensions are applied to Industrial Electricians at Level 3 of the German Qualifications Framework. It also describes Industrial Electricians' learning outcomes for each competence dimension (BMBF, 2013[13]).

Table 9.1. Competence requirements of the German Qualifications Framework at Levels 3 and 4

Level	Requirements	Professional competence		Personal competence		Education (example)
		Knowledge *Depth and breadth*	Skill *Instrumental and systemic skills; judgement*	Social competence *Team/ leadership skills, involvement, communication*	Autonomy *Autonomous responsibility; reflectiveness and learning competence*	
3	Being able to autonomously perform tasks that remain structured in some areas.	Having extended general or professional knowledge.	Having cognitive and practical skills for planning and processing tasks and evaluating them according to largely pre-determined criteria; transferring methods/results.	Working in a group and occasionally offering support; helping shape the working environment; presenting processes and results to the appropriate recipients.	Working autonomously and in less familiar contexts; appraising own actions and actions of others; requesting and selecting learning guidance.	VET (2 years)
4	Being able to autonomously plan and process tasks in fields that are subject to change.	Having deeper general or professional knowledge.	Having a broad spectrum of cognitive and practical skills to autonomously perform tasks and problem solving and to evaluate processes and results; considering alternative courses of action and effects with neighbouring areas; transferring methods/solutions.	Helping shape the work in a group and offer ongoing support; justifying processes and results; providing comprehensive communication on facts and circumstances.	Setting own learning and work objectives; reflecting on and assessing these objectives and taking responsibility for them.	VET (3-3.5 years)

Source: Adapted from the German Qualifications Framework (DQR, 2011[12]), pp. 7-15.

Table 9.2. Exemplary competence dimensions for Industrial Electricians (excerpt)

Level	Professional competence		Personal competence	
	Knowledge *Depth and breadth*	Skill *Instrumental and systemic skills; judgement*	Social competence *Team/Leadership skills, involvement, communication*	Autonomy *Autonomous responsibility; reflectiveness and learning competence*
3	Industrial electricians have an understanding of the interaction between mathematical and natural science contents and safety, economic and business administration, and environmental aspects. They have acquired extensive specialist knowledge particularly in electrical engineering, installation techniques, information technology, measuring and control technology. Industrial electricians are in possession of extended specialist knowledge allowing them to: • analyse electronic systems and test functions • plan and implement electrical installations • analyse and adapt control systems; roll out IT systems (…)."	Industrial electricians are in possession of cognitive and practical skills which enable them to process and connect mechanical components and equipment and to analyse electrical systems and test functions. They can assess work results and provide transfers of methods and solutions. They ... • process, assemble and connect mechanical components and electrical manufacturing resources • measure and analyse electrical functions and systems • assess the safety of electrical installations and manufacturing resources • install and configure IT systems (…)."	Industrial electricians are able to work in a team, provide mutual support, communicate correctly in technical language, help shape the learning and working environment, and present processes and results in a manner geared to their target group. They ... • work predominantly in a team and communicate, in their professional activity, using correct technical language both in house and externally with other people • apply work, time and learning planning methods • plan tasks within the team and co-ordinate them • research, procure and evaluate information • are able to present facts, draw up minutes, and use German and English technical terms (…)."	Industrial electricians are able to work independently. Particularly when dealing with current-carrying components they act responsibly and carefully. They... • plan work processes and subsidiary tasks, taking into account economic and scheduling guidelines • calculate and evaluate material and labour costs, record services performed • apply customary requirements and take advantage of training opportunities • recognise their own training requirements and take advantage of training opportunities (…)."

Source: German EQF-Referencing Report (BMBF, 2013[13]).

Applications of competence dimensions in practice

Comprehensive vocational competence dimensions, such as those in the German Qualifications Framework, must be broken into smaller sub-competences, both general and occupation-specific, to make them accessible for empirical assessment. General competences encompass a broad spectrum of basic skills related to mathematics, reading and writing or general problem solving, among others (Weinert, 2001[8]; Klotz and Winther, 2016[15]). Occupation-specific competences are pertinent to a specific occupation or occupational area, for example, specific rules, principles and action patterns (Klotz and Winther, 2016[15]).

Rüschoff (2019[6]) conducted a systematic review on the methods of competence assessment in German VET. This included both instruments used within VET examinations and those developed for research. The results show that most available methods and tools, roughly 60% of the instruments covered in the review, are focused on the assessment of occupation-specific professional competences. These include commercial knowledge and skills in industrial clerks (Klotz and Winther, 2015[16]) or problem-solving skills (e.g. error analysis/troubleshooting) in car mechatronics (Abele et al., 2016[17]; Gschwendtner, Abele and

Nickolaus, 2009[18]). Significantly fewer instruments, about 24%, were focused on general competences. Only 9% measured personal competences such as social or communicative competences (Rüschoff, 2019[6]).

Even if the professional competences and skills are clearly in the forefront, personal competences are equally relevant for most professions. The relative scarcity of instruments to assess social and communicative competences, then, does not indicate these competences are less important. Since these instruments are usually aligned with the framework curricula and training regulations, they typically emphasise professional aspects more than social and communicative aspects.

Yet, social and communicative skills will likely become more important for most professions, as employees will focus on more complex activities that require interdisciplinary exchange. This is especially the case with greater automation for routine tasks. The ability to communicate in a goal-oriented manner with professionals from a variety of backgrounds will hence be an important skill in the future.

Examinations in German vocational education and training

Structure and grading of examinations

Final examinations in (dual) VET are regulated by the Vocational Training Act or the Crafts Code (BBiG, §37 – §50/ §31-§40a HwO) and organised by the chambers for the respective occupations. The chambers appoint examination boards (*Prüfungsausschüsse*) and conduct the examinations. Examination boards consist of at least three people who are qualified experts in the profession. They must also include representatives of employers and employees (in equal numbers) and at least one vocational schoolteacher. The examination board may also obtain qualified expert opinions from third parties (e.g. vocational schools).

The learning fields covered in the examinations are aligned with the framework curricula and training regulations for an occupation. Such an alignment and the explicit inclusion of occupational experts ensure that assessed competences and the occupational scenarios in the examinations correspond to industry practices and reflect workplace conditions.

Examinations are structured in different examination areas. They must cover between three to five areas with “economics and social studies” being obligatory. Each area defines competences that apprentices must demonstrate and specifies certain instruments to assess them. Examinations are commonly designed to yield a total of 100 points across all tasks and examination areas. A very good performance or grade 1.0 in the German system is awarded 100 points (BIBB, 2020[19]).

Aggregated results for all final examinations held by the Chamber of Industry and Commerce (roughly 300 000 annually) are available in their examination statistics. These statistics contain information on the examination areas of the individual occupations, the pass rate of the examinations and the distribution of grades and average scores for individual examination areas (IHK, 2021[20]).

Development of examination tasks

The examination tasks are developed by one of several entities:

- the examination board
- a task development committee put in place for a certain profession by the responsible chambers
- a committee appointed by a chamber for a specific occupation in a federal state or region to develop the examination tasks, often together with a (supra-regional) development office for examination tasks

- development offices for examination questions, for example the PAL (www.stuttgart.ihk24.de/pal) for industrial and technical professions, the ZPA (www.ihk-zpa.de) or AkA (www.ihk-aka.de) for commercial professions and the ZFA (www.zfamedien.de) for professions in the printing and media industry.

Development offices develop the examination tasks and carry out the examinations for a variety of different occupations. For example, the PAL develops examination tasks for 133 of the currently 325 recognised occupations in German VET (PAL, 2020[21]).

The development offices usually hold the copyright and the exploitation rights to the examination tasks. For examination tasks still in use, secrecy regulations are strict. Examination tasks stemming from previous years can in some cases be acquired for inspection or exam preparation from the development offices.

Types of examination tasks

Table 9.3 provides an overview of the different types of tasks. Tasks can be written, oral or practical, and either closed (e.g. written multiple choice) or open (e.g. practical work assignments). The respective examination board decides which tasks are used; not all types of tasks are used in every examination. However, certain tasks are related in that the content of one requires the completion of a previous one. This is because many examinations are case-related. They have different tasks but refer to the same underlying work-related scenario and the results build on one another.

The case-related approach makes it possible to depict complete action sequences. This, in turn, permits a more representative picture of the apprentices' competences and skills in authentic work settings. For instance, apprentices might first complete a practical work assignment and then present and discuss the results. Table 9.4 depicts common and obligatory combinations of tasks.

Table 9.3. Types of assessment tasks in VET

Instruments	Description of the instruments	Evaluated dimension(s)
	Practical tasks	
Product sample	Apprentices craft a product typical for their profession, e.g. a wooden product.	• The final product
Work sample	Apprentices carry out a single professional activity (e.g. the provision of a service).	• Work procedure
Work task	Apprentices carry out a complex professional activity.	• Final result of the task • Work procedure
Work assignment	Apprentices carry out a typical assignment for their future profession (either in the company in which they receive their training or at the site of a customer).	• Work procedure and/or • Final result of the assignment
	Oral tasks	
Case-related expert discussion	Case-related expert discussions are based on a practical task that has either been completed beforehand or is described in detail as part of the examination. Apprentices may receive documents related to this practical task to prepare themselves and to be used during the discussion.	• Understanding of backgrounds and contexts • Methodical approach • Quality of solutions • Communication skills
Assignment-related expert discussion	Assignment-related expert discussions are based on a professional assignment (e.g. a work sample) that has been completed beforehand and discuss procedures, problem solving and solutions.	• Understanding of backgrounds and contexts • Methodical approach • Quality of solutions
Situational expert discussion	Situational expert discussions are based on a work sample or a work task and discuss apprentices' procedures, problem solving and solutions. Situational expert discussions are held during the completion of work samples or work tasks.	• Understanding of backgrounds and contexts • Methodical approach • Quality of solutions
Conversation simulations	Conversation simulations are a type of role play in which apprentices engage in conversations that are typical for their future professions. This can, for example, be conversations with customers or patients.	• Understanding of backgrounds and contexts • Methodical approach • Quality of solutions • Communication skills • Customer orientation
Presentations	Apprentices hold a presentation on, for example, a previously completed professional work task, a work assignment or a typical professional situation and respond to questions regarding their presentation.	• Methodical approach • Communication skills • Formal criteria of the presentation
	Written tasks	
Written tasks	Tasks to be completed in writing are typical professional activities for a certain profession. This might cover drafting a business letter in commercial professions or a circuit diagram in technical professions.	• Demonstration of professional knowledge • Understanding of backgrounds and contexts • Methodical approach • Quality of solutions
Documenting with practice-based documents	Documentations with practice-based documents are based on a previously completed professional work task, a created product or an assignment and documents the execution of this task in, for example, reports, work plans, operating instructions, etc.	Practice-based documentations are drawn on to support the evaluation of the work and/or the results

Source: Recommendation of the Main Committee of the Federal Institute for Vocational Education and Training (BIBB) on the structure and design of training regulations and examination requirements (BIBB, 2013[22]).

Table 9.4. Possible and obligatory combinations of tasks

Types of tasks	Can be combined with...	Must be combined with...
	Practical tasks	
Product sample	• Documenting with practice-based documents or • Presentation or • Assignment-related expert discussion	
Work sample	• Documenting with practice-based documents or • Assignment-related expert discussion or • Situational expert discussion	
Work task	• Written tasks or • Documenting with practice-based documents or • Presentation or • Assignment-related expert discussion or • Situational expert discussion	
Work assignment	• Presentation	• Documenting with practice-based documents • Assignment-related expert discussion
	Oral tasks	
Case-related expert discussion		
Assignment-related expert discussion		• Product sample or • Work sample or • Work task or • Work assignment
Situational expert discussion		• Work sample or • Work task
Conversation simulations		
Presentations		• Product sample or • Work sample or • Work task or • Work assignment
	Written tasks	
Written tasks	• Documenting with practice-based documents or • Situational expert discussion or • Work sample or • Work task or • Work assignment	
Documenting with practice-based documents		• Product sample or • Work sample or • Work task or • Work assignment

Source: Recommendation of the Main Committee of the Federal Institute for Vocational Education and Training (BIBB) on the structure and design of training regulations and examination requirements (BIBB, 2013[22]).

Psychometric properties of assessment instruments in vocational education and training

Regarding the psychometric quality of these examination tasks, one of the advantages lies in their validity. The validity of an assessment refers to the extent to which a method or instrument can depict the intended target construct (e.g. vocational competence). This can be established in a variety of ways. For example, content validity addresses how well the tasks of a test cover the range of all possible tasks. Hence, it addresses how well the test represents the underlying construct to be assessed. The examinations aim to ensure content validity in two ways. First, they align examinations to the framework curricula and training regulations. Second, they draw on experts in the field develop and select tasks to reflect the behaviour and knowledge needed by apprentices in their occupation.

A recent systematic review provides a comprehensive overview of the psychometric properties of the methods and instruments developed or in use in German VET (Rüschoff, 2019[6]), both in the context of VET examinations and in research contexts. Most instruments included in the review ensured the validity of the instruments through establishment of their content validity. They did this, for instance, by aligning the instruments with curricula and training regulations and drawing on expert ratings from the industry.

Roughly 22% of the instruments were further subjected to a construct validation, primarily through one of two methods. On the one hand, their convergent validity was assessed (i.e. the relationship of the instrument to other instruments measuring the same construct). On the other, their discriminant validity was assessed (i.e. ensuring the developed instrument does not correlate [strongly] with validated instruments measuring different constructs). The review identified a scarcity of results on the predictive validity of the instruments. It is unclear whether findings regarding the predictive validity are entirely lacking or just not publicly available. Most instruments in the review showed good or acceptable reliability (Rüschoff, 2019[6]).

An analysis of the psychometric properties of the final examinations in commercial professions in German VET studied the final examinations of 1768 apprentices (Klotz and Winther, 2012[23]; Winther and Klotz, 2013[24]). The results suggest the examinations are clearly aligned with the requirements of an occupation. They also align with the practical training apprentices received before their examination.

However, the relative weights of the examination areas sometimes deviate. For instance, a large part of the curriculum in commercial professions refers to the goods and services domain (roughly half of the curriculum and one-third of the practical training). Yet, only about 20% of the examination relates to this content (Winther, 2011[25]; Winther and Klotz, 2013[26]).

The analyses also indicated that reliabilities of the tasks used in commercial final examinations are sufficiently high. However, reliabilities differed by apprentices' ability levels. The examinations were especially reliable for test takers with average ability levels (with scores centred around the mean). They were somewhat less reliable for apprentices with either very high or very low ability levels (Winther and Klotz, 2013[26]). More detailed information on the psychometric properties of the examinations for individual professions is available to the respective chambers and is considered when tasks are to be reused.

Exemplary examination tasks for Advanced Manufacturing Technicians

Overall structure of the examination

The following examples provide an overview of the tasks used in an examination of Advanced Manufacturing Technicians at Level 4 of the German Qualifications Framework. The examples show excerpts of an examination held in the summer of 2017. All displayed tasks stem from the same examination. The tasks displayed were developed in Germany by the PAL (PAL, 2020[27]) and translated by the German American Chamber of Commerce of the Midwest (GACCMidwest, 2020[28]). To facilitate communication with an international readership, only the English translation is shown here.

The examination has a practical part, an oral part and a written part. As is common, the examination is case-related, meaning the different parts revolve around a typical professional scenario. In this case, they involve the construction and use of a disk separator. When completing the tasks, apprentices may use a mechanical and metal trades handbook, a collection of formulas, drawing tools and a non-programmable calculator with no communication with third parties. They are further provided with a set of technical drawings. Figure 9.1 depicts one of seven technical drawings that were part of the examination materials.

Figure 9.1. Example of a technical drawing forming part of the examination materials

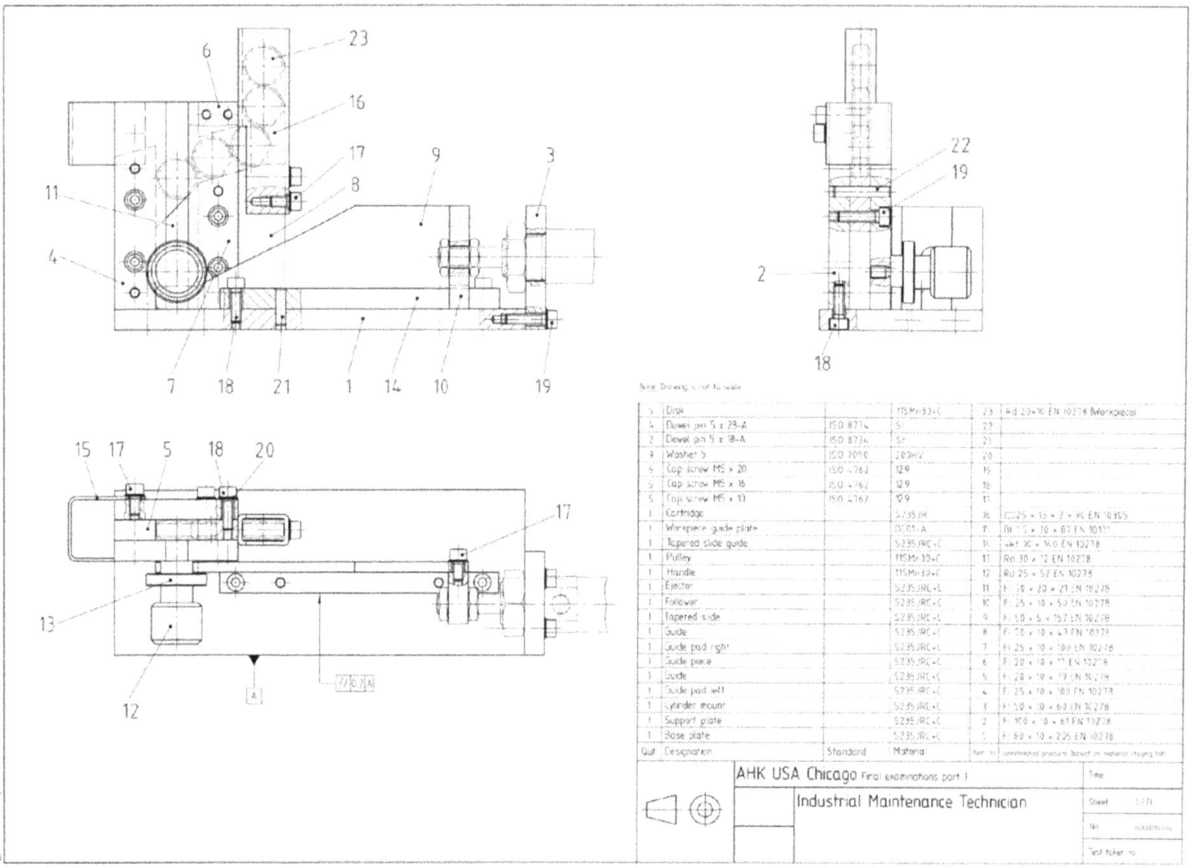

Practical tasks

In the practical component of the examination, apprentices demonstrate professional skills through the completion of a sequence of interrelated tasks (e.g. producing a functional module of a disk separator). Apprentices are provided with a functional description of the module to be produced, the setup requirements and the necessary drawings. They have 6.5 hours to complete the assignment. Their performance is evaluated according to pre-determined evaluation criteria on every sub-task. Figure 9.2 shows an excerpt of the assignment as presented to apprentices during the examination.

Oral tasks

The oral part of the examination consists of situational discussions. During the practical assignment (i.e. while producing a functional module of a disk separator), apprentices will be asked questions related to the completion of the assignment. They are expected to provide short answers that show they can explain professional issues and hold situational expert discussions.

Figure 9.2. Excerpt of a work assignment for Advanced Manufacturing Technicians

AHK Final examination, Part 1 – Summer 2017	
Work Assignment	**Advanced Manufacturing Technician**

1 **General Information**

In the final examination, Part 1, you will have to complete a work order.

2 **Specified time: 6.5 h**

3. **Examination documents that every test taker requires for the work assignment, in addition to this sheet:**

Drawings
"Inspection" work sheet, Page 1 of 4

4 **Identifying the examination documents**

At the location provided in the header of the examination documents, enter your first and last names and your test taker number.

5 **Situational discussion phases**

During the work assignment, the examination committee will have situational discussion phases with you. Answer the questions they ask you, whenever possible using short, proper answers.
In this process, show that you can explain issues and hold situational discussions.

6 **Functional description of the module**

The assembly with control technology function involves a device for separating disks.
By way of the cartridge (item no. 16), the disks (item no. 23) are supplied to the ejector (item no. 11). The vertical motion of the ejector (item no. 11) occurs by way of the tapered slide (item no. 9), which is connected to the double-acting cylinder 1A. With each advance stroke of the piston rod of cylinder 1A, a disk is separated and ejected from the mechanical assembly.

7 **Performance phase**

Your assignment is to produce a functional module with control technology on your assembly panel according to the examination drawings and examination documents. Always be sure to follow safety procedures. Your assignment includes the following:

- produce individual parts through forming and cutting processes
- mark all components
- assemble of the individual parts according to drawing
- ensure quality standards are followed
- set-up, check and adjust all assembly components
- inspect for proper function

Continued on next page

Written tasks

The written part of the examination has two parts. Part A consists of 23 multiple-choice questions of which 6 are mandatory and 3 are optional. Apprentices can decide which three questions they wish to skip. Part B consists of eight short open questions that all need to be answered. Both parts A and B refer to the technical drawings provided beforehand (see Figure 9.1). Apprentices have 90 minutes to complete the two parts. Figure 9.3 provides examples of multiple-choice questions in part A. Figure 9.4 provides examples of short open questions in part B.

Figure 9.3. Examples of multiple-choice questions used in examinations for Advanced Manufacturing Technicians

AHK German American Chambers of Commerce
Deutsch-Amerikanische Handelskammern

German American Chamber of Commerce of the Midwest, Inc.
321 North Clark Street, Suite 1425
Chicago, Illinois 60654-0714
Phone: (312) 644-2662 | Fax: (312) 644-0738
www.gaccmidwest.org

Questions 1 to 23 relate to separating disks using a tapered slide and associated topics, such as evaluation of technical documents, planning for material and tools, defining technical parameters, planning and coordinating workflows, manufacturing components using manual and automated processes, using test methods and test equipment, using accident prevention regulations and observing environmental protection.

1

To which material group can the material of the handle (Item No. 12) be assigned?

1 Case-hardened steels

2 Tempered steel

3 Free-cutting steel

4 Spring steel

5 Tool steel

2

Handle (Item No. 12)
The abbreviation for the material includes the additional symbol +C. What does this symbol mean in the designation system?

1 It is a coarse-grained steel.

2 It is the code letter for carbon.

3 It is a hot-rolled steel.

4 It is a drawn bright steel.

5 The +C indicates that carbon must be added for tempering.

Figure 9.4. Example of a short open question used in examinations for Advanced Manufacturing Technicians

AHK German American Chambers of Commerce
Deutsch-Amerikanische Handelskammern

German American Chamber of Commerce
of the Midwest, Inc.
321 North Clark Street, Suite 1425
Chicago, Illinois 60654-0714
Phone: (312) 644-2662 | Fax: (312) 644-0738
www.gaccmidwest.org

Description of examination task

Rating scale (10 to 0 points)

The disk separator with tapered slide and with control system function shown in the general drawing, page 1(7), is to be manufactured.
Work on the following questions. Answer them with short sentences, wherever possible.

U1

Describe the function of the mechanical module by using the specified names of the components and their item numbers.

Task solution:

Result U1

Evaluation and scoring of the results

The practical assignment is scored in three areas:

- performance (testing the function of the model, control technology, visual inspection of the mechanical module and dimensional inspection)
- inspection (based on a test protocol)

- the oral situational discussion (which is deemed part of the work assignment as it is held during the assignment).

Each of the three areas can yield a maximum of 100 points (with points already awarded and weighted in intermediate steps). In the final evaluation of the work assignment, a maximum of 100 points can be attained. The breakdown of points awarded is 85% for the performance phase; 10% for the test protocol; and 5% for the discussion phase. In the written part of the examination, each multiple-choice question yields 1 point, resulting in a maximum number of 20 points for part A (20 of 23 multiple-choice questions must be answered). Part B consists of eight short-answer questions that must all be answered. Each short-answer question yields 10 points, resulting in a maximum number of 80 points for part B. The divisors of parts A and B are 0.4 and 1.6, respectively (i.e. a maximum of 20/0.4 = 50 points for part A and a maximum of 80/1.6 = 50 points for part B). This gives parts A and B each 50 points or a weighting of 50% each. Together, apprentices can get 100 points for the written part of the examination. The practical and written parts of the examination each count for half of the final grade.

Suitability of contemporary vocational education and training tests to assess capabilities of artificial intelligence and robotics

Advantages of performance-based assessments

The above review of the psychometric properties of the examination tasks in VET suggests these tasks are suitable for assessing vocational competences *in humans*. Are they equally suitable to assess vocational competences *in machines*? Addressing this question requires first understanding what the tasks reviewed above do and do not measure and how they differ from other assessment methods and instruments.

The tasks presented above are action-oriented and inherently practical, which are advantages. With work-related tasks, examiners can observe performance on the same behaviours (or very similar behaviours) that apprentices will need to demonstrate in their subsequent professions. Performance on these tasks is thus likely a direct predictor of later job performance.

Although these tasks draw on observable task-related behaviour (e.g. crafting a certain product), they indicate more than task performance. The examinations aim to use action-oriented tasks to draw inferences on an underlying characteristic of the apprentices such as the competences (e.g. knowledge or skills) that enable apprentices to exert this task-related behaviour. Yet, in action-oriented examinations, the tasks themselves already have considerable meaning. They are not only abstract indicators of the underlying construct but also demonstrate the target behaviour. This distinguishes the action-oriented tasks used in VET from other tests seeking to assess underlying characteristics or constructs such as tests of general mental ability. Tests of general mental ability are likewise used to predict job performance and occupational attainments in humans (Schmidt and Hunter, 2004[29]; Bertua, Anderson and Salgado, 2005[30]). However, they do not draw on work-related tasks to make inferences about future professional performance. Rather, they aim to assess general mental abilities that are expected to manifest in various domains, including future job performance. For example, individuals who score high on a general mental ability test have proven to be able to solve complex test items (observable behaviour). This observable behaviour is theoretically related to the concept of general mental ability. This, in turn, is assumed to relate to performance on other complex problem-solving tasks in real-life situations.

Advantages of performance-based assessments in artificial intelligence and robotics

The results that hold for humans may not apply in the assessment of the capabilities of AI and robotics. Questions in a general mental ability test are only indicators of general intelligence. The resulting score, then, is not an end in itself. For example, individuals will likely attain a high score on a general mental ability test if they memorise the answers first. However, this score will not measure underlying general mental ability. Rather, it will represent their ability to memorise answers to a test. As the result would no longer measure general intelligence, it would also not predict outcomes beyond the immediate test. Hence, it would be unrelated to outcomes usually associated with higher general mental ability.

In humans, the practice described above would be considered cheating on a test. When training an AI, algorithms are trained by knowing the answers in advance, at least for specialised/weak AI. Through repeated exposures to certain tasks or challenges, the AI is trained to solve tasks with increasing accuracy and efficiency. As a result, the AI's performance will be a measure of its ability to learn the answers (or answering logics) to this type of test rather than of the presumed underlying characteristics (e.g. mental ability). Like an individual learning the answers to a test by heart, the resulting score will not be a measure of the intended underlying construct. Therefore, it will most likely no longer predict other outcomes typically associated with general mental ability because such measures gain their predictive value through links with other complex problem-solving skills.

In humans, research has consistently demonstrated these links. However, to successfully apply their general mental ability to other problem-solving tasks, humans likely also draw on other sources of information. These could be basic knowledge about material objects and their properties, events and their effects, or beliefs and desires – what computer science calls "common sense knowledge" (McCarthy, 1989[31]; Miller, 2019[32]). Whereas it can be assumed that humans possess this common sense knowledge, this is not (yet) the case in AI. Consequently, an AI trained on such tests might only be able to solve the questions on this type of test without allowing for inferences about their (job) performance beyond the tests.

The tasks used in VET assess apprentices' underlying competences through authentic professional tasks and requirements. In professional contexts, such work-related tasks are widely recognised as diagnostic instruments and are used successfully for assessment and selection (Hunter and Hunter, 1984[33]; Roth, Bobko and McFarland, 2005[34]; Lievens and Patterson, 2011[35]). Although practical work-related tasks also intend to infer underlying competences from task performance, the tasks themselves are directly related to vocation-specific requirements. In this way, they assess not only underlying competences but also vocation-specific task performance.

Thus, training an AI to learn and reproduce specific procedures to complete a professional task or to solve a work-related problem would train it to perform vocation-specific tasks efficiently. However, performance on one task cannot or can hardly be generalised to performance on a different task or an unknown context. The generalisability of measures of this type of task performance are thus narrow. Still, it would be certain that a person (or AI) can successfully complete this specific task.

Conclusions

Instruments used in German VET offer several advantages that could be useful for training and assessing the capabilities of AI and robotics:

- Reference to labour market requirements

Measurements used in VET are aligned with the framework curricula and training regulations, and fall under the responsibility of the respective chambers. They are thus developed together with industry and reflect authentic labour market requirements.

- Performance-based measurements

Measurements used in VET are performance-based. Performance-based measures capture competences through concrete professional behaviours. The assessed target behaviour has immediate relevance to exercise of the profession.

- Interpretability

Competence assessments draw strongly on psychometric procedures and aim to capture underlying competences. However, assessing performance on concrete vocational tasks relies far less on theoretical assumptions regarding underlying psychological constructs than other assessment approaches. Thus, in performance-based assessments, a "successfully completed" task is meaningful in itself and there is no need to draw inference on any underlying competences that enabled its completion.

- Ongoing development process and topicality

The respective chambers constantly develop and evaluate examination tasks through ongoing examinations. Hence, they exist for a large number of different occupations.

In sum, training and assessing AI and robotics based on concrete work-related tasks might yield scientifically interesting results. Moreover, due to the strong practical orientation of the tasks, it can provide relevant insights for the industry when anticipating the capabilities of AI and robotics. Industry can then change the educational and skill requirements of their future workforce accordingly.

Recommendations

When seeking to select tasks to train and assess the capabilities of AI and robotics, the following points should be considered in this order of sequence:

- **Choose occupations that are subject to similar regulatory structures**

In German VET, there are different responsible authorities and regulations for different types of occupations and training. It is advisable to start with occupations that fall under similar legislation, so that the requirements for the occupations and the examination structures used are largely comparable. The majority of occupations are regulated by the Vocational Training Act (BBiG) or the Crafts Code (HwO).

- **Choose occupational areas that represent a broad spectrum of the working world**

Choose both industrial-technical and commercial professions, as well as craft trades and possibly health professions to represent as broad a spectrum of the working world as possible. Within these areas, a variety of occupations should be chosen that require different competences.

- **Choose common occupations**

The objective is to explore the future role of AI and robotics in the labour market. The most meaningful and relevant information will therefore be obtained by looking at occupations prevalent in today's labour market. In addition, access to testing instruments will be easier for more common occupations than for less common ones. Statistics on the number of apprentices per occupation and year are publicly available in Germany (BIBB, 2019[36]).

Against the background of representing a broad spectrum of the working world, other occupations with more diverse competence requirements can be selected if there is too much overlap between the most common occupations.

- **Cover different competence domains**

Initially, one competence domain (e.g. skills or knowledge) can be chosen as a starting point. Ultimately, all relevant domains (i.e. skills, knowledge, social competences and autonomy) should be covered to give a comprehensive picture of AI and robotics capabilities.

- **Consider different types of tasks**

All types of tasks (i.e. practical tasks, written tasks and oral tasks) should be covered to gain a comprehensive picture of the capabilities of AI and robotics. Initially, only one type of task could be chosen as a starting point. The different types of tasks in the assessment of relevant competences are complementary. Consequently, all types of tasks should ultimately be considered to provide a comprehensive impression of the capabilities of AI and robotics.

- **Select individual tasks to be used**

As the last step, individual tasks to be used must be selected. The chambers, or the bodies commissioned by the chambers to develop the examination tasks, hold the rights to the examination materials. Therefore, materials should be obtained through the chambers.

As some occupations stemming from the same occupational areas are also somewhat similar, they may overlap in their test instruments (e.g. in different sales occupations or different types of mechanic occupations). In such cases, tasks used across occupations within a field should be prioritised.

References

Abele, S. et al. (2016), "Berufsfachliche Kompetenzen von Kfz-Mechatronikern – Messverfahren, Kompetenzdimensionen und erzielte Leistungen (KOKO Kfz) ["Occupational Competencies of Mechatronics Technicians – Measurement, Dimensions and Achievement Goals"]", *Technologiebasierte Kompetenzmessung in der beruflichen Bildung: Ergebnisse aus der BMBF-Förderinitiative ASCOT*, Bertelsmann. [17]

Baumgarten, C. (2020), *Vocational Education and Training: Overview of the Professional Qualification Opportunites in Germany*, German Office for International Cooperation in Vocational Education and Training, Bonn, https://www.govet.international/dokumente/pdf/vocational_education_and_training_web.pdf. [2]

Bertua, C., N. Anderson and J. Salgado (2005), "The predictive validity of cognitive ability tests: A UK meta-analysis", *Journal of Occupational and Organizational Psychology*, Vol. 3, pp. 387-409, http://dx.doi.org/10.1348/096317905X26994. [30]

BIBB (2020), *Richtlinie des Hauptausschusses des Bundesinstituts für Berufsbildung: Musterprüfungsordnung für die Durchführung von Abschluss- und Umschulungsprüfungen*, ["Guidelines of the Committee of the Federal Institute for Vocational Education and Training: Recommended Regulations for Final Exams and Exams for Entering Retraining"], Bundesanzeiger Amtlicher Teil (BAnz AT 21.12.2020 S3), https://www.bibb.de/dokumente/pdf/HA120.pdf. [19]

BIBB (2020), *Verzeichnis der anerkannten Ausbildungsberufe 2020*, ["Directory of Recognised Occupations in Vocational Education and Training"], https://www.bibb.de/veroeffentlichungen/de/publication/show/16754. [1]

BIBB (2019), *Rangliste 2019 der Ausbildungsberufe nach Anzahl der Neuabschlüsse*, ["Ranking of Apprenticeships sccording to the Number of New Degrees 2019"], https://www.bibb.de/de/103962.php. [36]

BIBB (2013), *Empfehlung des Hauptausschusses des Bundesinstituts für Berufsbildung (BIBB) zur Struktur und Gestaltung von Ausbildungsordnungen (HA 158)*, ["Recommendations of the Committee of the Federal Institute for Vocational Education and Training for the Structure and Design and Training Regulation Frameworks"], Bundesanzeiger Amtlicher Teil (BAnz AT 13.01.2014 S1), http://www.bibb.de/de/32327.htm. [22]

BMBF (2019), *Berufsbildungsgesetz - BBiG (The new Vocational Training Act)*, Bundesministerium für Bildung und Forschung/Federal Ministry of Education and Research, Bonn, https://www.bmbf.de/upload_filestore/pub/The_new_Vocational_Training_Act.pdf. [3]

BMBF (2019), *Report on Vocational Education and Training 2019*, Bundesministerium für Bildung und Forschung/Federal Ministry of Education and Research, Bonn, https://www.bmbf.de/upload_filestore/pub/Berufsbildungsbericht_2019_englisch.pdf. [37]

BMBF (2013), *Deutscher Qualifikationsrahmen für lebenslanges Lernen [German EQR Referencing Report]*, Bundesministerium für Bildung und Forschung/Federal Ministry of Education and Research, Bonn, https://www.dqr.de/media/content/German_EQF_Referencing_Report.pdf (accessed on 24 August 2020). [13]

Cedefop (2020), "European qualifications framework. Initial vocational education and training: Focus on qualifications at levels 3 and 4", *Research Paper*, No. 77, Publications Office of the European Union, Luxembourg, http://data.europa.eu/doi/10.2801/114528. [14]

DQR (2011), *Original Title in German [The German Qualifications Framework for Lifelong Learning adopted by the "German Qualifications Framework Working Group"]*, Willkommen auf der Internetseite zum Deutschen Qualifikationsrahmen, http://www.dqr.de/media/content/The_German_Qualifications_Framework_for_Lifelong_Learning.pdf (accessed on 24 August 2020). [12]

GACCMidwest (2020), "German American Chamber of Commerce in the Midwest", webpage, https://www.gaccmidwest.org/ (accessed on 8 September 2020). [28]

Gschwendtner, T., S. Abele and R. Nickolaus (2009), "Computersimulierte Arbeitsproben: Eine Validierungsstudie am Beispiel der Fehlerdiagnoseleistungen von Kfz-Mechatronikern [Computer-simulated Work Samples: A Validation Study using the Example of Failure Detection by Mechatronics Technicians]", *Zeitschrift für Berufs- und Wirtschaftspädagogik*, Vol. 4, pp. 557–578. [18]

Hunter, J. and R. Hunter (1984), "Validity and utility of alternative predictors of job performance", *Psychological Bulletin*, Vol. 96/1, pp. 72–98, https://doi.org/10.1037/0033-2909.96.1.72. [33]

IHK (2021), *Prüfungsstatistik der Industrie- und Handelskammer*, http://pes.ihk.de/Berufsauswahl.cfm. [20]

Klieme, E. and D. Leutner (2006), "Kompetenzmodelle zur Erfassung individueller Lernergebnisse und zur Bilanzierung von Bildungsprozessen. Beschreibung eines neu eingerichteten Schwerpunktprogramms der DFG", *Zeitschrift für Pädagogik*, ["Competence Models for Measuring Individual Learning Outcomes and Educational Processes. Description of a Programme of the German Research Foundation"], pp. 876–903. [7]

Klotz, V. and E. Winther (2016), "Zur Entwicklung domänenverbundenerund domänenspezifischer Kompetenz im Ausbildungsverlauf: Eine Analyse für die kaufmännische Domäne", *Zeitschrift für Erzeihungswissenschaft*, ["On the Development of Domain-related and Domain-specific Competences in Vocational Education. Analysis in the Commercial Domain"], pp. 765-782, http://dx.doi.org/10.1007/s11618-016-0687-1. [15]

Klotz, V. and E. Winther (2015), "Kaufmännische Kompetenz im Ausbildungsverlauf – Befunde einer pseudolängsschnittlichen Studie ["Commercial Competence in the Course of Vocational Training - Findings of a Pseudo-longitudinal Study"]", *Empirische Pädagogik*, Vol. 29/1, pp. 61-83. [16]

Klotz, V. and E. Winther (2012), "Kompetenzmessung in der kaufmännischen Berufsausbildung: Zwischen Prozessorientierung und Fachbezug: Eine Analyse der aktuellen Prüfungspraxis", *Berufs- und Wirtschaftspädagogik*, ["Competence Measurement in Vocational Training in Commerce. Between Process Orientation and Field Relevance: Analysis of Current Examination Procedures"], pp. 1-16, http://www.bwpat.de/ausgabe22/klotz_winther_bwpat22.pd. [23]

KMK (1996, 2007, 2011, 2017), "Handreichung für die Erarbeitung von Rahmenlehrplänen der KMK für den berufsbezogenen Unterricht in der Berufsschule und ihre Abstimmung mit Ausbildungsordnungen des Bundes für anerkannte Ausbildungsberufe", *Ständige Konferenz der Kultusminister der Länder in der Bundesrepublik Deutschland*, ["Handout for the Development of Framework Curricula of the KMK for Job-related Teaching in Vocational Schools and their Coordination with the Federal Training Regulations for Recognised Training Occupations"]. [10]

Koeppen, K. et al. (2008), "Current issues in competence modeling and assessment", *Zeitschrift Für Psychologie*, Vol. 2, pp. 61-73, http://dx.doi.org/10.1027/0044-3409.216.2.61. [4]

Lievens, F. and F. Patterson (2011), "The validity and incremental validity of knowledge tests, low-fidelity simulations, and high-fidelity simulations for predicting job performance in advanced-level high-stakes selection", *Journal of Applied Psychology*, Vol. 96/5, pp. 927-940, https://doi.org/10.1037/a0023496. [35]

McCarthy, J. (1989), "Artificial intelligence, logic and formalizing common sense", in Thomason, R. (ed.), *Philosophical Logic and Artificial Intelligence*, Springer, Dordrecht, https://doi.org/10.1007/978-94-009-2448-2_6. [31]

Miller, T. (2019), "Explanation in artificial intelligence: Insights from the social sciences", *Artificial Intelligence*, Vol. 267, pp. 1-38, http://dx.doi.org/doi:10.1016/j.artint.2018.07.007. [32]

Nickolaus, R. (2018), "Kompetenzmodellierung in der beruflichen Bildung – eine Zwischenbilanz (Competency modeling in vocational training - an interim balance)", in Schlicht, J. and U. Moschner (eds.), *Berufliche Bildung an der Grenze zwischen Wirtschaft und Pädagogik*, Springer VS, https://doi.org/10.1007/978-3-658-18548-0_14. [9]

PAL (2020), *Jahresbericht 2019 ["Annual Report 2019"]*, Prüfungsaufgaben- und Lehrmittelentwicklungsstelle der IHK Region Stuttgart, https://www.stuttgart.ihk24.de/blueprint/servlet/resource/blob/4391022/cd49a986c985f428a9716ea48bb06c08/pal-2019-jahresbericht-data.pdf. [21]

PAL (2020), "PAL-Prüfungsbücher ["PAL-Examinations"]", webpage, https://www.stuttgart.ihk24.de/pal/publikationen/buecher/pal-pruefungsbuecher-liste-3811836 (accessed on 8 September 2020). [27]

Roth, H. (1971), *Pädagogische Anthropologie – Band 2. Entwicklung und Erziehung*, ["Educational Anthropology - Volume 2. Development and Nurture"], Hermann Schroedel Verlag, Braunschweig. [11]

Roth, P., P. Bobko and L. McFarland (2005), "A meta-analytic analysis of work sample test validity: Updating and integrating some classic literature", *Personnel Psychology*, Vol. 58/4, pp. 1009-1037, http://dx.doi.org/10.1111/j.1744-6570.2005.00714.x. [34]

Rüschoff, B. (2019), *Methoden der Kompetenzerfassung in der beruflichen Erstausbildung in Deutschland: Eine systematische Überblicksstudie*, ["Competency Assessment Methods in Initial Vocational Training in Germany: A Systematic Overview Study"], Bundesinstitut für Berufsbildung, City. [6]

Schmidt, F. and J. Hunter (2004), "General mental ability in the world of work: Occupational attainment", *Journal of Personality and Social Psychology*, Vol. 1, pp. 162–173, http://dx.doi.org/10.1037/0022-3514.86.1.162. [29]

Weinert, F. (2001), *Vergleichende Leistungsmessung in Schulen – eine umstrittene Selbstverständlichkeit ["Comparative Performance Measurement in Schools - a Controversial Matter of Course"]*, Weinert, F.E. (ed.), Publisher, City. [8]

Winther, E. (2011), "Kompetenzorientierte Assessments in der beruflichen Bildung–Am Beispiel der Ausbildung vonIndustriekaufleuten ["Competence-oriented Assessments in Vocational Training - Using the Example of the Training of Industrial Clerks"]", *Zeitschrift für Berufs- und Wirtschaftspädagogik*, Vol. 107/1, pp. 33–54. [25]

Winther, E. and V. Klotz (2013), *Measurement of vocational competences: An analysis of the structure and reliability of current assessment practices in economic domains*, http://www.ervet-journal.com/content/5/1/2. [24]

Winther, E. and V. Klotz (2013), "Measurement of vocational competences: An analysis of the structure and reliability of current assessment practices in economic domains", *Empirical Research in Vocational Education & Training*, Vol. 5/2, http://www.ervet-journal.com/content/5/1/2. [26]

Zlatkin-Troitschanskaia, O. and J. Seidel (2011), "Kompetenz und ihre Erfassung – das neue 'Theorie-Empirie Problem' der empirischen Bildungsforschung? ["Competences and their Assessment – a New Problem between Theory and Empiricism in Empirical Education Research"]", in Zlatkin-Troitschanskaia, O. and J. Seidel (eds.), *Stationen empirischer Bildungsforschung: Traditionslinien und Perspektiven*, Springer, VS Verlag für Sozialwissenschaften. [5]

Notes

[1] Other forms of training include full-time school-based training. For a more detailed overview of the overall German VET system, see the German Federal Ministry of Education and Research (BMBF, 2019[37]) or the German Office for International Cooperation in Vocational Education and Training (GOVET; www.govet.international).

10. An occupational taxonomic approach to assessing AI capabilities

David Dorsey, Human Resources Research Organization (HumRRO)

Scott Oppler, Human Resources Research Organization (HumRRO)

This chapter proposes an approach for comparing capabilities of human and artificial intelligence (AI) based on comprehensive occupational taxonomies. After a summary of general methodological recommendations, it describes four major steps of the proposed approach. As the first step, it discusses the identification of an occupational taxonomy and its requirements. Second, it proposes a strategy for sampling occupations from the taxonomy. Third, it provides guidance on collecting expert judgement on AI capabilities with regard to the selected occupations. Fourth, it considers the implications of data analysis from expert interviews.

Introduction

This chapter proposes a four-step approach for comparing capabilities of human and artificial intelligence (AI) based on comprehensive occupational taxonomies. To that end, it looks to identify an appropriate taxonomy of human capabilities to assess and to identify available human assessments. After considering assumptions about design, it proposes a four-step approach to evaluate the capabilities of AI and robotics with regard to the world of work.

"Work," "skills," "tasks," etc. are often used in fundamentally diverse ways. Assessment developers typically start with a specific definition of work, usually via a job or practice analysis. This allows understanding of the essential underlying tasks (specific work behaviours) that comprise an occupation, job or position (Figure 10.1). Based on the tasks, one infers the underlying knowledge, skills, abilities and other key characteristics (KSAOs, or worker characteristics) needed to perform these tasks. These KSAOs then serve as the direct targets for assessment development.

Assessment results can only be "valid" to the extent that they support inferences or linkages back to performing actual job-relevant behaviours. Therefore, developers use a variety of validation techniques to gather and form data-based validity arguments. In this way, they complete the inference chain back to "work".

Figure 10.1. An understanding of the work-related assessment development inference chain

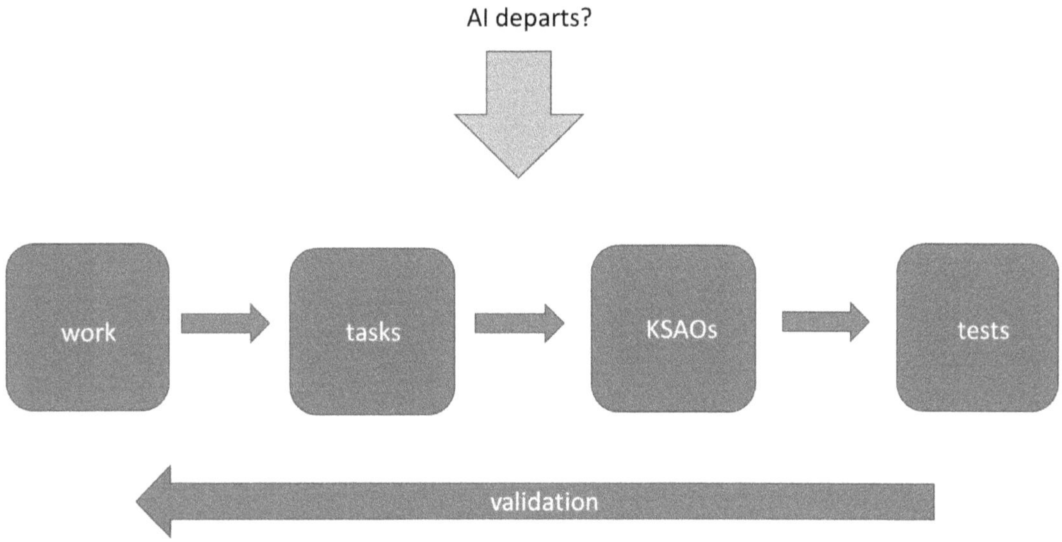

To build a regular programme of assessments that track development of AI capabilities and compare them to the distribution of human capabilities, over time, two related problems need to be solved: identifying an appropriate taxonomy of human capabilities to assess; and identifying available human assessments. It will also be necessary to sample occupations and work descriptors (rather than assessing all of them) and sample assessments from a limited set of occupations (based on the availability of relevant assessments).

Assumptions and method design considerations

There are several assumptions and key design considerations.

Subject matter experts will be required

Subject matter experts (SMEs) are needed to judge whether AI can successfully perform a given work activity and/or correctly answer a particular assessment item. Sets of SMEs could include a combination of computer scientists, industrial-organisational psychologists and incumbents in the occupations to evaluate the capabilities of AI systems with regard to these elements.

There is an implicit major "levels of analysis" question

A major "levels of analysis" question is implicit in the methodology. Both human capabilities and AI technologies change in meaning and specificity as descriptions move from the abstract to the concrete. In the United States, for example, the Department of Labor's Occupational Information Network (O*NET)[1] is the official system for describing work at the national level. It describes the same type of work activity in general terms, somewhat more specific terms or very job-specific terms (Figure 10.2). Specific constructs implied in the activity can change at various levels of description.

This same level of analysis question also applies to other work- and worker-oriented descriptors such as skills, knowledge and even aspects such as personality. At possibly the most specific level of analysis are individual assessment items (such as those shown in Appendix A). These are used to assess competence for hiring, advancement, credentialing, etc. within occupations.

Within occupations, there are various career stages or career levels, often described as entry, journeyman, full performance and expert. Thus, the requirements of occupations change across career levels. The measurement of career levels and "career paths" is in and of itself a discrete area of practice and science [e.g. (Carter, Cook and Dorsey, 2011[1])].

The capabilities of artificial intelligence can be judged against many different work-related descriptors

In addition to determining the level of work description used for this research, work can be described in terms of a variety of descriptor types, including:

- job characteristics (e.g. tasks, activities)
- worker characteristics (e.g. abilities, skills)
- occupational assessments (e.g. items on credentialing exams).

Within the study of occupations, pre-employment conditions or requirements might correlate with "lower-level skills". For example, the US Social Security Administration uses a set of "activities of daily living" (ADLs) to determine whether someone in a disability status can work. These include common activities that any self-sufficient person may be expected to perform, such as grocery shopping, dressing and going to work. There are various measurement approaches for determining the requirements of work in terms of such ADLs.

Figure 10.2. Levels of analysis in work descriptors

O*NET Generalized Work Activity Example

- Establishing and Maintaining Interpersonal Relationships — Developing constructive and cooperative working relationships with others and maintaining them over time.

O*NET Detailed Work Activity Example

- Liaise between departments or other groups to improve function or communication.

O*NET Job Task Example

- Represent organization at personnel-related hearings and investigations.

Source: www.onetonline.org/.

Significant trade-offs exist between "fidelity" and "generalisability" across occupations

Within this research challenge, there is a trade-off between "fidelity" and "generalisability". Specifically, the more fine-grained the information about a given occupation is, the larger the amount of this information gets, but it becomes harder to generalise to other occupations. For example, items on occupational credentialing exams tend to be specific with respect to occupations. This makes such exams an intuitive and presumably well-grounded basis for judging work-related capabilities pertaining to an occupation.

However, the judgements associated with these exams have limited application to other occupations. In contrast, abilities and skills tend to be less specific to occupations. Therefore, judgements associated with them apply more readily to other occupations, given the required level of abilities and skills in the other occupations is known.

As discussed above, job characteristics can range from specific tasks to more generally defined work activities. The latter allows judgements to apply across a wider range of occupations, assuming the required level in the other occupations is known. Further, the more occupationally specific the elements included in a judgement task are, the more judgements are required for a given occupation.

Conversely, the less occupationally specific the elements included in the judgement task are, the fewer judgements are required for a given occupation. The challenge, then, is to determine the minimum level of occupational specificity needed to produce useful information about AI capabilities for a given occupation, while maximising generalisability.

It is appealing to assume the "right" level of granularity of information for making decisions can be determined in advance. However, collective experience suggests this level depends heavily on the purpose and context of the decisions, and the ultimate use of the information. With respect to testing AI, there is limited prior research, calling for different approaches, via pilot testing, cognitive lab testing, etc. These are discussed below.

Worker-oriented characteristics (and related job performance) can also be divided into "can do" aspects. These are typically based on knowledge, skill or ability, whereas "will do" aspects focus on non-cognitive aspects such as motivation, personality, stress tolerance, etc. (Borman et al., 1991[2]).

To put a fine point on the trade-offs mentioned above, specifically around the number of judgements needed, consider that O*NET currently contains information regarding requirements for nearly 1 000 occupations. Taking as a prototypical credentialing exam, a given form of the widely known Society for Human Resource Management certification exam presents examinees with 160 items, 96 knowledge items and 64 situational judgement items.[2]

Thus, across all occupations, credentialing exams of this type would reach an astronomical number of judgements (approximately 1 000 x 160 = 160 000), if they were used to assess all possible AI capabilities. Each of these approximately 160 000 items are likely to be specific to the occupation in question, affording little generalisability to other occupations.

One alternative would be indirect generalisability via careful sampling of occupations. In this case, the relevance of results to other occupations could be inferred. However, if the goal is to make inferences about an entire economy, claims would be stronger through a direct form of generalisability evidence.

Higher-level work descriptors, such as those on O*NET, could be a direct mechanism for generalising judgements about capabilities across occupations. Such an approach would not negate the use of specific credentialing assessment items. Instead, a methodology with various levels of information and judgements could gauge what can be gleaned from different levels of analysis. Moreover, using this methodology, specific assessment items can be linked to higher level descriptors as is often done in various forms of content validation.

Proposed approach

Given the trade-offs noted above, a four-step approach is proposed to evaluate the capabilities of AI and robotics with regard to the requirements of the world of work.

Step 1: Identify an occupational information system that includes work-related descriptors or elements representing a range of levels of occupational specificity

Identify an occupational information system. This needs to specify the work and worker requirements for a wide range of occupations. It should also specify the descriptors at different levels of occupational specificity, ranging from the specific to the general. O*NET could be a useful "content model" as it is made up of several different taxonomies regarding occupational characteristics and worker requirements.

O*NET has two characteristics that make it particularly relevant. First, it describes occupations in terms of the knowledge, skills and abilities required of workers in those occupations. Second, it describes how the work is performed in terms of both occupationally specific tasks and work activities at three different levels of specificity (Detailed, Intermediate and Generalised Work Activities). O*NET also has links to Europe's European Skills, Competences, Qualifications and Occupations (ESCO).[3]

Step 2: Identify a sample of occupations representing a range of job families (e.g. manufacturing, health care)

Identify a sample of occupations to include in the SME judgement workshops (described in Step 3). As noted previously, the number of occupationally specific judgements required for a single occupation can be quite large. This is especially true if the rated stimuli are individual test items on an occupational credentialing exam (which are often in the range of 150 items or more). Therefore, the number of occupations that can be feasibly included in the research is limited.

The proposed approach selects occupations as exemplars of broader job families, allowing most of the workforce to be represented, at least to some extent. For example, the near 1 000 occupations in the O*NET database are classified into 23 job families, as well as 16 broader career clusters. Therefore, one occupation could represent each job family (or career cluster) to help cover the range of occupations in today's workforce. It would select occupations with existing (and available) credentialing exams. They would then be included in the expert judgement tasks as part of Step 3.

Even working with only 16 or 23 occupations (depending on whether they have been sampled from career clusters or job families) would require a sizeable number of judgements on the credentialing exam items (i.e. 16 x 160 = 2 560). However, this is certainly more manageable than attempting to collect such judgements for all occupations in the workforce. Of course, if necessary, this number could be further reduced by grouping the job families or career clusters into a smaller number.

Step 3: Convene subject matter experts to judge the capabilities of AI with respect to different sets of descriptors ranging in degree of occupational specificity

The third step concentrates most of the effort. It collects judgements from SMEs regarding the capabilities of AI with respect to the various sets of descriptors in the research. There are two subcomponents to this step: determining which sets of descriptors and other stimuli (e.g. credentialing exam items) to collect judgements about; and collecting the judgements.

Determining descriptors and stimuli

The collection of judgements should represent a range of levels of occupational specificity to evaluate the trade-off between fidelity and generalisability. To that end, inferences supported by descriptors at different levels of occupational specificity should be compared to identify the level(s) that best manage the trade-off (i.e. providing the greatest fidelity, while still enabling generalisability of judgements across the greatest numbers of occupations).

Using O*NET, for example, SMEs could judge AI capabilities with respect to occupational-specific tasks, as well as at the progressively less occupationally-specific work activities (Detailed, Intermediate and Generalised). There are over 20 000 occupationally specific tasks in the O*NET system. However, only those associated with the occupations identified for the sample (as identified in Step 2) would be included in the proposed data collection. In comparison, there are approximately 2 000 Detailed Work Activities, 300 Intermediate Work Activities and 23 Generalised Work Activities, all of which would be included in the proposed research.

In addition to collecting judgements for items on occupational-specific credentialing exams, it would be possible to collect judgements with regard to O*NET's cross-occupational Abilities, Skills and Knowledge descriptors. The job requirement scales for each of these descriptors are defined in terms of tangible work behaviours (see Appendix A).

Altogether, 52 abilities, 35 skills and 33 knowledge areas can be included in the data collection. Finally, the research should consider inclusion of descriptors representing ADLs discussed earlier. Although not included in O*NET, taxonomies and descriptors for ADLs exist in other sources (Edemekong et al., 2019[3]).

Collecting judgements

Multidisciplinary teams of SMEs should collaborate in judgements concerning AI capabilities with respect to the selected descriptors. These teams should comprise computer scientists with expertise in AI and robotics, industrial-organisational psychologists with expertise in job analysis and human performance, and job incumbents employed in occupations identified in Step 2. Each group would bring different, and important, perspectives regarding the judgements being gathered.

Judgement tasks could vary depending on the descriptors being considered by the SMEs. For example, judgements regarding the capabilities of AI systems to respond correctly to items on the occupational-specific credentialing exams might use an approach similar to that used to assess computer capabilities to respond correctly to items on the Survey of Adult Skills (PIAAC) (Elliott, 2017[4]).

Alternative procedures might be used for judging the capabilities of AI with regard to the work-related tasks and activity descriptors or the abilities, skills and knowledge areas in the O*NET Content Model. For instance, SMEs could identify the maximum point on each O*NET scale to represent the level of activities at which an AI system could be expected to perform, given a specified amount of resources required for development.

These judgements would be similar to those collected in standard-setting studies using the Bookmark method (Karantonis and Sireci, 2006[5]). In that case, SMEs are asked to identify the most difficult item on an assessment that a minimally qualified examinee would be likely to answer correctly.

These judgements would not be tied to specific occupations. Therefore, they would not necessarily need to be collected for each individual occupation included in the sample. That said, job incumbents from a variety of occupations should be included in the teams. They would provide the judgements for these cross-occupational descriptors. They could also collect judgements regarding these descriptors from different groups of SMEs (each focusing on different occupations) to evaluate the extent to which they generalise across occupations.

Regarding the occupationally specific tasks associated with occupations in the sample, SMEs could estimate the level of effort required to develop an AI system to replace (and/or potentially assist) human workers in performing a given task, activity or work behaviour.

The various approaches to eliciting these judgements (using some combination of cognitive laboratories and pilot testing) should be evaluated before embarking on any full-scale effort to collect data.

Step 4: Compare results associated with the different sets of descriptors (or combinations thereof)

The final step would be to analyse the data collected from the SMEs to achieve two goals.

Determine occupations primed to be replaced or aided by AI

First, the specific occupations included in the sample would be evaluated. This would examine the extent to which all or portions of each occupation are primed to be replaced and/or aided by AI technology.

Results would not likely suggest that AI could completely replace workers in any of the occupations in the research. However, they might point to particular types of tasks and activities that may no longer require human workers. More likely perhaps, they could identify the skills and abilities needed by human workers, given the assistance that AI might provide.

Comparing results across the admittedly limited sample of occupations may lead to some valuable inferences regarding the prevalence of these potential changes. Identifying specific "job components" that could be done by AI would allow a search for these job components across a database like O*NET. This, in turn, could possibly generalise the results to new occupations. This is similar to how "job components" are used in synthetic approaches to validation [e.g. (Johnson and Carter, 2010[6])].

Evaluate judgements of AI capabilities

Second, and perhaps more importantly, a data analysis would evaluate the extent to which the different sets of judgements make similar conclusions about AI capabilities.

On the one hand, results associated with the more general descriptors could lead to similar conclusions as the more specific descriptions. In other words, descriptors such as skills; detailed and intermediate work activities could have the same conclusions as occupational-specific tasks and items on occupational-specific credentialing exams. This would suggest that more general judgements could be applied to the other occupations.

On the other hand, results may indicate that inferences associated with descriptors representing different levels of occupational specificity are not sufficiently similar to one another. This would suggest the continued need to collect judgements with regard to occupationally specific descriptors (job-specific tasks; items on credentialing exams) for those occupations of interest not included in Step 3.

The psychometric quality of SME judgements is an additional consideration. The four steps do not include a separate validation of the SME judgements regarding the capabilities of AI to perform various tasks and activities.

However, the data collection could be used to estimate the extent to which different SMEs (or different groups of SMEs) provide similar judgements about different sets of descriptors. For example, judgements regarding the more occupationally specific descriptors may demonstrate greater levels of inter-rater agreement than do the more general descriptors.

Alternatively, the level of agreement among the ratings for the more general descriptors may be relatively high among SMEs in the context of a given occupation (or family of occupations). Yet these judgements may differ for the same descriptors collected in the context of different occupations. The design for data collection strategies needs to consider these possibilities.

Consistency of judgements is a necessary aspect of psychometric quality but not enough to establish validity. To truly establish validity, these judgements must be compared with how AI systems carry out the activities that are the subject of the judgement task. Moreover, this should take place in the context of the specific occupations for which the judgements are being made.

This process could potentially be expensive for occupations new to AI systems. However, it could be possible to include several occupations in the sample for which automated systems have begun to proliferate. Judgements about these occupations and systems could then be compared to judgements about occupations without automation.

References

Borman, W. et al. (1991), "Models of supervisor job performance ratings", *Journal of Applied Psychology*, Vol. 76, pp. 863-872. [2]

Carter, G., K. Cook and D. Dorsey (2011), *Career Paths: Charting Courses to Success for Organizations and their Employees*, Wiley-Blackwell. [1]

Edemekong, P. et al. (2019), "Activities of daily living (ADLs)", *StatsPearls [Internet]*, https://www.ncbi.nlm.nih.gov/books/NBK470404/. [3]

Elliott, S. (2017), *Computers and the Future of Skill Demand*, Educational Research and Innovation, OECD Publishing, Paris, https://dx.doi.org/10.1787/9789264284395-en. [4]

Johnson, J. and G. Carter (2010), "Validating synthetic validation: Comparing traditional and synthetic validity coefficients", *Personnel Psychology*, Vol. 63/3, pp. 755-795. [6]

Karantonis, A. and S. Sireci (2006), "The Bookmark Standard-Setting Method: A literature review", *Educational Measurement: Issues and Practice*, Vol. 25, pp. 4-12, https://doi.org/10.1111/j.1745-3992.2006.00047.x. [5]

Annex 10.A. Example rating scales and assessment items

Figure 10.A.1. O*NET GWA rating

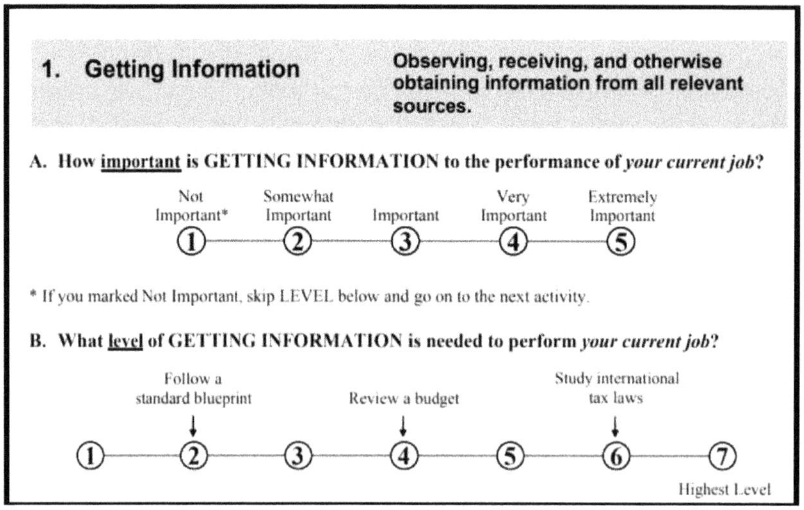

Source: http://www.onetonline.org/.

Figure 10.A.2. O*NET skill rating

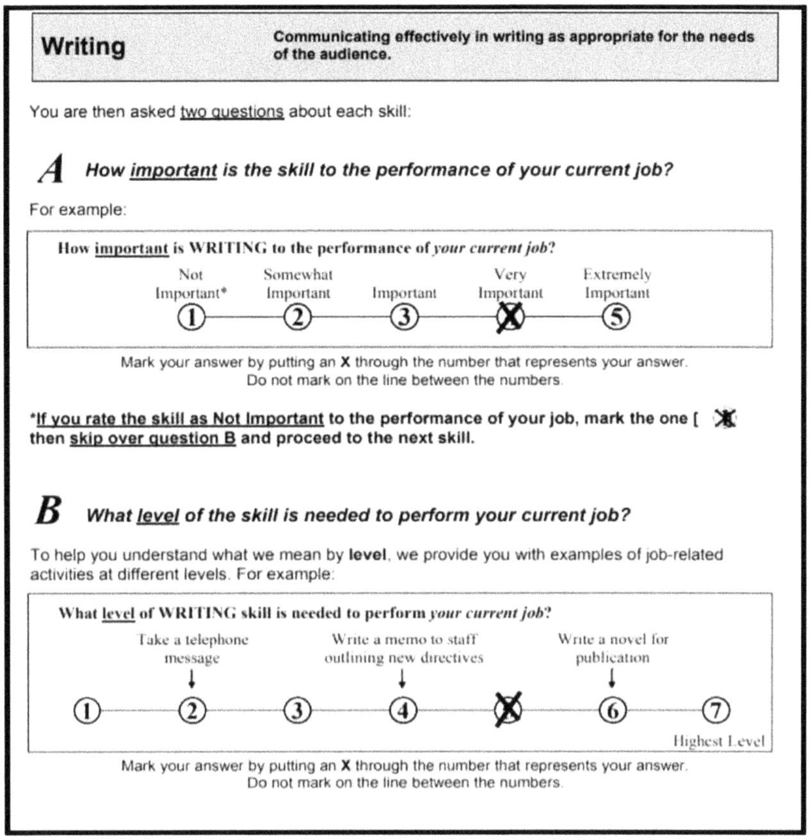

Writing — **Communicating effectively in writing as appropriate for the needs of the audience.**

You are then asked two questions about each skill:

***A* How important is the skill to the performance of your current job?**

For example:

How important is WRITING to the performance of *your current job*?

Not Important* ① — Somewhat Important ② — Important ③ — Very Important ④ (marked X) — Extremely Important ⑤

Mark your answer by putting an **X** through the number that represents your answer.
Do not mark on the line between the numbers.

***If you rate the skill as Not Important** to the performance of your job, mark the one [X] **then skip over question B and proceed to the next skill.**

***B* What level of the skill is needed to perform your current job?**

To help you understand what we mean by **level**, we provide you with examples of job-related activities at different levels. For example:

What level of WRITING skill is needed to perform *your current job*?

① — ② (Take a telephone message) — ③ — ④ (Write a memo to staff outlining new directives) — ⑤ (marked X) — ⑥ (Write a novel for publication) — ⑦ Highest Level

Mark your answer by putting an **X** through the number that represents your answer.
Do not mark on the line between the numbers.

Source: www.onetonline.org/.

Box 10.A.1. SHRM-CP practice question

A small start-up software company realises the technology skill sets of newly hired programmers are more advanced than the existing programmers' skillsets. Recognising the constant business need for these evolving, state-of-the-art skillsets, which is the best workforce development strategy to implement?

- Perform a job redesign for the existing employees that will not require new, updated skills.
- Design a rigorous in-house training programme to get longer-tenured programmers up to speed with the newer programmers.
- Partner with a local community college to offer programmers the opportunity to update their skill sets.
- Offer new hires shorter-term contracts to allow for a continual hiring of programmers with the most up-to-date skills.

Note: SHRM = The Society for Human Resource Management

Box 10.A.2. ASE Mechanic engine repair practice question

Which of the following creates a flapping sound near the front of the engine?

- timing belt tension too tight
- drive belt too tight
- drive belt too loose
- timing belt tension too loose.

Notes

[1] For an overview of O*NET, please see O*NET online at www.onetonline.org/

[2] www.shrm.org/certification/about/descriptions-of-exams/Pages/default.aspx. The full distribution of types of credentialing tests and their tasks in the United States is not known because no comprehensive inventory exists. A recent report from Credential Engine (https://credentialengine.org/counting-credentials-2021/) suggests as many as 967 734 unique credentials in the United States. The report defines "credentials" in a different way than credentialing assessments; yet it does provide a sense of the variety. It highlights "occupational licences" and "occupational certifications", which numbered about 20 000 in 2020.a

[3] More information regarding the O*NET Content Model is available on the O*NET website at www.dol.gov/agencies/eta/onet.

Part III. Artificial intelligence capabilities and their measures

11. Identifying artificial intelligence capabilities: What and how to test

José Hernández-Orallo, Universitat Politècnica de València

Evaluating the capabilities of artificial intelligence (AI) has enormous implications in many areas, especially for the future of work and education. The context is also changing rapidly: the capabilities of humans and AI co-evolve, with scenarios of replacement, displacement or enhancement. Beginning with a review of several taxonomies from human evaluation and AI, this chapter presents a 14-ability taxonomy to identify abilities as potentially disassociated clusters to characterise AI systems. It explores a range of human tests used for decades in recruitment and education, contrasting them with the increasing trend towards basing AI evaluation on benchmarks. The chapter reviews the challenges of bringing human tests to evaluate AI, identifying guidelines to devise reliable tests to compare the capabilities of humans and AI.

Introduction

This chapter analyses how the evaluation of artificial evaluation (AI) systems differs from that of other hardware and software systems, and how it diverges from the evaluation of human cognition – from abilities to skills. It covers common skill taxonomies used for humans as opposed to those used in subfields in AI. However, it proposes a taxonomy based on abilities since skills, knowledge and task performance, if not programmed specifically, must ultimately develop from abilities.[1]

This chapter discusses a recently introduced taxonomy, including 14 abilities, that could be useful for both humans and AI. It examines measurement problems of tests for human evaluation (psychometric, educational and professional) and the bevy of AI evaluation platforms. Subsequently, it argues that human tests cannot be directly used as measurement instruments for an ability-oriented evaluation of AI.

Nonetheless, it identifies the elements of tests that should be abandoned and those that could be reused for working adaptations or newly created tests. It also outlines pragmatic solutions of mapping human tests and AI benchmarks through the ability taxonomy. It ends with recommendations to evaluate AI systems more precisely to determine what they can really do.

Future artificial intelligence roles

Displacement vs. replacement

Some narratives envision that AI will "replace" humans[2], taking over highly skilled and enjoyable activities from engineering to the arts. However, with AI being around for a long time, history suggests that AI will first transform work, creating new kinds of occupations and tasks for humans to do on their own and to share with machines. For instance, the portfolio of a handful of clients handled by a human has been transformed into a more sophisticated framework dealing with massive portfolios of clients. Such transformation was made possible by machine learning models and other AI techniques such as planning, optimisation, natural language processing and sentiment analysis. These AI components now talk with customers, anticipate their behaviour, promote cross-selling and even ensure their overall satisfaction with the company.

Even in cases where the whole task is seemingly automated, a deeper analysis shows that humans usually assume subsidiary new subtasks so the system can work. This phenomenon is not new, as when repurposed workers had to oil or repair a newly introduced machine. However, this kind of "incomplete automation" is easier to hide with cognitive tasks, which humans now do (even unpaid, such as when a customer orders with their phone at a "restaurant").

Variants of this phenomenon have been referred to as *fauxtomation* (Jackson, 2019[1]) or simply "human computation" (Von Ahn, 2008[2]; Taylor, 2018[3]). Figure 11.1 shows how the narrative of AI increasingly taking on more tasks than humans today is incomplete. While some new tasks are created for and only done by machines, humans are doing new tasks, too.

Figure 11.1. New tasks are done by humans, by AI and by both together

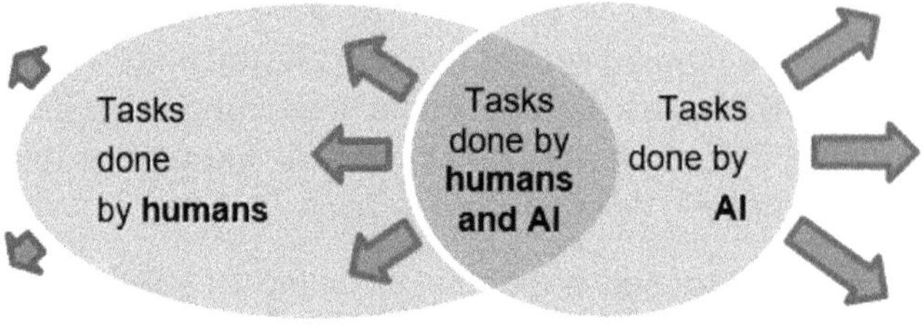

In this transformation, there is displacement rather than just replacement. Human labour has not disappeared in areas where AI has had an important impact. Humans do different tasks, and many of them are new. Typically, these tasks relate to setting goals and targets, monitoring robots and other AI systems, adapting and integrating their decisions, and curating "training" material for AI systems.

This narrative of displacement vs. replacement will determine the path of AI research. The prevalence of responsible and human-centred AI today is introducing a culture based on three factors. First, humans must always be in the loop. Second, humans and machines should collaborate and not compete. Third, machines should always be subordinate to humans.

This vision suggests the separation between what AI will be able to do from what it will be allowed to do. For instance, one day it will be possible to automate most of a physician's tasks. However, this may never happen because of ethical issues or mainstream social pressure. This is not new or particular to AI; the "human touch" is considered a special value in some domains that can already be automated (e.g. hand-made delicatessen). However, ethical and political decisions more than technology may dominate decisions about what AI can do and what is *reserved* for humans in the future.

Human-machine collaboration through externalisation and extension

Another important factor is the different forms in which humans and machine collaborate, or create new behaviour (Rahwan, 2019[4]). In basic "externalisation", a human delegates a task to a machine, giving it the instructions, goals or input. However, humans are also integrating AI capabilities in a more coupled way through "AI extenders" (Hernández-Orallo and Vold, 2019[5]). For instance, most humans use navigational tools in their phones to go to a new address. While the tool shows the position and optimal routes, the human still navigates the city and ultimately decides where to go and how. In this way, the full cognitive process of going from position A to B is not externalised but extended.

In general, machines complement or extend humans, rather than replace them. Humans then adapt to the new situation, developing new "digital" skills (e.g. using new AI apps). This means that human skills are changing significantly. Young people text extremely fast because they use the predictive hints given by their instant messaging app.

This phenomenon – which goes beyond skills – is changing abilities as well. For instance, factual memory capabilities are falling because people can check any fact on the Internet. This is known as the "Google effect" (Bohannon, 2011[6]). Provided the technology does not fail, the new situation (and the associated atrophy) should be seen as empowering. When coupled with their AI extenders, humans should thus be considered more capable overall.

Accordingly, as humans can do more things using tools, they should be evaluated with those tools. As calculators are allowed in many math exams, AI extenders should be allowed for writing, editing, drawing or speaking in more proficient and creative ways. Almost no one writes today without spell check or online access to Wikipedia. Limiting access to AI tools and extenders in tests would thus measure an unrealistic situation. The implications of all this for the following sections are important:

- Tasks humans do are changing more rapidly.
- Human skills, and even some specific abilities, are changing.
- Human evaluation tests are (or should be) changing.

In a rapidly changing world fuelled by AI, tasks are going to change faster, and so will skills and knowledge. This affects humans but also AI. For AI to become economically advantageous over human labour, it needs general abilities rather than specialised skills. In this way, AI systems can learn new skills efficiently and autonomously, adapting faster than humans to new tasks and procedures.

From skill lists to ability taxonomies

Evaluating artificial intelligence through its abilities rather than task completion

Evaluating what AI does in terms of tasks would be short-sighted, unlikely to be comprehensive and prone to overfitting.[3] An AI or robotic system showing performance at a particular task gives poor indications of what other things AI can do. For instance, while computer chess reached superhuman level in the late 1990s (Weber, 1997[7]; Campbell, Hoane and Hsu, 2002[8]), it took AI 20 years to reach human-level performance on other "similar" games such as Go (Silver et al., 2016[9]).

Quite remarkably, both the chess and Go milestones used completely different approaches. Similarly, floor robotic cleaners have been in residential homes for quite some time, but no robot can yet clean a table full of objects properly. For these and other reasons, an ability-oriented[4] evaluation in AI should be preferred over a task-oriented evaluation (Hernández-Orallo, 2017[10]; Hernández-Orallo, 2017[11]).

While still unusual in AI, the standard approach to evaluate humans is by abilities and skills rather than through tasks. Specific tasks for a job, such as driving a lorry, are evaluated only occasionally. Moreover, this evaluation is usually accompanied with verification of some other skills, knowledge and abilities, while many others are taken for granted (e.g. being able to understand an order from the boss).

Defining abilities and skills

Before looking at taxonomies of abilities and skills, the terms should be clarified. Abilities and skills are sometimes used interchangeably, but skills and knowledge are more properly used for effective competences. For example, solving differential equations, programming in Python, playing the guitar or working in teams is a skill. Abilities usually refer to potential capacities related, for example, to verbal, spatial or metacognitive domains.

Abilities allow humans to acquire and deploy skills and knowledge (Schuelke and Day, 2012[12]). This interpretation goes beyond clustering tasks into skills, and skills into abilities. Rather, it means that abilities allow individuals to develop skills and knowledge, which leads to further skills and, ultimately, the capacity to do tasks. Also, abilities in humans are stable for a significant part of the adult life. Conversely, skills are learnt, improved or even forgotten at different moments in the life of an individual.

There are many taxonomies for human cognitive abilities, and some differ significantly. The Cattell-Horn-Carroll (CHC) taxonomy (Carroll, 1993[13]) characterises humans hierarchically (Figure 11.2).

Figure 11.2. Carroll's three-stratum model

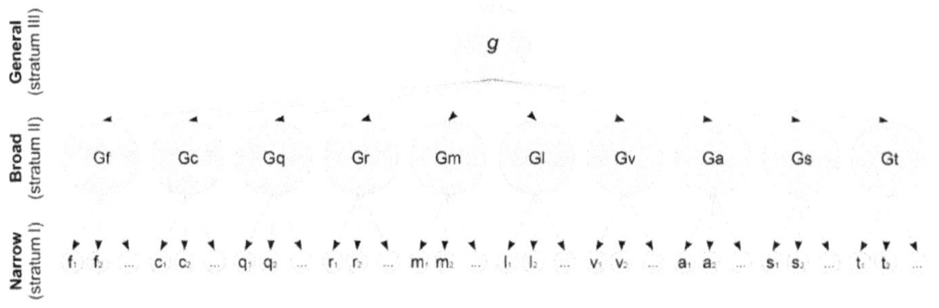

Note: This model is usually known as Cattell-Horn-Carroll taxonomy. The g factor is on the top (third) level, ten broad abilities are at the second level, with the bottom (first) level including many more "narrow" abilities.

The Cattell-Horn-Carroll taxonomy

Table 11.1 describes the ten factors in stratum II of the CHC taxonomy. Some, like processing speed, are not really "abilities". Rather, they are "factors" identified through techniques such as factor analysis. Their interpretation requires examination of their loadings, connecting with underlying tests and the lower stratum.

Nonetheless, these factors capture some areas of cognition and intelligence for which humans involve different intrinsic mechanisms. For example, inductive inference is associated with Gf and Gc. Meanwhile, deductive inference is associated with Gq or neurological substrate (e.g. Gv and Ga).

Table 11.1. Broad abilities at stratum II of Cattell-Horn-Carroll taxonomy

Factor	Ability
Gf	Fluid intelligence
Gc	Crystallised intelligence
Gq	Quantitative reasoning
Gr	Reading and writing ability
Gm	Short-term memory
Gl	Long-term storage and retrieval
Gv	Visual processing
Ga	Auditory processing
Gs	Processing speed
Gt	Reaction time/decision speed

Developmental perspective

Cognitive development offers a more dynamic take on what an individual can do. Piaget introduced four main stages in child development (Piaget, 1936[14]; Piaget, 1964[15]): sensorimotor (0-2 years), preoperational (2-7 years), concrete-operational (7-12 years) and formal-operational (12 and more years). Many variants and extensions have followed, sometimes referred to as neoPiagetian models (Morra et al., 2012[16]).

The developmental perspective studies how skills are built over other skills. From the perspective of education, the focus is on skills, as many can be acquired in some degree at any age. Conversely, abilities have a greater innate component and consolidate in adolescence. However, ignoring abilities altogether leads to poor understanding of why some individuals develop some skills better than others.

Taxonomies for job analysis

The use of skills and knowledge is more common in the analysis of the workplace. Taxonomies for job analysis such as O*NET-SOC, ISCO and ESCO. O*NET-SOC[5] includes more than 1 000 "occupational titles" classified into 23 major groups[6], which are mostly sector-oriented (e.g. health, construction, etc.). ISCO[7] includes more than 1 500 categories of occupations, organised into ten major groups (Table 11.2). ESCO[8] covers about 3 000 occupations and around 13 500 skills. It is partially hierarchical. Apart from a sector-oriented hierarchy, it has a top hierarchy of skills that is similar but different from ISCO, including eight categories (Table 11.3).

Table 11.2. Ten major groups in the ISCO occupation categories

Code	Category
1	Managers
2	Professionals
3	Technicians and associate professionals
4	Clerical support workers
5	Service and sales workers
6	Skilled agricultural, forestry and fishery workers
7	Craft and related trades workers
8	Plant and machine operators, and assemblers
9	Elementary occupations
0	Armed forces occupations

Table 11.3. Eight major groups in the ESCO occupational skills

Code	Category
S1	Communication, collaboration and creativity
S2	Information skills
S3	Assisting and caring
S4	Management skills
S5	Working with computers
S6	Handling and moving
S7	Constructing
S8	Working with machinery and specialised equipment

The cognitive ability taxonomies, developmental models and job competence classifications are derived from and aimed at humans. When taxonomies are derived from human populations, some elements that are essentially different may fall into the same human ability, simply because they are correlated in the human species. As well, similar elements may be separated because they are handled by different modules or genes in humans.

In cognitive development, some stages seem more universal than others, but even for some animals the stages may differ significantly. Foals and other precocial animals can stand and run in a few hours, for example. Similarly, the competences and skills used for labour depend on the society and the economy at a particular time and culture. Neither of these human taxonomies is immediately applicable to AI.

Domains in AI are strongly linked to underlying techniques, which have varied significantly in a few decades. AI typically organises its functionalities with terms such as learning, planning, recognition, inference, etc. To date, there is no standard taxonomy of AI skills or abilities. Areas and domains are typically used for textbooks and conferences or bibliometric research (Machado et al., 2018[17]; Frank, Wang and Cebrian, 2019[18]).

A taxonomy of cognitive abilities

Taking all this into account, Hernández-Orallo and Vold (2019[5]) introduce a taxonomy of cognitive abilities, merging several categorisations in psychology, animal cognition and AI. To be comprehensive about all cognitive abilities, the methodology started with elements from all these disciplines, distilling from different sources:

- Thurstone's primary mental abilities according to factors from CHC hierarchical model (stratum II, Figure 11.2)
- areas of animal-cognition research according to Wasserman and Zental (2006[19])
- main areas in AI according to the *AI Journal* (as per 2017)
- "competency" areas in AGI according to Adams et al. (2012[20])
- I-athlon "events" from Adams, Banavar and Campbell (2016[21]).[9]

These different lists were integrated by matching synonyms and related terms, and trying to keep a manageable number of broad capabilities. There were tensions between both distinctiveness and comprehensiveness against the number of abilities. The main criterion for distinguishing between two abilities A and B (and not merging them) was the understanding that a system or component (either natural or artificial) could *conceivably* master one of them while failing at the other. The compromise for completeness was easier to find. Some elements (such as processing or decision speed in the CHC) are not proper abilities. In addition, some abilities related to multimodality were not explicitly included in the final list of 14 (e.g. olfactory processing). The current version only covers "visual" and "auditory" processing, being the two most representative sensory modalities.

The taxonomy is further developed into a rubric in Martínez-Plumed et al. (2020[22]). The 14 cognitive abilities are shown in Table 11.4 and a number of principles behind their development are presented in Box 11.1.

Table 11.4. Cognitive abilities applicable to both humans and AI systems

Ability	Description
MP: Memory processes	Storage of information in an appropriate medium to be recovered at will according to some keys, queries or mnemonics. This covers long-term memory and episodic memory.
SI: Sensorimotor interaction	Perception of things, recognising patterns and manipulating them in physical or virtual environments with parts of the body (limbs) or other actuators, through various sensory and actuator modalities, and representations.
VP: Visual processing	Processing of visual information, recognising objects and symbols in images and videos, movement and content in the image, with robustness to noise and different angles and transformations.
AP: Auditory processing	Processing of auditory information, such as speech and music, in noisy environments and at different frequencies.
AS: Attention and search	Focusing attention on the relevant parts of a stream of information in any kind of modality, by ignoring irrelevant objects, parts, patterns, etc. Similarly, seeking those elements that meet some criteria in the incoming information.
PA: Planning, sequential decision making and acting	Anticipating the consequences of actions, understanding causality and calculating the best course of actions given a situation.
CE: Comprehension and compositional expression	Understanding natural language, other kinds of semantic representations in different modalities, extracting or summarising their meaning, as well as generating and expressing ideas, stories and positions.
CO: Communication	Exchanging information with peers, understanding what the content of the message needs for a given effect, following different protocols and channels of informal and formal communication.
EC: Emotion and self-control	Understanding the emotions of other agents, how they affect their behaviour and also recognising their own emotions and controlling them and other basic impulses depending on the situation.
NV: Navigation	Moving objects or oneself between different positions, through appropriate, safe routes and in the presence of other objects or agents, and changes in the routes.
CL: Conceptualisation, learning and abstraction	Generalising from examples, receiving instructions, learning from demonstrations and accumulating knowledge at different levels of abstraction.
QL: Quantitative and logical reasoning	Representation of quantitative or logical information that is intrinsic to the task, and the inference of new information from them that solves the task, including probabilities, counterfactuals and other kinds of analytical reasoning.
MS: Mind modelling and social interaction	Creation of models of other agents to understand their beliefs, desires and intentions, and anticipate the actions and interests of other agents.
MC: Metacognition and confidence assessment	Evaluation of their own capabilities, reliability and limitations, self-assessing the probability of success, the effort and risks of own actions.

Source: Hernández-Orallo and Vold (2019[5]).

Box 11.1. Principles behind the cognitive ability taxonomy

There are some principles behind the taxonomy in Table 11.4:

- First, the clusters should not be informed by the categories in human or AI taxonomies only. Abilities should be identified as different when, with the current knowledge, they are thought to conceivably rely on different mechanisms (e.g. deductive and inductive inference).
- Second, it is convenient to associate taxonomies with rubrics that determine whether a task or skill requires the ability. This can be understood as a representational definition and understanding for each ability, and not simply as a cryptical latent "factor" or a meaningless construct.
- Third, developing a test that only measures one ability for every kind of subject (natural or artificial) is complicated. It is more practical to think of many-to-many quantitative connections between abilities and tests, with the advantage of reusing the results of existing tests and benchmarks.
- Finally, skills must always be connected with abilities in the context of development. For instance, the progression of an AI system in the acquisition of elemental skills can be a good way to ensure the system has the abilities needed to develop the skills.

Source: Hernández-Orallo and Vold (2019[5]).

The above taxonomy is not static. However, it serves as a stable source to do mappings between other AI/human taxonomies (Martínez-Plumed et al., 2020[22]; Tolan et al., 2021[23]) and especially tests, as explored in the following section.

Tests: Caveats and pathways

Evaluating humans

Abilities[10] are usually latent variables or constructs, with tests being instruments for measuring them. During the 20th century, a plethora of tests was developed for evaluating humans:[11]

- Psychometric tests for general abilities

These notably include those related to IQ tests. Example: Wechsler Adult Intelligence Scale, with tests aggregated into four categories: verbal comprehension, perceptual reasoning, working memory and processing speed.[12]

- Developmental tests

These cover a series of stages for different purposes (e.g. detecting disabilities). Example: the Bayley scales (Bayley, 1993[24]) evaluate children from ages to 3.5 years with items in three categories: mental scale, motor scale and behaviour rating scale.

- Tests for consolidated knowledge or general education skills

These explore "attainment" or "achievement". For instance, military psychometric tests (such as Armed Services Vocational Aptitude Battery) and college entrance exams (such as ACT and SAT) cover a mixture of abilities and skills (English, mathematics, reading, writing and science). The Bennett Mechanical Comprehension Test covers more specific abilities and skills.

- Personnel selection and certificates

This combines psychometric tests, interviews and practical demonstrations to certify certain abilities, attitudes, knowledge and skills. In many countries, for instance, a driver's licence test evaluates reaction time, visual acuity, knowledge and ability to judge traffic signs and rules. It combines these tests with a practical exam with a real car and sometimes a simulator.

Evaluating machines

Given the range and diversity of evaluations for humans, what can be used to evaluate machines? AI evaluation differs from the evaluation of many other software and hardware systems. For AI, there is usually no formal or procedural description of how the system must solve a goal (otherwise, the solutions would be programmed). Experimental evaluation then becomes more relevant in AI (from learning to planning) than in other areas of computer science.

Apart from informal and subjective assessments (e.g. the Turing test), AI has rubrics and benchmarks.

Rubrics are generally based on human assessment about the capability of the system. Unlike open evaluations such as the Turing test, rubrics are systematic. For instance, (Brynjolffson and Mitchell, 2017[25]; Brynjolffson, Mitchell and Rock, 2018[26]) present a series of questions about a task (the rubric), giving a score that represents whether machine learning could automate the task. A recent OECD project (Elliott, 2017[27]) relies on subject matter experts to assess AI capabilities for three areas in the Programme for the International Assessment of Adult Competencies. A more general approach is based on technology readiness levels. Here, the rubric distinguishes different levels, from research ideas in the lab to viable products (Martínez-Plumed, Gómez and Hernández-Orallo, 2020[28]; Martínez-Plumed, Gómez and Hernández-Orallo, 2021[29]).

Benchmarks are repositories of instances of a task (or collections of tasks) that serve as challenges for AI to improve on several metrics of *performance*. Benchmarks are undoubtedly fuelling the progress of the field (Hernández-Orallo et al., 2016[30]) but are still limited as valid measurement instruments. AI systems commonly reach superhuman performance on a benchmark but do not display the associated capability; the systems usually fail beyond the conditions and distribution of the benchmarks. Accordingly, many benchmarks are soon replaced, entering a "challenge-solve-and-replace" (Schlangen, 2019[31]) or a "dataset-solve-and-patch" (Zellers et al., 2019[32]) dynamic.[13]

Assessing artificial intelligence with tests designed for humans

Given these validity problems in AI evaluation, and the wide range of valid tests for humans, using tests designed for humans might seem a good idea for AI. There are several reasons why this is not advisable:

1. Tests are devised as measuring instruments for a particular population. Human tests lack measurement invariance beyond the human population (even beyond adults).
2. Humans are embodied agents. Many AI systems do not take the form of an agent, and sometimes not even the form of a system. Instead, they appear as cognitive components or modules.
3. A single human can perform well for many tasks and tests. When AI is said to solve A and B, for example, this typically means that one AI system solves A and another AI system solves B.
4. AI systems and components can be built on purpose for a task. The designers put a lot of specific knowledge, bias or curated training data for the particular benchmark.
5. The behavioural traits of humans and AI overlap. AI may "conquer" more human abilities in the future, but AI is introducing many other new abilities (see Figure 11.1).
6. Humans and AI differ on the resources used (e.g. data, compute, sensors) or ignore/ban associated human cognitive labour (e.g. labelling data, delegation to human computation).

The six reasons are exemplified by the use of IQ and other human intelligence tests as benchmarks for AI. Whenever a type of IQ problem or a battery of intelligence tests is made available for AI researchers, less and less time, but increasingly more computational resources are needed for a new AI system to excel at the tests (Hernández-Orallo et al., 2016[33]). However, this AI system can do nothing else beyond the particular IQ tests. Remarkably, such tests are not about knowledge and specialised skills but rather about *human* core reasoning capabilities and abstract problems. They include letter series, number series, Raven's progressive matrices, odd-one-out problems, vocabulary analogies, geometric analogies, etc. The success of AI systems on these intelligence tests has not shown real progress in AI. It cannot be used as evidence that AI systems have general intelligence (Hernández-Orallo et al., 2016[30]).

While intelligence tests are just a kind of cognitive test for humans, other human tests are also problematic when applied to AI (Dowe and Hernández-Orallo, 2012[34]; Hernández-Orallo, 2017[11]). For instance, AI challenges have used questions from educational exams, including diagrams, geometry and mathematics from 4th grade science exams (Clark and Etzioni, 2016[35]). The bad results were reassuring: "no system [came] even close to passing a full 4th grade science exam". Good results from the Aristo Project and other (ensembles of) language models just three years later were labelled as "a significant milestone" (Clark et al., 2019[36]).

However, the new AI solutions are not really "general question-answering" systems and cannot compare to a human with a similar result. Massive language models certainly solve many questions – more than the average human – but a closer look reveals the system is not robust to minor variations of the questions. Ultimately, it does not really understand the questions. In other words, the positive test results for AI on human tests, when compared to humans, are hugely overestimated because of overfitting.

More promising avenues

To avoid the recurrent problem of overfitting, some new benchmarks for AI are taking inspiration from human tests but have been profoundly modified or reconstructed. The tests aim to be easy for humans but challenging for state-of-the-art AI. However, they should not contain hidden statistical patterns or other artefacts that AI systems could exploit to circumvent what the inspirational tests are supposed to measure in humans.

Winograd and Winogrande

The Winograd Schema Challenge has been one of the most important attempts in this direction. It was presented as a collection of text comprehension questions using pronouns that must be disambiguated (Levesque, Davis and Morgenstern, 2012[37]). Levesque initially sought to design questions whose answer would show a high level of common sense reasoning around the elements appearing in the question. However, several AI systems have recently shown excellent performance by exploiting some statistical artefacts in the way the questions are generated. These systems use "clever tricks involving word order or other features of words or groups of words". However, they do not really display the capabilities for referential disambiguation that the test is assumed to be measuring (Kocijan et al., 2020[38]).

Winogrande is a much larger version meant to replace Winograd's schemas. However, new language models have quickly reached good performance too, while still being far from general language understanding. This happens in all areas of AI, from natural language to machine vision. It is sometimes called the Clever Hans phenomenon, as AI finds alternative cues and tricks to solve the task in the same way as a celebrated 19th century horse did to amaze spectators (Lapuschkin et al., 2019[39]; Hernández-Orallo, 2019[40]). Due to the Clever Hans phenomenon, the validity of many tests for AI systems is constantly questioned. This, in turn, provokes the "challenge-solve-and-replace" (Schlangen, 2019[31]) dynamics mentioned earlier. At its heart it reveals an *adversarial* game between AI developers (and their systems) and the evaluators (Hernández-Orallo, 2020[41]). Such an adversarial philosophy is intrinsic to evaluation and should be incorporated in the design of benchmarks and evaluation procedures.

New benchmarks from natural language processing

There are several good examples of these new benchmarks in natural language processing.[14] MOSAIC, for example, includes the adversarial generation of examples found in SWAG (Zellers et al., 2018[42]) or DynaBench.[15] An adversarial example is modified slightly such that humans are not significantly affected, while AI systems fail catastrophically. Understanding how these examples must be generated for different kinds of AI systems can help improve the systems.

Ultimately, if the only possible examples that make an AI system fail also make humans fail, the AI system may really be better than humans. At this point, the question of what the test measures, and all the variations of the instances that become part of the measure, can be considered. As AI systems become designed for the test, an adversarial mindset is needed more than for the evaluation of humans. Training to the test also happens for humans but to a lesser extent.

Non-human animal tests and sandbox evaluation

A less anthropocentric stance to the evaluation of AI looks at non-human animals. In the 1990s, for instance, the Cognitive Decathlon was built for DARPA's Biologically Inspired Cognitive Architecture programme, based on developmental tests (Mueller, 2010[43]). The battery was discontinued around ten years ago (Mueller et al., 2007[44]). However, in a related approach, the animal-AI environment builds on animal tests rather than human tests (Crosby et al., 2019[45]; Crosby et al., 2020[46]). If the adaptation of these tests become more common in the future, the main risk would still be overfitting the AI system to the test distribution, instead of really solving the constructs the test was supposed to measure.

Sandbox evaluation provides a possible solution to overfitting. In these cases, rather than training instances, AI developers are provided with a "sandbox environment" to create different curricula for the AI system. Only when the system has been "raised" in the environment can evaluators disclose the tasks, and test the system, without further training or adaptation. The idea is to encourage the construction of systems that can master a domain, rather than mastering tasks from a distribution. The Animal-AI Olympics followed this philosophy (Crosby et al., 2019[45]; Crosby et al., 2020[46]).

Expert judgement

More immediate options to compare human and AI capabilities are needed until new tests inspired by human or animal evaluation or based on more fundamental principles are developed (Hernández-Orallo, 2017[11]). A pragmatic alternative to tests is the use of human experts to assess AI capabilities through rubrics or other kind of questionnaires.

Indirect mapping between job market and AI benchmarks

Martínez-Plumed et al. (2020[22]) explore a hybrid solution that performs a mapping between AI benchmarks and the 14 cognitive abilities in Table 11.4. Through expert questionnaires and other methodologies (e.g. Delphi), a matrix of many-to-many correspondences is established between benchmarks and abilities. As the performance for different AI benchmarks is usually incommensurate, Martínez-Plumed et al. measure the intensity of research in terms of number of related papers or media articles. This measure, in turn, is mapped to intermediate abilities. By doing a similar mapping from abilities to other elements, such as occupational tasks, AI benchmarks can be linked with occupations (Figure 11.3).

Figure 11.3. Bidirectional and indirect mapping between job market (ISCO-3 specifications) and AI benchmarks

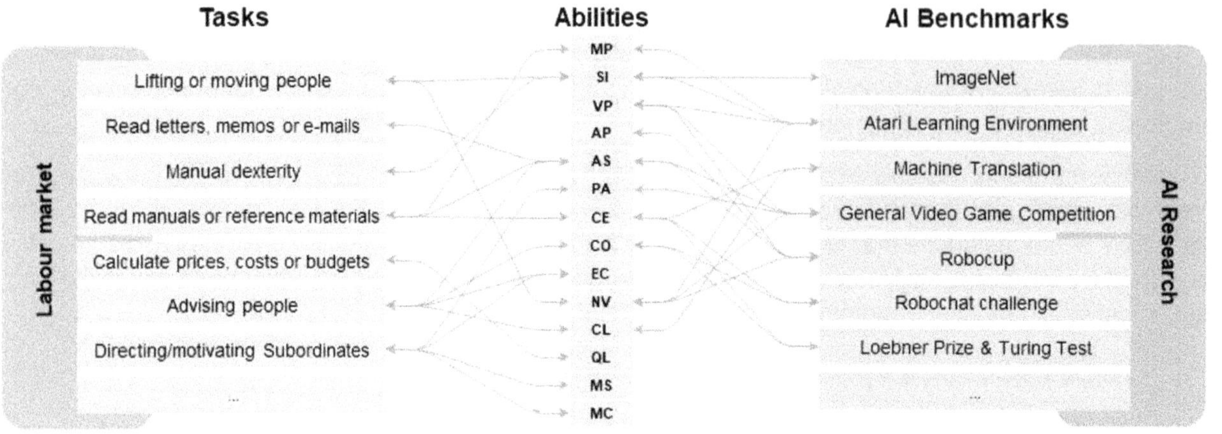

Source: (Martínez-Plumed et al., 2020[22]).

Bidirectional and indirect mapping is a promising approach but must be used cautiously for quantifying abilities. The "latent" intermediate abilities could, in principle, be mapped to other elements, such as human test results. However, the performance results of different benchmarks should not be aggregated; research intensities can be calculated for the ability but not the magnitude of the ability. This is because results at the leader boards for all benchmarks are about different AI systems. The systems that "solve" ImageNet are different from those used for Robocup.

More importantly, even if the same general AI system is used for many of the benchmarks (e.g. a language model), the magnitudes of performance (the scales) are different (Hernández-Orallo and Vold, 2019[5]; Hernández-Orallo, 2020[41]). Indeed, 90% success in ImageNet cannot be averaged with 72% correct answers in Winogrande or a score of 570 points in Pacman.

Normalising results against a human population is possible. However, the transformation should be based on *percentiles* over the human population rather than using human average performance. Moreover, this does not solve the scaling problem, as the relevance of each test would depend on the variance of the human population for that test.

AI Collaboratory

Some initiatives are exploring better mappings and aggregations of evaluation results for AI and humans to allow for meaningful comparison. One such initiative is the AI Collaboratory (Martínez-Plumed, Gómez and Hernández-Orallo, 2020[47]; Martínez-Plumed, Gómez and Hernández-Orallo, 2020[48]; Martínez-Plumed, Gómez and Hernández-Orallo, 2020[49]). As part of the AI WATCH programme (Martínez-Plumed, Gómez and Hernández-Orallo, 2020[47]) of the European Commission, it collects and structures evaluation results for AI and humans, and building mappings and hierarchies.

The AI Collaboratory is structured with a multidimensional schema. It contains information about the facts (the measurements) and satellite information about *who* is measured (the intelligent systems), *what* is measured (the tests) and *how* it is measured (the procedures). Each dimension is hierarchical. For instance, in the "who" dimension, systems can be aggregated into populations, populations into families, etc. In the "what" dimension, examples can be aggregated into tests, tests into batteries, etc. A taxonomy of abilities, as those seen in Figure 11.2 or Table 11.4, could be easily defined in the "what" dimension. Despite all the caveats for comparing human results with AI results, data-driven tools and meta-analyses are essential to understand how measurements relate to each other.

Recommendations

With the dominance of the machine learning paradigm, and skills changing more rapidly, AI systems will likely become less specialised for particular skills and tasks to be profitable. Consequently, a taxonomy of abilities, such as the one shown in Table 11.4, using the principles in Box 11.1, serves as a foundation for the evaluation of more adaptable AI systems. There will be exceptions like testing standardised skills or tasks, such as driving, but more general, ability-oriented AI will be able to adapt to evolving tasks and skills required at home and in the workplace.

- **Develop new testing protocols and share detailed results for data-driven exploration**

It is not enough to think in terms of abilities rather than tasks, or to build new benchmarks that cover an ability rather than a task. Instead, new testing protocols in AI are needed that go beyond training-test, avoiding learning specialisation or even AI systems built for the test. Also, evaluations should consider all the resources and costs involved in the solution (data, compute, human computation, etc.) (Martínez-Plumed et al., 2018[50]). Evaluation must become more iterative, more adversarial to avoid being gamed by AI researchers (willingly or not) (Hernández-Orallo, 2020[41]). Finally, there is a lack of meta-analysis and data-driven exploration of AI capabilities.

- **Use an intrinsic scale for new test designs**

Lack of common categories, and especially of commensurate scales, is a critical concern when reusing results from different AI benchmarks, and especially when comparing them with human tests results. New test designs should use an intrinsic scale, independent of the population to be tested. As an ultimate resource, scores could be normalised according to the distribution of human results rather than as a single individual or population average. AI often sets this average as a misleading threshold for "human-level performance".

- **Learn from human evaluation**

Use of human tests for evaluating AI *directly* is not feasible for a number of reasons. However, ignoring human evaluation, its tests and its associated techniques would be a mistake. Such approaches offer lessons to learn from. They may imply a cognitive overhaul of evaluation in AI. Exploring mappings and meaningful aggregations that capitalise on the information from human tests and AI benchmarks is a worthwhile initiative. AI and robotics – and human hybridisations yet to come – deserve more than simple performance-based, task-oriented evaluation.

References

Adams, S. et al. (2012), "Mapping the landscape of human-level artificial general intelligence", *AI Magazine*, Vol. 33/1, pp. 25-42. [20]

Adams, S., G. Banavar and M. Campbell (2016), "I-athlon: Towards a multidimensional Turing test", *AI Magazine*, Vol. 33/1, pp. 25-42. [21]

Bohannon, J. (2011), "Searching for the Google effect on people's memory", *Science*, Vol. 335/15 July, p. 277. [6]

Brynjolffson, E. and T. Mitchell (2017), "What can machine learning do? Workforce implications", *Science*, Vol. 358/6370, pp. 1530-1534. [25]

Brynjolffson, E., T. Mitchell and D. Rock (2018), "What can machines learn and what does it mean for occupations and the economy?", *AEA Papers and Proceedings*, Vol. 108/May, pp. 43-47. [26]

Campbell, M., A. Hoane and F. Hsu (2002), "Deep blue", *Artificial Intelligence*, Vol. 134, pp. 55-83. [8]

Carroll, J. (1993), *Human Cognitive Abilities: A Survey of Factor-analytic Studies*, Cambridge University Press, New York. [13]

Clark, P. and O. Etzioni (2016), "My computer is an honor student–But how intelligent is it? Standardized tests as a meausre of AI", *AI Magazine*, Vol. 37/1, pp. 5-12. [35]

Clark, P. et al. (2019), "From 'F' to 'A' on the N.Y. Regents Science Exams: An overview of the Aristo Project", *arXiv*, Vol. 1909.10958. [36]

Crosby, M. et al. (2019), "Translating from animal cognition to AI", *NeurIPS 2019 Competition and Demonstration Track, PMLR*, pp. 164-176. [45]

Crosby, M. et al. (2020), "The animal-AI testbed and competition", *NeurIPS 2019 Competititon and Demonstration Track, PLMR*, pp. 166-176. [46]

Dowe, D. and J. Hernández-Orallo (2012), "IQ tests are not for machines, yet", *Intelligence*, Vol. 40/2, pp. 77-81. [34]

Elliott, S. (2017), *Computers and the Future of Skill Demand*, Educational Research and Innovation, OECD Publishing, Paris, https://dx.doi.org/10.1787/9789264284395-en. [27]

Firestone, C. (2020), "Performance vs. competence in human-machine comparisons", *Proceedings of the National Academy of Sciences*, Vol. 117/43, pp. 26562-26571. [51]

Fleishman, E. (1972), "On the relation between abilities, learning and human performance", *The American Psychologist*, Vol. 27/11, pp. 1017-1032. [52]

Frank, M., D. Wang and M. Cebrian (2019), "The evolution of citation graphs in artificial intelligence research", *Nature Machine Intelligence*, Vol. 1/2, pp. 79-85. [18]

Hamilton, E., J. Rosenberg and M. Akcaoglu (2016), "The Substitution Augmentation Modification Redefinition (SAMR) model: A critical review and suggestions for its use", *TechTrends*, Vol. 60, pp. 433-441. [53]

Hernández-Orallo, J. (2020), "Twenty years beyond the Turing test: Moving beyond the human judges too", *Minds & Machines*, Vol. 30, pp. 533-562. [41]

Hernández-Orallo, J. (2019), "Gazing into Clever Hans machines", *Nature Machine Intelligence*, Vol. 1/4, pp. 172-174. [40]

Hernández-Orallo, J. (2017), "Evaluation in artificial intelligence: From task-oriented to ability-oriented measurement", *Artificial Intelligence Review*, Vol. 48/3, pp. 398-447. [10]

Hernández-Orallo, J. (2017), *The Measure of All Minds: Evaluating Natural and Artificial Intelligence*, Cambridge University Press, New York. [11]

Hernández-Orallo, J. et al. (2016), "A new AI evaluation cosmos: Ready to play the game", *AI Magazine*, Vol. 38/3, pp. 66-69. [30]

Hernández-Orallo, J. et al. (2016), "Computer models solving intelligence test problems: Progress and implications", *Artificial Intelligence*, Vol. 230, pp. 74-107. [33]

Hernández-Orallo, J. and K. Vold (2019), "AI extenders: The ethical and societal implications of humans cognitively extended by AI", *Proceedings of the 2019 AAAI/ACM Conference on AI, Ethics, and Society*, Association for Computing Machinery, New York. [5]

Jackson, G. (2019), "Why the rise of the robots hasn't happened just yet", 23 January, The Financial Times, https://www.ft.com/content/ec2f65c8-1e61-11e9-b2f7-97e4dbd3580d. [1]

Kocijan, V. et al. (2020), "A review of Winograd schema challenge datasets and approaches", *arXiv preprint arXiv:2004*, Vol. 13831. [38]

Krizhevsky, A. (2009), *Learning Multiple Layers of Features from Tiny Images*, University of Toronto, https://www.cs.toronto.edu/~kriz/learning-features-2009-TR.pdf. [54]

Lapuschkin, S. et al. (2019), "Unmasking Clever Hans predictors and assessing what machines really learn", *Nature Communications*, Vol. 10/1, pp. 1-8. [39]

Levesque, H., E. Davis and L. Morgenstern (2012), "The Winograd schema challenge", presentation, Thirteenth International Conference on the Principles of Knowledge Representation and Reasoning. [37]

Machado, M. et al. (2018), "Revisting the arcade learning environment: Evaluation protocols and open problems for general agents", *Journal of Artificial Intelligence Research*, Vol. 61, pp. 523-562. [17]

Martínez-Plumed, F. et al. (2018), "Accounting for the neglected dimensions of AI progress", *arXiv preprint arXiv*, Vol. 1806.00610. [50]

Martínez-Plumed, F., E. Gómez and J. Hernández-Orallo (2021), "Futures of artificial intelligence through technology readiness levels", *Telematics & Informatics*, Vol. 58/101525. [29]

Martínez-Plumed, F., E. Gómez and J. Hernández-Orallo (2020), "AI Watch: Methodology to monitor the evaluation of AI technologies", *JRC Working Papers*, No. 120090, Joint Research Centre. [49]

Martínez-Plumed, F., E. Gómez and J. Hernández-Orallo (2020), *Assessing Technology Readiness Levels of Artificial Intelligence*, Joint Research Centre Report, AI Watch, European Commission, Brussels. [28]

Martínez-Plumed, F., E. Gómez and J. Hernández-Orallo (2020), "Tracking AI: The capability is (not) near", presentation, 24th European Conference on Artificial Intelligence, Santiago de Compostela, Spain. [48]

Martínez-Plumed, F., E. Gómez and J. Hernández-Orallo (2020), "Tracking the impact and evolution of AI: The AIcollaboratory", Evaluating progress in AI, First International workshop, European Conference on Artificial Intelligence, Santiago de Compostela, Spain. [47]

Martínez-Plumed, F. et al. (2020), "Does AI qualify for the job: A Bidirectional model mapping labour and AI intensities", *Proceedings of the AAAI/ACM Conference on AI, Ethics and Society*, pp. 94-100. [22]

Morra, S. et al. (2012), *Cognitive Development: Neo-Piagetian Perspectives*, Psychology Press, Hove, UK. [16]

Mueller, S. (2010), "A partial implementation of the BICA cognitive decathlon using the Psychology Experiment Building Language (PEBL)", *International Journal of Machine Consciousness*, Vol. 2/2, pp. 273-288. [43]

Mueller, S. et al. (2007), "The BICA cognitive decathlon: A test suite for biologically-inspired cognitive agents", *Proceedings of behavior representation in modeling and simulation conference*. [44]

Piaget, J. (1964), "Cognitive development in children", *Journal of Research in Science Teaching*, Vol. 2/3, pp. 176-186. [15]

Piaget, J. (1936), *La naissance de l'intelligence chez l'enfant*, Delachaux et Niestlé, Lonay, Switzerland. [14]

Puentedura, R. (2006), "Transformation, technology and education", presentation, Strengthening Your District Through Technology workshop, 18 August, Maine School Superintendents Association, http://hippasus.com/resources/tte/. [55]

Purves, C., C. Cangea and P. Veličković (2019), "The PlayStation reinforcement learning environment (PSXLE)", *arXiv preprint arXiv*, Vol. 1912.06101. [56]

Rahwan, I. (2019), "Machine behaviour", *Nature*, Vol. 568/7753, pp. 477-486. [4]

San Antonio, T. (ed.) (1993), *Bayley Scales of Infant Development*. [24]

Schlangen, D. (2019), "Language tasks and language games: On methodology in current natural language processing research", *arXiv prepreint1398arXiviv*, Vol. 1908.10747. [31]

Schuelke, M. and E. Day (2012), "Ability determinants of complex skill acquisition", *Encyclopedia of the Sciences of Learning*, Springer. [12]

Seel, D. (2012), "Skill", *Encyclopedia of the Sciences of Learning*, Springer. [57]

Silver, D. et al. (2016), "Mastering the game of Go with deep neural networks and tree search", *Nature*, Vol. 529/7587, pp. 484-489. [9]

Taylor, A. (2018), "The Automation Charade", *Logic*, Vol. 5/1 August, https://logicmag.io/failure/the-automation-charade/. [3]

Tolan, S. et al. (2021), "Measuring the occupational impact of AI: tasks, cognitive abilities and AI benchmarks", *Journal of Artificial Intelligence Research*, Vol. 71, pp. 191-236. [23]

Vinyals, O. et al. (2017), "Starcraft II: A new challenge for reinforcement learning", *arXiv preprint arViv*, Vol. 1708.04782. [58]

Von Ahn, L. (2008), *Human computation*, presentation to IEEE 24th International Conference on Data Engineering, 7-12 April, Cancun. [2]

Wang, A. et al. (2019), "Superglue: A stickier benchmark for general-purpose language understanding systems", *arXiv preprint arXiv*, Vol. 1905.00537. [59]

Wasserman, E. and T. Zentall (2006), *Comparative Cognition: Experimental Explorations of Animal Intelligence*, Oxford University Press. [19]

Weber, B. (1997), “Computer defeats Kasparov, stunning the chess experts”, 5 May, New York Times. [7]

Zellers, R. et al. (2018), “Swag: A large-scale adversarial dataset for grounded commonsense inference”, *arXiv preprint arXiv*, Vol. 1808:05236. [42]

Zellers, R. et al. (2019), “Hellaswag: Can a machine really finish your sentence?”, *arXiv preprint arXiv*, Vol. 1905.078301400. [32]

Notes

[1] The terms capability and capacity will be used more broadly, while the terms skill, knowledge and ability have more precise uses as follows. A skill is the "overlearned behavioural routine resulting from practice" (Seel, 2012[57]), which is represented by “the level of proficiency on specific tasks. It is the learned capability of an individual to achieve desired performance outcomes (Fleishman, 1972[52]). Thus, skills can be improved via practice and instruction” (Schuelke and Day, 2012[12]). Knowledge is typically used in a similar sense as skills but assumes a more theoretical or conceptual nature as opposed to the practical or actionable nature of skills. An ability "refers to a general trait, reflecting the relatively enduring capacity to learn tasks. Although fairly stable, ability may change over time primarily in childhood and adolescence through the contributions of genetic and developmental factors" (Schuelke and Day, 2012[12]). New skills are acquired using cognitive abilities and can build on previous skills and knowledge. See also section 3.

[2] The “Substitution, Augmentation, Modification, and Redefinition” (SAMR) model is a popular taxonomy covering the ways in which technology may affect tasks (Puentedura, 2006[55]; Hamilton, Rosenberg and Akcaoglu, 2016[53]), but the opposition of enhancement vs transformation does not work well for AI, especially as enhancement based on AI is usually coupled with significant modification and redefinition of tasks.

[3] A model or system overfits when it shows good performance for the examples seen during training (or examples from the training distribution) but generalises poorly for examples that are different from those seen during training.

[4] This has recently been rephrased as performance versus competence (Firestone, 2020[51]).

[5] www.onetcenter.org/taxonomy.html

[6] www.bls.gov/soc/2018/major_groups.htm

[7] www.ilo.org/public/english/bureau/stat/isco/

[8] https://ec.europa.eu/esco

[9] Summaries of these sources can be found in several tables and figures in Chapters 3-5 of (Hernández-Orallo, 2017[11]).

[10] This analysis excludes non-cognitive behavioural features, such as personality traits, from the analysis, as their extrapolation to AI systems is even more farfetched. This does not mean that personality in machines has not been studied or even tried to be measured. For more information about non-cognitive behavioural features in machines, see (Hernández-Orallo, 2017[11]).

[11] For a summary of intelligence tests, developmental tests and attainment tests for humans, see Chapters 3 and 12 of Hernández-Orallo (2017[11]).

[12] Most of these tests are not freely available. This is partly for commercial reasons and partly because having the questions in advance would lead to specialisation: humans would prepare for the test.

[13] This has happened from CIFAR10 (image classification) to CIFAR100 (Krizhevsky, 2009[54]), SQuAD1.1 (Q&A) to SQuAD2.0, GLUE (language understanding) to SUPERGLUE (Wang et al., 2019[59]), Starcraft (real-time strategy) to Starcraft II (Vinyals et al., 2017[58]) and the Atari Learning Environment (ALE) (Machado et al., 2018[17]) to the PlayStation Reinforcement Learning Environment (PSXLE) (Purves, Cangea and Veličković, 2019[56]).

[14] https://mosaic.allenai.org/projects/mosaic-commonsense-benchmarks

[15] https://dynabench.org/

12. Using human skills taxonomies and tests as measures of artificial intelligence

Ernest Davis, New York University

This chapter looks at using human skills taxonomies and tests as measures of artificial intelligence (AI). It examines the strong points of computers, such as their ability to store enormous memories and access them reliably and quickly. It also reflects on weaknesses of AI systems compared to humans related to vision and manipulation, and use of natural language. It pays special attention to the limited capacity of AI to use common sense reasoning and world knowledge. In addition, the chapter looks at the ability of AI to detect subtle patterns in data as a double-edged sword. With all this in mind, the chapter proposes four consequences for testing, and looks ahead to building trustworthy AI systems.

Introduction

It is tempting to use human tests to measure the power of artificial intelligence (AI). The body of educational and vocational tests developed evaluating human beings is extensive. These tests have been developed, studied and validated meticulously. Moreover, the science of human testing is powerful and deep, grounded in both psychology and practical experience.

However, using human tests to measure AI has many potential pitfalls that can lead to misleading results. Human tests have been developed to distinguish between human beings, or to evaluate the fitness of a human test taker for a given task. Their designs therefore take for granted that test takers share basic features of human intelligence. They are not designed to evaluate intelligences that are radically different, such as AIs.

Moreover, an AI system, or a computer system generally, may play a different role than a human in any kind of undertaking. When a new technology is introduced in an application, for example, it does not simply do the same thing people did. Often, it transforms the whole process. Hence, measuring the human-level abilities of AI may be entirely irrelevant to predicting its usefulness or to what extent it may replace human workers.

The abilities and limitations of humans and computers are dramatically different. This banal truth lies at the heart of the problems in applying human tests to AI, and often gets overlooked. This chapter thus begins enumerating the obvious advantages of computers and their significance before discussing their more subtle limitations at greater length.

Where computers shine

The strong points of computers are obvious. They can carry out complicated calculations extremely quickly. They have enormous memories that can be accessed reliably and extremely quickly. Conversely, the working memory of humans is extremely small and their long-term memory is limited and unreliable. Moreover, computers do not suffer from fatigue or distractions, and therefore do not make careless errors. Additionally, data or programs can be rapidly and accurately copied from one computer to another. To a large extent, data in one computer can be conveyed to another in minutes or seconds at essentially zero cost.[1]

Even in tasks where computers excel, they may carry their abilities over to similar tasks, or even to components of the same task, differently than people do. Computer programs are, inevitably and properly, designed to take maximal advantage of the computer's strengths. A program may thus achieve a human or superhuman level of skill at some particular task. However, in so doing it may have mastered only one aspect of a problem.

Humans could remain superior at mastering multiple aspects of a problem. For example, a web search engine can record and retrieve web pages on a scale unimaginable for a person. Yet humans are still enormously superior in judging whether a particular webpage includes the answer to a specific question. Games-playing programs are enormously superior to humans in terms of tactics; in terms of strategy, humans still sometimes have the advantage.

Where computers fall short

In many respects, AI systems fall far short of humans, suffering from weaknesses that are quite unlike anything seen in people.

Unique weaknesses of artificial intelligence systems

Vision and manipulation

Computer vision and manipulation are not comparable to humans or, indeed, many animals. AI systems can match or even exceed human abilities in certain narrowly defined tasks. These include recognising digits or identifying particular medical situations or well-defined categories. However, they are nowhere near human abilities in identifying activities or relation between objects in still photos, and still worse in videos.

Natural language

The abilities of AI technology at natural language tasks are uneven. Speech transcription is quite reliable under certain conditions. These include when the system has been trained on the speaker; when the speaker does not have a strong accent; when there is a single speaker; and when the background is quiet. However, even under less restrictive conditions, AI transcription is generally good.

In addition, machine translation between the major European languages, Japanese and Chinese is generally fine. Web search is generally successful at finding relevant documents. However, while intelligent assistants such as Siri and Alexa can be useful, their quality is uneven (Dahl and Doran, 2020[1]). Developing systems that can actually understand an utterance or text remains a distant goal.

Computers are mostly not embodied

A human child learns the basics of the physical and social worlds by interacting with them. Research in "developmental robotics" has attempted to do likewise. In this approach, researchers allow the robot to interact with the real world or with a realistic virtual world (Asada et al., 2009[2]; Shanahan et al., 2020[3]).

However, the overwhelming majority of AI programs are just passive observers of a large dataset with almost no inherent structure. Vision programs are trained on a collection of millions of images labelled by category. The labels are the only connection between one image and another. Neither the images nor the labels have any connection to a larger context. Language programs are trained on immense text corpora entirely ungrounded in the real world. Most robots are built to carry out limited tasks in controlled worlds with limited perceptual feedback.

This gap has obvious implications for robots that have to engage with a complex open world but also for the depth of the understanding of natural language. Humans' understanding of natural language is fundamentally grounded in their experience of the external world and their interactions with it (Lake and Murphy, 2020[4]). It is not clear whether, ultimately, any amount of textual data or advance in learning technology can make up for this fundamental difference.[2]

Common sense reasoning and world knowledge

AI programs remain limited in their basic understanding of the world (Davis and Marcus, 2015[5]; Levesque, 2017[6]). A particularly important stumbling block is a poor understanding of time. Many AI programs exist in a timeless present with only a superficial ability to deal with past, future and change. Other domains such as spatial reasoning, intuitive physics, folk psychology and folk sociology are even less incorporated in AI programs; their absence likewise severely limits the depth of understanding that an AI program can achieve.

Meta-reasoning

AI programs are generally unaware of their own reasoning processes and characteristics. Only a minority of question-answering systems know when to say, "I don't know" or "I need more information". Many AI systems are configured to give what they somehow judge to be the best answer. However, these systems do not consider whether their judgement is reliable, whether they have enough information to answer the question in principle or even whether the question is meaningful.

Generation vs. understanding

For humans, it is almost always easier to understand something than to create it. For example, it is generally much easier to understand a foreign language than to speak or write it and much easier to interpret an image than to create it.

For computer programs, the reverse often holds. It is much easier to construct computer graphics that can create complex images of any form (photorealistic, line drawing, schematic, etc.) than a computer vision program that can interpret them. (The use of CAPTCHAs to distinguish humans from bots depends on this; the CAPTCHA program can easily generate an image with distractors that a human can easily see through but that will confuse a bot.)

The GPT-3 system can generate long articles that, on surface inspection, are plausible imitations of journalists, philosophers, essayists, poets and so on. However, on closer inspection, it does not understand its own writing. Chapter 9 discusses a skill test that asks the test taker to identify a side view of a disk separator. It would be unreasonable to ask a human being to draw the side view instead, but this task might well be easier for a computer program.

This difference is the source of some common misunderstandings and exaggerated views of AI. It is natural to suppose that, if AI can draw a picture, surely it can see a picture. Similarly, if it can write a text, surely it can understand the text it has written. However, although that conclusion holds for humans, it does not at all hold for AIs.

Extreme specialisation

Some of the most prominent AI successes are specialised to an extraordinary degree. A beginner Go player with a 20x20 board instead of the standard 19x19 will probably play the entire game without noticing. A good or champion-level player might be disconcerted when they notice the 20x20 board but will still likely play well. However, a champion AI Go player will not play well on a 20x20 board. Indeed, it will be unable to even conceive such a board could exist. It is hard-wired to understand the world consists of 361 positions that may be white, black or empty and that only 362 moves (the 361 squares and "pass") are possible. With a 20x20 board, they are as much at a loss as a human transported to a 17-dimensional universe.

Similar limits appear in other areas as well. Any human who can easily translate from English to high-quality French can presumably answer questions posed in French. However, a program that can translate from English to French cannot answer any questions at all in French. A chess-playing program can choose a superlatively fine move but may not be able to do anything else chess-related. For instance, it probably cannot say anything about a game it has just played. It can certainly not offer an opinion about the skill of its opponent.

Combining skills

AI programs cannot always combine skills. A person with two skills can normally carry out tasks that require both of them. However, an AI program that can tell whether a picture contains a cat or a dog may not be able to say if it contains both a cat and a dog. As of 12 May 2020, Google search can answer the questions, "What is the US state with the largest area?" and "What is the population of Alaska?" but not "What is the population of the US state with the largest area?"

Certainly, if one AI program has ability A and another has ability B, it does not follow that another program could do tasks requiring both A and B. For example, systems trained for pronoun reference resolution can attain near-perfect behaviour. However, question-answering and translation programs may stumble when confronted with the same kinds of problems.

Grotesque errors

AI programs often give answers that are not merely wrong but, by human standards, bizarre or grotesque[3]. The GPT-3 system can generate page after page of text that is superficially well-written and appears coherent. However, in one experiment Marcus and Davis (2020[7]) gave a prompt to GPT-3 that produced a nonsensical continuation:

The prompt was:

> At the party, I poured myself a glass of lemonade, but it turned out to be too sour, so I added a little sugar. I didn't see a spoon handy, so I stirred it with a cigarette. But that turned out to be a bad idea because

GPT-3 produced the continuation:

> it kept falling on the floor. That's when he decided to start the Cremation Association of North America, which has become a major cremation provider with 145 locations.

In another experiment, Metz (2020[8]) prompted GPT-3 to write a love story. It managed relatively well until the last sentence, which started "We went out for dinner and drinks and dinner and drinks and dinner and drinks ..." It repeated the phrase "and dinner and drinks" 55 times before running out of steam. In another instance, a medical Chabot powered by GPT-3 recommended suicide to an imaginary patient (Rousseau, Bauderlaire and Riviera, 27 October, 2020[9]).

Self-driving cars often do the correct thing. However, in one case a self-driving car mistook a truck for a billboard above the road, causing a fatal crash (Evarts, 11 August 2016[10]). The humorous, absurd or embarrassing mistakes produced by the "autocorrect" feature of text messaging systems are legion.

This aspect of computers dates back to the early days of computer technology. In the 1960s and 1970s, there were endless horror stories about computer systems that sent bills for millions of dollars or repeatedly sent bills for zero dollars. There was nothing to be done because "it's the computer". No one at the company knew how, or was able, to fix the third-party software. That kind of problem has become much less frequent over the decades. Computer systems have not become any more aware of the absurdity of these bills. However, this kind of software has gradually been debugged. In addition, interfaces and protocols for the humans using the software have probably improved.

The problem now is fundamentally the same, although it takes much more sophisticated forms. For example, AI programs based on machine learning are often effectively impossible to debug. They can only be retrained.

Finding subtle patterns in data

Most AI successes in the last 20 years have been based on applying machine learning techniques to large datasets. A large data corpus relevant to a particular task is assembled by some means or other. For example, it may be collected from resources such as the web. It could also be generated by human labour for this purpose or synthesised by another computer program. Finally, it could be assembled using some combination of these approaches. The machine learning technology then finds patterns, generally extremely complex, in the dataset and uses them to carry out the task.

The patterns found in machine learning are generally not ones that humans have found or could find. For patterns that humans could find, it is usually more effective to use conventional programming rather than machine learning. Indeed, even once the AI has found the patterns, it is usually beyond human abilities to explain why they are effective or even to describe them in any meaningful way. This ability to find complex obscure patterns in data underlies AI successes in specialised tasks, such as economic modelling, scientific research and sociological studies; and in basic human abilities, such as vision and language.

However, this reliance on complex patterns is a double-edged sword. Since the patterns cannot be explained or understood, empirical tests are the only way to judge their reliability or where they break down. If they work reliably on well-chosen test examples, then presumably, or hopefully, they will continue to work well in the future.

This, in turn, leads to a further danger. Generally, for testing, the corpus of examples is divided randomly into a "training set" for input to the learning component and a "test set" to evaluate the quality of the system. Patterns may apply to the corpus as a whole because of the way it was assembled but not to examples outside the corpus. In these cases, the learning module will find those patterns in the training set. Consequently, the program using the patterns will work properly on all the examples in the test set. However, it will fail on new examples that do not conform to these patterns.

For example, the SNLI dataset contains pairs of English sentences A and B characterised in terms of their logical relations: B is a consequence of A, B contradicts A, or B is neutral with respect to A (i.e. B could be either true or false). Programs based on machine learning trained on a training set from this dataset achieved a significant measure of success when tested on a test set; this was taken as a sign of progress towards understanding the logical significance of texts. It was later discovered the relation could be identified by looking only at sentence B. The dataset had been constructed by giving crowd workers sentence A and asking them to construct a sentence B with the target relation. It turned out that the crowd workers had used a few simple strategies in constructing their examples. For instance, they had often constructed an entailed sentence by leaving out gender or number information, a neutral sentence by adding a motivation that might be true or false, and a contradiction by adding a negation (Table 12.1). The program then categorised the relation between the sentences, with fair accuracy, purely on the basis of these features in sentence B (Gururangan et al., 2018[11]).

Table 12.1. Annotation artefacts in a corpus of sentence entailment

Category	Example
Premise	A woman is selling bamboo sticks talking to two men on a loading dock
Entailment	There are **at least** three **people** on the loading dock.
Neutral	A woman is selling bamboo sticks **to help provide for her family**.
Contradiction	A woman is **not** taking money for any of her sticks.

Source: Gururangan (2018[11]).

As another consequence to constructing AI programs in this way, the programs can be vulnerable to variations in input that seem entirely inconsequential to humans or even to other computer programs. Vision programs can be fooled by small changes invisible to the human eye (Figure 12.1). This kind of problem in AI programs has been demonstrated innumerable times; indeed, the construction of "adversarial examples" that break AI programs is at this point a significant subfield of AI research.

Figure 12.1. Pig or airliner: Changes imperceptible to the human eye can lead to misclassifications

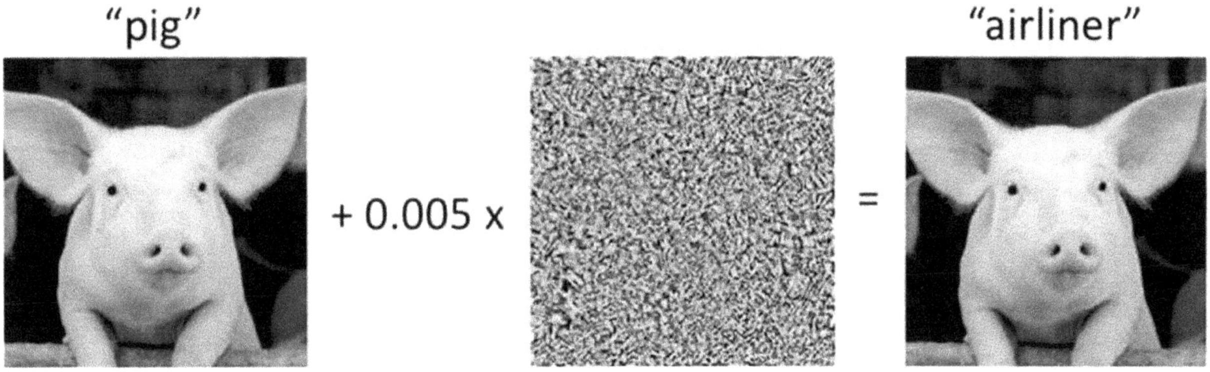

Source: Madry and Schmidt (Image created by Logan Engstrom) (6 July 2018[12]).

Another experiment with the GPT-3 program accidentally left a blank space at the end of a line (Marcus and Davis, 2020[7]). This trivial error was invisible to the human eye and unproblematic for most programs that deal with natural language text. However, it caused GPT-3 to flounder on a test that otherwise it would have gotten right. Similarly, machine translation systems can be bewildered by a misspelled word that would never confuse a human reader. Typically, programs equipped with spelling correction, such as Google search, can handle these challenges easily.

Consequences for testing

Four consequences of limitations call for caution in using human-oriented tests on artificial intelligence

Due to the four consequences of all these limitations and idiosyncrasies noted below, extreme caution is needed in evaluating the significance of the success of an AI program on a human-oriented test. When humans do well on a test, one can conclude, with some degree of confidence, that they know the material or can carry out the tasks the test was designed to measure. If an AI system does well on the same test, that conclusion is altogether unreliable.

Humans and artificial intelligence systems have different notions of difficulty

What is easy for a human is often difficult for a computer, and vice versa. It may be much easier for an AI program to search the entire web for an answer than to find the answer in a text provided with the test. As mentioned, it is often easier for a computer to draw a picture than to recognise a picture. Yes/no questions are particularly difficult for AI programs (Clark et al., 2019[13]).

The questions in standardised human tests are carefully calibrated in terms of difficulty, but this calibration does not apply to AI programs taking the same test. AI programs that succeed on standardised science tests (Clark et al., 2019[14]) may be at a loss when asked a simple question about the physical world that is not the kind that appears on tests (Davis, 2016[15]). For example, a program may be able to answer the question, "How are the particles in a block of iron affected when the block is melted?" from the eighth grade NY State Regents exam, but not the question, "Is it possible to fold a watermelon?"

AI systems are sensitive to "inconsequential" changes

AI programs may be extraordinarily sensitive to changes in question format that, to a human, would be inconsequential or even invisible. This kind of sensitivity becomes likely if the dataset used to train or "fine-tune" the program is in some way related to the source of the test questions.

Human abilities cannot be taken for granted in AIs

Important human abilities are taken for granted and so not tested but these same abilities cannot be taken for granted in AIs. No test, for instance, rewards a test taker for answering "I don't know". At most, the scoring system is set up so there is, on average, no value in guessing randomly.

In real-world settings, it is often important that a person, or an AI program, realises the limits of its knowledge. Either it must try to find out what it needs to know or proceed with suitable caution. Only a small fraction of AI programs even attempt this approach (Davis, 2020[16]). Likewise, basic common sense knowledge is almost never explicitly tested in human-oriented tests because it can be assumed in people.

Grotesque errors of artificial intelligence can lead to disaster

Finally, the tendency of AI programs to fall into grotesque errors raises concerns that generally do not arise with humans. If these are fairly rare but catastrophic, then, like so many computer bugs, they may avoid detection during controlled testing. However, they can emerge disastrously when the system is deployed in the world at large.

Towards trustworthy artificial intelligence systems

Even if an AI system suffers from the kind of limitations described above, it still may be useful in a practical setting. An AI program may be specialised, sensitive to small changes and adversarial examples, lack common sense and make grotesque errors. Nonetheless, it might still be placed in a work environment where it can remain within its area of specialisation. It could only be given inputs it can handle and not required to use common sense. Finally, its grotesque errors could be either avoided or caught.

Humans should be able to adapt to the idiosyncrasies of AI. Collectively, people have some 60 years of experience of dealing with computer programs. They are used to word processing or spreadsheets doing things that would be bizarre and unacceptable in a human secretary or accountant.

It is critical, however, to understand the scope of AI programs and their limitations. Human-oriented tests are a poor way of determining the scope and limitations of AI programs. An AI program should not be considered as "an unusual human being". Consequently, it should not be evaluated using the same yardsticks applied to human job applicants. Rather, it should be seen as a potentially powerful but poorly understood piece of software engineering.

Above all, the "attribution error" should be avoided. When a human succeeds at a task, it understands what has been achieved. This is not the case for computers. Methods, insights and cautions for avoiding these errors developed through studying the cognitive psychology of animals are relevant here (see also Chapter 17) (Shanahan et al., 2020[3]).

Ideally, programs used in critical tasks should come with the kinds of product reliability information that accompanies dangerous physical tools or medications. This would permit users to say with confidence that, under normal circumstances, when used properly, the programs are reliable. At the same time, this would alert users to risk if programs were used in unusual circumstances or in some strange way (Marcus and Davis, 2019[17]).

Historically, computer technology (with the exceptions of hardware and cybersecurity), and AI in particular, has not provided these kinds of guarantees. The addition of new features has been typically prioritised over ensuring that existing features worked reliably. However, with more reliance on computers for critical activities, the technology increasingly needs to be trustworthy.

In this regard, systems like the GPT series are a step in the wrong direction for a variety of reasons (Marcus, 2020[18]). They have no specified purpose. They carry out an ill-defined category of tasks, often impressively, sometimes absurdly, with no demarcation or predictability. Finally, they are sold to the AI community and to the world at large as a tonic medicine "good for what ails you".

The AI research community has become increasingly aware of these issues, particularly in the last two years. The development of evaluation strategies for AI technology and the careful analysis of its capacities is a major and urgent area of research (Dodgen et al., 2019[19]; Heinzerling, 21 July 2019[20]; Pineau, 2020[21]). The problems of determining what an AI system can do and how it can most productively be used in practical settings are major challenges.

Predicting the impact of artificial intelligence

A computer system does not have to emulate or achieve the abilities of a human worker to revolutionise the workplace (Shneiderman, 2020[22]). Word processing programs, for example, have largely displaced the role of typists. Yet the personal computer revolution did not require computers to learn skills like feeding paper or changing a ribbon. At the same time, typists have skills that surpass word processing technology. A typist, for example, can give a frank opinion on the quality of a letter.

Like a word processor, AI will remove some frustrations and introduce new ones. This dynamic has implications for testing. The tests that are used to compare human typists – mostly speed and errors per page – are irrelevant for word processing technology. In fact, there is no useful way to measure the quality of a word processing software other than user satisfaction (which is nebulous) and profitability (which depends on many factors other than inherent quality).

In any given field, an enormous impact cannot be tied to any specific ability, let alone any human ability, let alone an ability that is addressed in human tests.

Recommendations

It is difficult to design tests for AI that are meaningful and reliable. AI technology is evolving at breath-taking speed. Moreover, AI developers are generally more concerned with creating products and capacities than with evaluating them.

AI technology has become more sophisticated and ubiquitous. Given limited and inadequate understanding of AI, it is both increasingly urgent and difficult to evaluating AI and predict its impact. With this in mind, some guidelines for the design of tests are presented:

- **Keep in mind the differences between AI and humans**

The most reliable tests will measure how well an AI carries out a well-defined task in a particular workplace. For instance, how well can the program detect conditions of a specified type in medical data or images of some kind? Even in such a limited setting, tests should reflect the limitations of AI and the differences between AI and humans. The tests should determine robustness of AI in relation to flawed data (e.g. misspellings in a text or imaging anomalies) and to potential situations (e.g. a patient with some other, unusual condition). If these have not been tested, then the human interpreting the results needs to keep these concerns in mind as potential sources of error.

- **Stay flexible**

AIs used directly by the public at large – that are embedded in a commercial product or put on a website – require special attention. Tests should try to guard against all the possible ways that things can go wrong with users who are often careless, impatient and occasionally malicious. One must also expect that users will find ways to make things go wrong that the tests did not anticipate. Designers should thus stay flexible so they can respond adequately to these issues.

- **Look for strengths and weaknesses rather than a specific score**

Designing tests to evaluate the capabilities of state-of-the-art AI against those of humans on a broadly defined, open-ended task is much more challenging than on a simple task. In general, it is more useful and more meaningful to probe the strengths and weaknesses of the system rather than assigning a score between 0 and 100.

- Test problems that seem easy; problems posed in a variety of forms and that require a variety of kinds of answers; and problems collected or generated from a variety of sources.
- Include questions that pinpoint each of the system's individual capacities and problems that test its ability to use two of its capacities in combination.
- Test data on another source (if the source of the training data is known).
- Collect the problems from some natural source if possible rather than generating problems to serve as a test set.
- Test the system against adversarial examples and against anomalous examples. Success on any narrowly defined task should not be considered an adequate measure of the system's ability at a much broader set of tasks.

References

Asada, M. et al. (2009), "Cognitive developmental robotics: A survey", *IEEE transactions on Autonomous Mental Development*, Vol. 1/1, pp. 12-34, https://ieeexplore.ieee.org/abstract/document/4895715. [2]

Chomsky, C. (1986), "Analytic study of the Tadoma method: Language abilities of three deaf-blind subjects", *Journal of Speech, Language, and Hearing Research*, Vol. 29/3, pp. 332-347, https://doi.org/10.1044/jshr.2903.347. [23]

Clark, C. et al. (2019), "BoolQ: Exploring the surprising difficulty of natural yes/no questions", *arXiv*, Vol. 1905.10044, https://arxiv.org/abs/1905.10044. [13]

Clark, P. et al. (2019), "From 'F' to 'A' on the NY Regents Science Exams: An overview of the Aristo project", *arXiv*, Vol. 1909.01958, https://arxiv.org/abs/1909.01958. [14]

Dahl, D. and C. Doran (2020), "Does your intelligent assistant really understand you?", 6 October, Speech Technology, https://www.speechtechmag.com/Articles/Editorial/Industry-Voices/Does-Your-Intelligent-Assistant-Really-Understand-You-143235.aspx. [1]

Davis, E. (2020), "Unanswerable questions about images and texts", *Frontiers in Artificial Intelligence: Language and Computation* 29 July, https://doi.org/10.3389/frai.2020.00051. [16]

Davis, E. (2016), "How to write science questions that are easy for people but hard for computers", *AI Magazine* Spring, https://cs.nyu.edu/faculty/davise/papers/squabu.pdf. [15]

Davis, E. and G. Marcus (2015), "Commonsense reasoning and commonsense knowledge in artificial intelligence", *Communications of the ACM*, Vol. 58/8, pp. 92-105, http://dx.doi.org/10.1145/2701413. [5]

Dodgen, J. et al. (2019), "Show your work: Improved reporting of experimental results", *arXiv preprint*, Vol. 1909.03004, https://arxiv.org/abs/1909.03004. [19]

Evarts, E. (11 August 2016), "Why Tesla's Autopilot isn't really autopilot", Car Buying Tips, News and Features blog, https://cars.usnews.com/cars-trucks/best-cars-blog/2016/08/why-teslas-autopilot-isnt-really-autopilot. [10]

Gururangan, S. et al. (2018), "Annotation artifacts in natural language inference data", *Proceedings of the 2018 Conference of the NorthAmerican Chapter of the Association for Computational Linguistics: Human Language Technologies*, Vol. 2/Short Papers, http://dx.doi.org/10.18653/v1/N18-2017. [11]

Heinzerling, B. (21 July 2019), "NLP's Clever Hans moment has arrived", Benjamin Heinzerling blog, https://bheinzerling.github.io/post/clever-hans/. [20]

Lake, B. and G. Murphy (2020), "Word meaning in minds and machines", *arXvi preprint*, Vol. 2008.01766, https://arxiv.org/abs/2008.01766. [4]

Levesque, H. (2017), *Common Sense, the Turing Test and the Quest for Real AI*, MIT Press, Cambridge, MA. [6]

Madry, A. and L. Schmidt (6 July 2018), "A brief introduction to adversarial examples", Gradient Science blog, https://gradientscience.org/intro_adversarial/. [12]

Marcus, G. (2020), "GPT-2 and the nature of intelligence", *The Gradient,* 25 January, https://thegradient.pub/gpt2-and-the-nature-of-intelligence/. [18]

Marcus, G. and E. Davis (2020), "GPT-3, Bloviator: OpenAI's language generator has no idea what it's talking about", *Technology Review,* 22 August, https://www.technologyreview.com/2020/08/22/1007539/gpt3-openai-language-generator-artificial-intelligence-ai-opinion/. [7]

Marcus, G. and E. Davis (2019), *Rebooting AI: Building Artifificial Intelligence We Can Trust*, Pantheon Press, New York, http://rebooting.ai. [17]

Metz, C. (2020), "When AI falls in love", 24 November, The New York Times, https://www.nytimes.com/2020/11/24/science/artificial-intelligence-gpt3-writing-love.html. [8]

Pineau, J. (2020), "Building reproducible, reusable, and robust machine learning software", *DEBS '20: Proceedings of the 14th ACM International Conference on Distributed and Event-based Systems*, http://dx.doi.org/10.1145/3401025.3407941. [21]

Rousseau, A., C. Bauderlaire and K. Riviera (27 October, 2020), "Doctor GPT-3: Hype or reality?", Nabla blog, https://www.nabla.com/blog/gpt-3/. [9]

Shanahan, M. et al. (2020), "Artificial intelligence and the common sense of animals", *Trends in Cognitive Science*, Vol. 24/11, pp. 862-872, http://dx.doi.org/10.1016/j.tics.2020.09.002. [3]

Shneiderman, B. (2020), "Design lessons from AI's two grand goals: Human emulation and useful application", *IEEE Transactions on Technology and Society*, Vol. 1/2, pp. 73-82, http://dx.doi.org/10.1109/TTS.2020.2992669. [22]

Notes

[1] This, perhaps, has been as important and as disruptive in both positive and negative respects, as any of the other features of computers. Legal, social, commercial and even conceptual frameworks for intellectual property are still grappling with its consequences.

[2] The nature of that experience in humans can vary widely without making much difference to the language. The language use of the deaf-blind, such as Helen Keller, who are limited to tactile interactions with the world is not significantly different from the hearing and sighted, although there are measurable differences in the learning process (Chomsky, 1986[23]).

[3] This *mot juste* is due to Mark Steedman, who used it at the online meeting of the *AI and the Future of Skills* project held on 5-6 October 2020.

13. The encroachment of artificial intelligence: Timing and prospects

Richard Granger, Dartmouth College

This chapter describes several instances of artificial intelligence (AI), artificial neural network and machine learning systems that are judged to be highly successful. It also highlights the shortcomings of these systems to explain their limitations. Through examples such as so-called self-driving cars, image recognition, handwriting analysis and digital virtual assistants like Siri, the chapter explores the ways in which AI is both like and unlike human intelligence. It clarifies ways in which AI will and will not be useful in various workplaces. It also examines capabilities of humans that are likely to outpace AI for some time, and therefore may remain critical factors in employment practice.

Introduction

Artificial intelligence (AI) systems are steadily growing as tools to accomplish specified tasks, increasingly at the expense of jobs that would otherwise be carried out by human workers. How can people stay relevant and employed?

This chapter focuses on shortfalls of AI, explaining limitations likely to persist for many years. It clarifies ways in which AI will and will not be useful in various workplaces. It also examines capabilities of humans that are likely to outpace AI for some time, and therefore may remain critical factors in employment practice.

To that end, it focuses on ways in which AI – and related advances in machine learning and artificial neural networks (ANN) – is like, and quite unlike, human intelligence. How are the architectures – the innards – of ANNs like, and unlike, actual networks of neurons, i.e. brain circuits? What behavioural or computational abilities arise from ANNs, and how are they like and unlike human behavioural and computational abilities?

Similar and different underlying architectures

Artificial and natural neural networks

How are artificial "neural networks" similar to, and different from, actual brain circuitry? AI systems have shown an impressive ability to play difficult games, perform language translation, find patterns in complex data and many other seemingly human-level behaviours. These systems are often designated as "brain-like", or "brain-inspired", i.e. based on or derived from design principles of human brains.

Given the impressive accomplishments of these systems, it is perhaps understandable, even flattering, to connect them to human brain designs. Where do such claims come from? What characteristics of brains do these artificial systems actually exhibit?

Brains differ from standard (non-AI) computers in myriad ways. Among other things, brains learn and retain information over decades, whereas computers do not. Further, computers are predominantly serial whereas brains exhibit large-scale parallelism.

Surprisingly few characteristics of brains that might have become standard ANN properties have been adopted. These include long-term learning and parallelism, as well as the neural property of only producing simple computations such as addition and multiplication. In this way, the engines of the brain (neurons) contrast with computers that have extensive mathematical abilities.

Today, ANNs comprise much (though not all) of what is often more broadly termed machine learning and AI. They are composed of simple computing units, acting in parallel and learning from data. This indeed makes ANNs far more "brain-like" than standard computers.

However, these points of similarity between artificial networks versus actual brains represent a surprisingly small fraction of brain characteristics. These neural and brain-circuit properties (e.g. location and temporal differences between excitatory and inhibitory cells) can seem insignificant. Yet the capabilities of ANNs versus real brains are vastly different. It is not yet known what cognitive abilities arise from what features of our brains. Many AI and ANN researchers have shown that incorporation of additional actual brain properties may substantially, even drastically, alter and enhance the capabilities of ANNs.

Many AI researchers are studying how to overcome the shortfalls of AI systems compared to human abilities. Although AIs outperform many human abilities, these strongly centre on circumscribed tasks such as game-playing and online product recommendations. Interestingly, these closely correspond to the type of task for which computers have always outperformed humans: large-scale numeric calculation and data analysis.

AIs excel at tasks that can be reduced to pre-specified outcomes such as either winning or losing a game of chess or Go. It remains unclear whether more open-ended intelligent behaviours will be susceptible to similar approaches. Correspondingly, AIs still are markedly inferior to humans in reasoning, lifelong learning, common sense and much more.

Brain algorithms are intrinsically parallel; other algorithms typically are not

Unlike many algorithms, brain algorithms are intrinsically parallel. Computational steps can only be carried out in parallel if later steps in a process do not depend on previous steps. This avoidance of "serial dependency" is at the core of parallelism.

In a large computational set of operations, any given step in the process (say, step 146) may depend on prior steps (e.g. steps 26, 91 and 108) to have produced their partial outputs. The essence of "serial" computation is the dependency of later operations on earlier operations. All standard computers are intrinsically serial.

Can these serial dependencies somehow be turned into independent parallel operations? This is a present-day, much-researched problem that remains unsolved. Despite many specialised approaches, there is no general method for taking the dependencies in serial operations and somehow "parallelising" them.

A first step in understanding brains is to understand their already-parallel methods – their parallel algorithms – rather than assuming that computational serial methods can be rendered parallel.

Many hypothetical models of brain systems have been characterised in terms of their parallel and serial components. A prominent example includes "unsupervised" categorisation methods, which are predominantly intrinsically parallel – not "parallelised" from serial steps; these have been proposed as partial models of some cortical operations. Another such example is "reinforcement learning" (RL) systems, which include highly parallel operations. These have been widely proposed as partial models of the basal ganglia structures in brains.

Some additional related computational operations are also intrinsically parallel. For instance, most of what a search engine does is parallel: I type in "antiviral" to Google. It can parcel out the task to millions of independent computers to search through separate, independent, parts of the Internet.

After all those independent jobs return their findings, a non-parallel job must put them all together and then order them from first to last. However, most of the task is intrinsically parallel: each separate location in the Internet can be looked at separately during the search. Since none of them depends on others, all these separate searches can be done simultaneously and independently with enough separate computers to assign in parallel to this task.

Parallel computers are becoming increasingly important and prevalent. Such computer hardware (such as supercomputers, clusters, GPUs) contain substantial numbers of computing units (think neurons). These can carry out their operations independently of each other, and therefore can operate simultaneously. Many parallel methods (such as search) are typically implemented onto parallel hardware to speed up their operation.

Remarkably, most ANN operations, especially the still-prevalent "supervised" systems such as "backpropagation" (and "multi-layer perceptrons", more generally), are not entirely parallel. Consequently, they still require expensive hardware to run.

Interestingly, vast swaths of land around major hydroelectric dam sites are owned by the few major tech firms. Most of these firms specialise in AI, including Google (Alphabet), Apple, Amazon, Facebook, Netflix, Microsoft and IBM.

Why this strange connection between computer tech companies and hydroelectric dams? Their "server farms", i.e. stations of large numbers of computer clusters, run so hot that their power and cooling costs require plentiful power sources. In other words, AI is not cheap or prevalent as yet: much of it still requires extreme resource usage just to run at all (Jones, 2018[1]; Pearce, 2018[2]).

Similar and different behaviours and "cognitive" capabilities

How is someone's intelligence or aptitude assessed for a task? Typically, specific tests designed to identify traits and abilities have been found to correlate with humans who previously performed well in the task or job of interest.

Tests of this kind omit enormous sets of actual worker characteristics desired by an employer. This is because the worker is assumed to be human with standard human capabilities. These capabilities are taken for granted, and not explicitly tested.

In other words, humans are ostensibly tested for their necessary skills. However, many crucial skills are simply assumed since all humans possess them. One such set of skills is the ability to not commit catastrophic and unexpected errors. The four examples below – i) self-driving cars, ii) image recognition, iii) handwriting analysis and iv) digital virtual assistants – explore these concerns.

"Self-driving" cars

The term "self-driving" car is a misnomer. The numeric multi-tier "levels" scale, outlined below, roughly describes increasingly difficult abilities rather than levels of autonomy (NHTSA, 2018[3]; SAE, 2018[4]):

0. No automation: i.e. normal cars; all driving responsibilities resting with human drivers.
1. Driver assistance: human drivers are assumed to be responsible for all tasks, and may be assisted by a system that partially performs steering, *or* acceleration/deceleration.
2. Partial automation: similar to "1", humans are assumed to be responsible for all tasks, and may be assisted by a system that partially performs steering *and* acceleration/deceleration.
3. Conditional automation: the artificial system takes over most aspects of driving, including steering and acceleration/deceleration. Yet the human driver is expected to "respond appropriately" to a request to take over. In other words, ongoing vigilance is still required on the part of the human driver.
4. High automation: the artificial system takes over most aspects of driving. If the human driver does not respond appropriately to a request to take over, the system can pull over to stop the vehicle.
5. Full automation: the artificial system performs any and all driving tasks of a human driver, with no intervention required from the human.

No current products go beyond "Level 3". In other words, they all depend on the human driver to take over whenever called to do so by the AI system. Higher-level products have often been predicted, but many experts explain they remain a long way off.

When "Level 3" cars have accidents, the accidents themselves are not the primary issue. (Humans, too, have accidents.) Rather, it is the novelty of Level 3 accidents that is striking. The accidents involve things that, again, no one even thought to test for because they are so far outside the realm of human experience. In one case, for instance, a Level 3 car drove at high speed into the side of a truck. To the AI system, the truck somehow looked like the sky (Tian et al., 2018[5]; Boudette, 2019[6]; TKS Lab, 2019[7]).

There are multitudes of such potential unexpected and catastrophic errors lurking in AIs. These errors are not tested for in advance, and current tests consistently fail to anticipate them. Humans would not, for instance, accidentally insert an iPhone into a coffee machine or ignite their office chair.

Image recognition

Equally compelling examples of catastrophic errors occur in image recognition. The consequences are typically (though not always) less dramatic than Level 3 car accidents. However, the errors are every bit as illuminating. Other things besides images that fool humans can fool AI image experts. Figure 13.1 shows a set of "adversarial" images (Szegedy et al., 2014[8]; Kurakin, Goodfellow and Bengio, 2016[9]) that are judged, by these expert AI systems, to be the object labelled below them. For example, one pixilated multicolour image was identified by the AI as a cheetah. Another such pixilated image was identified as an armadillo. To humans, these images do not remotely resemble animals or objects.

Figure 13.1. Adversarial images showcase judgement errors of AI systems

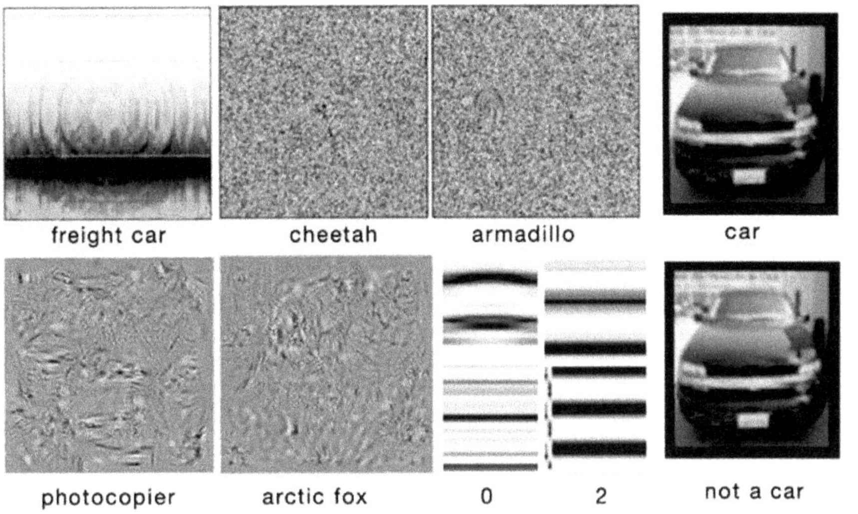

Source: (Nguyen, Yosinski and Clune, 2015[10]).

The designers of these systems did not think to test for such errors, which illustrates the range of the problem. It appears not to be practicable to anticipate all such errors; the range of misreads is so broad and unexpected that it cannot readily be predicted.

Why do these strangely inhuman errors arise at all? If AI systems performed somewhat like humans do, they might fall short in certain ways, much as some humans exhibit capabilities that other humans may lack. However, AI systems are not carrying out the kinds of operations that humans do.

Although AI systems are touted as "human-like" and "brain-like", there is overwhelming evidence by AI experts indicating this is not so. For example, the widely used "supervised" learning systems of most ANNs have sometimes been suggested as relating to operations of the human neocortex. However, brain projections do not contain "error correction" signals of the kind required by supervised systems.[1]

A retrospective case: Handwriting recognition

Building on two examples of the non-human nature of AIs (operating cars and recognising images), this section considers the "retrospective" example of handwriting recognition. It illustrates the perspective of time. When errors occur, how long does it take the field to recover from them?

Early focus on handwriting in the 1990s

In the 1990s, handwritten text recognition presented a major challenge to AI. Major conferences were held, focusing solely on handwriting recognition. In the United States, entire funding programmes of the Defense Advanced Research Projects Agency supported the work, and entire divisions of the National Institute of Standards and Technology developed datasets for it.

By the late 1990s, the field believed it was succeeding (LeCun, Bottou and Haffner, 1998[11]; Von Ahn et al., 2003[12]). Publications routinely highlighted their "correct hit rate" achievements, with suspiciously precise-seeming values such as 91.4% correct. These assessments assumed that the phrase "handwriting recognition" was self-explanatory.

Development of CATCHAs in the 2000s

However, in the early 2000s, new data appeared in the form of "Completely Automated Public Turing test to tell Computers and Humans Apart" (CAPTCHAs) (Von Ahn et al., 2003[12]). Websites began using these forms of distorted letters (now ubiquitous) to determine whether a purported user was a human or an AI. Humans could see the letters and numbers in CAPTCHAs effortlessly.

However, handwriting-recognition AI systems performed abysmally on CAPTCHAs. A few papers described promising recognition successes on limited databases but failed to generalise to other datasets. The supposedly highly successful field of handwriting recognition went almost entirely dry for more than ten years. Dozens of claims were made that CAPTCHAs had been cracked; these were repeatedly refuted.

Finally, 14 years later, a publication showed an approach that could reliably address many popular CAPTCHA systems (Bursztein et al., 2014[13]). It took additional years for other such reports to emerge; some were still considered to be sufficiently noteworthy that they appeared in prestigious journals (George et al., 2017[14]).

In sum, a problem introduced in 2000, into a supposedly successful field, essentially crushed that field until the mid to late 2010s. All of this took multiple lifetimes in terms of typical technology cycles.

The need for broader assessment of systems

Today, the stakes are similar. The predominant focus of researchers on the statistics of huge data, such as games and shopping recommendations, artfully alters the metric for success. They are not addressing open-ended problems, i.e. the problems that humans typically face. Rather, these approaches aim at achieving known metrics such as game wins.

These closed-ended tasks are being won by massive memorisation and processing of millions of instances (Serre, 2019[15]). The claims of game-playing systems, for instance, refer to software trained on the equivalent of the entire lifetimes of imagined hundreds of thousands of human players. By contrast, humans perform highly complex tasks of recognition, retrieval, decision and inference, after learning on comparatively miniscule quantities of data, many orders of magnitude less than the artificial systems require.

Without the specifications of actual human behaviour, it can be all too easy to imagine researchers are formally addressing a task such as handwriting recognition, or chess or Go.

The lesson is not being learnt. Tasks are carefully steered away from far-reaching human abilities, focusing instead on data memorisation mixed with slight generalisation. By this nostrum, a stream of attractive successes are toted up, largely disregarding failures and shortfalls (Serre, 2019[15]).

Many in the field are mindful of the situation. Researchers are striving for applicable metrics that could assess systems more broadly, seeking "feasible and reasonable" tests to which a system could be subjected. A range of views on this approach is worth further pursuit (Legg and Hutter, 2007[16]; Dowe and Hernandez-Orallo, 2012[17]; Hernandez-Orallo, Dowe and Hernandez-Lloreda, 2014[18]).

"Digital virtual assistants"

This section turns to a set of examples perhaps somewhat closer to direct experience: the performance of Siri (and corresponding conversational AIs, such as Alexa, etc.).

Siri and its cousins, now many years old, are still revolutionary and impressive accomplishments. Their recognition of spoken English and several other languages is still among the more skilful achievements of commercial AIs. Yet Siri makes jarring errors that even a young child would not make. Why is this so? Why do Siri's errors not match those that people might produce?

Indeed, how does one know when Siri has made an error at all? One way to address this is to envision a rigorous system for recognising and cataloguing such errors. After all, if humans cannot catch the errors, how could they possibly design systems that produce fewer of them?

Importantly, it is not possible, even in principle, to construct any rigorous system for recognising Siri's errors. This is because the measure of "when something is an error" is solely empirical: a human must judge that the response (somehow) does not make sense. By Siri's internal logic, of course, the response was somehow the correct output computed from the input she received.

Empirically, then, a human evaluator is required to judge whether or when Siri errs. This is radically different from the corresponding circumstances of other systems used in offices. If a corporation purchases a photocopying machine or an assembly line system, they come with specifications: careful, relatively precise descriptions of what the system will do. Deviations from these specs are errors. Even software, including complex software, comes with specifications. There are explicit instructions for its use, and careful characterisations of its corresponding prescribed behaviours.

Siri comes with no such specifications. In fact, its creators cannot, with any precision at all, produce such specifications. Errors, then, are not deviations from (non-existent) specifications. What counts as an error? Only what humans notice, after the fact.

This is the crucial difference between all AI systems and their hybrids, compared to all other contemplated office or workplace systems. With few or no specifications, errors cannot be reliably predicted in advance (Granger, 2020[19]).

Conclusions

AI system problems are real, and sometimes nefarious

It is a time of substantial upheaval in the fields of AI, machine learning and neural networks. The normally well-regarded journal *Nature* recently published a report co-authored by researchers at Google, describing testing of an AI system for medical screening of breast cancer mammograms (McKinney et al., 2020[20]). It then published a critique by a group of AI experts, who argued "the absence of sufficiently documented methods and computer code underlying the study effectively undermines its scientific value" (Haibe-Kains et al., 2020[21]).

The original authors then issued a brief counter-reply, asserting they would not release all of the code or data used to obtain the results. They argued that "much of the remaining code … [is] of scant scientific value and limited utility to researchers outside our organisation." They further stated that releasing the system could risk it being treated as "medical device software" and "could lead to its misuse" (McKinney et al., 2020[22]).

These claims seem transparently spurious for two reasons. First, the requested code is indeed of direct interest to the researchers attempting to evaluate and replicate the medical claims. Second, the code could be issued solely to AI researchers and clearly labelled as experimental and expressly not for medical use.

Corporations are similarly reluctant to release data around the pursuit of autonomous driving, medical diagnostics, and many other systems with potentially widespread and dangerous impacts. This continues to make it impossible in most cases for researchers to evaluate the claims of such systems. When the methods are hidden, they cannot in any way be seriously evaluated. In such cases, they simply remain unsubstantiated claims, and should be treated as such.

Even supposedly non-profit research organisations have refused to make their research code and materials available for evaluation by scientific researchers. The perhaps inaptly named "Open AI" switched from being a non-profit to a "capped profit". This turns out to mean a for-profit corporation that says it sets a 100x limit on investors' returns. For comparison, early investors in Google have received roughly a 20x return.

OpenAI publicised GPT-2 and later GPT-3 that can write impressive-seeming text stories. These initially were presented as literally "understanding human language". However, the code producing these purported wonders was secret. OpenAI, despite its name, would not release the code (Gershgorn, 2020[23]).

Even well-respected academic researchers could not access the code to test the company's claims. Eventually, some researchers did (not via OpenAI itself), and published findings that drastically call into question the code's "understanding" of language. A sample failed story, for example, left most readers befuddled at what was being said[2] (Marcus and Davis, 2020[24]; Marcus and Davis, 2020[25]). Extended discussions of other failures of common sense in present-day AI are well presented in Marcus and Davis (2019[26]).

A statement may be syntactically correct and seemingly coherent, and yet not make any sense. Yet, like Siri, this is not an example of a system failing in the typical sense. These systems have opaque designs, such that their failures cannot be rigorously predicted. Even after the fact, the failures cannot be cogently systematised.

How can we be more informed consumers and testers of proposed AI systems?

What can be done to ensure that humans continue to participate and thrive in work environments targeted by AI systems? Despite enormous publicity and fanfare, AIs do not appear ready to take over jobs that require human language usage and common sense. AIs applying for such jobs will continue to be blindsided by many real-world situations, like trucks on the highway. They are even more vulnerable to directed attacks such as adversarial inputs.

Recognise shortcomings of AI systems

It remains unclear whether specific examples of the kind described here could be generalised in a way to thoroughly identify likely AI errors. A systematic effort to collect known errors can assist in individual assessments of proposed AI systems (by checking for known errors). However, it is not known whether these will lead to systems that are so improved that such errors no longer occur.

Some companies are working to develop products aimed directly at overcoming these AI shortcomings. A "hybrid" approach, for example, incorporates higher-level symbol-manipulation operations with lower level ANN systems for statistical handling of big data (Marcus and Davis, 2019[26]). This approach includes symbol-manipulation systems, such as "expert systems" from past successful AI efforts. As such, they allow specific "common sense" rules, such as "if this, then that", which systems can use to evaluate possible actions.

Hybrid systems may not be the answer, but new approaches of some kind must be brought to bear. At the very least, AI "experts" could increasingly be trained to assess proposed systems for known errors. They could also acknowledge the lack of generalisation that is a reliable hallmark of extant systems.

Train AI experts and customers to look for known errors

Customers of hybrid systems could perhaps usefully be trained to query experts with these examples of widespread AI shortcomings in mind. If potential AI customers are presented with sample errors such as these, they may become increasingly well-informed users. As a result, they may become qualified to ask AI providers and experts about the precise limits of a candidate system under specific conditions. With the intended tasks of the system specified, experts can then be asked to consider an expanded set of tasks. The aim would be to clarify which generalisations of the AI system can and cannot be counted on.

These conclusions indicate how difficult it may be to define any remotely "complete" set of tasks that a buyer, or an AI expert, could use to identify all potentially relevant AI errors. It is far from obvious how to pin down even a small representative set of such tests.

The four categories of examples described in this manuscript suggest initial starting points for these common areas of supposed expertise (self-driving cars; image recognition; handwriting; digital virtual assistants). These and many other available instances of AI hard limitations should be made available to entities looking to acquire and use AI systems.

It will likely be beneficial for the pre-planned domain of the intended AI system to draw on these starting points to formulate examples. For instance, in each example, a few themes appear to emerge. Limitations to AI systems are described in terms that appear technical and detailed but in practice are vague with respect to what a user can expect (as in self-driving cars, handwriting, image recognition). Operations not explicitly tested are not highlighted for their possible divergence from reasonable behaviour (all categories display this shortcoming).

Use concrete vignettes to avoid incorrect inferences

In addition, AI systems are presented such that an intended user may infer abilities that a human would exhibit in a job, but that the AI system is not at all guaranteed to do. For example, the differences between "Level 2" and "Level 3" in self-driving cars are described by long, and highly detailed and technical descriptions. However, these can readily mislead readers into making inferences not promised in the specification.

The leap between Level 2 and Level 3 is supposedly separated by a system in which "the human monitors the driving environment" (Level 2) vs. "the automated system monitors the driving environment" (Level 3). Yet, despite the addition of multiple sensors in the latter system, the human is nonetheless still required to continuously monitor all conditions, and be prepared to take control at a moment's notice.

How might users detect such potential misreads in advance? One approach is to take the specification of the task, and identify specific, concrete vignettes in which a difficulty might arise. These difficulties could be either for the human in a job, or for the AI system supposedly performing that job. Ask specific questions about what the human might be expected to do, if, say, a shipment does not arrive on schedule. AI system descriptions do not typically volunteer such examples.

In each case, the more scenarios that are specifically inquired about, the more likely that system errors may be identified. This approach may appear too cumbersome or piecemeal, but it is exactly this characteristic that may make it useful.

The approach may seem piecemeal because it appears to lack an undergirding principle connecting different proposed vignettes. Indeed, AI systems do not yet have anything resembling complete principles for behaviours in complex settings. This is why a car may crash at full speed into a wall, or an image that looks like white noise may be labelled as an armadillo, with high (but erroneous) confidence.

A human would handle edge effects with "common sense", but the AI system may prove erratic. Indeed, it may be erratic in any untested situation. Testers would do well to think of tasks where a human might wince or express concern, but finally react sensibly. In these situations, AI may well fail the test.

AI systems yet to be

An oft-told adage is that present-day AI approaches are like climbing a tree, or even flying a helicopter, with the aim of reaching the moon. They can cite measurable ongoing progress: they do indeed keep climbing higher, closer to their goal. Yet it will require utterly distinct understandings and methods to go beyond the current achievements.

In the long run, there is no principled reason why artificial systems cannot duplicate, and exceed, human abilities. Indeed, current human abilities already arise from physical machines – our brains; they simply are made of meat rather than metal. In describing the original purpose of AI, John McCarthy said, "Every aspect of learning or any other feature of intelligence can be so precisely described that a machine can be made to simulate it" [McCarthy et al., 1955 in (McCarthy et al., 2006[27])].

The field of AI is not currently "precisely describing" the features of intelligence. Quite the opposite: such features are still highly elusive. This definition of AI does not emphasise analysing large datasets or commercial products per se. Rather, it focuses on the scientific objective of understanding the mind, and the brain that produces the mind. This may appear to be a too-theoretical pursuit, but it is the most pragmatic path to follow: to outperform humans, one must first equal them.

AIs will eventually outperform humans but are nowhere close to doing so. AIs currently outperform humans in much the way that computers have always outperformed humans: in large-scale numeric calculation and data analysis. Correspondingly, AIs still wildly underperform humans in reasoning, lifelong learning, common sense and much more.

As the "precise descriptions" of these human abilities are achieved, the inception of true intelligences will also come closer. Crucial information will come from neuroscience, psychology, computer science, mathematics and other related fields. Just as flying machines were based on principles of aerodynamics used by flying animals, intelligent machines will arise from understanding the principles of intelligence. Great advances have been made towards this aim, and more will come. In the meanwhile, human workers significantly outpace AIs in their judgements and practicality.

When AIs do come to verge on human common sense, the questions of their industrial utility will be even more urgent. Jobholders have been repeatedly infringed on in the past, in times of economic and societal reorganisation. All such previous upheavals have entailed humans taking the jobs of other humans, but the resulting instabilities were nonetheless real. As AIs advance beyond their current limitations, continued threats to stable human employment will recur. Much work is yet to be done to address the future livelihoods of humans in an increasingly artificially intelligent world.

References

Boudette, N. (2019), "Despite high hopes, self-driving cars are 'Way in the future'", 17 July, New York Times. [6]

Bursztein, E. et al. (2014), "The end is nigh: Generic solving of text-based CAPTCHAs", presented at WOOT 14, San Diego, CA, 19 August, https://www.usenix.org/conference/woot14/workshop-program/presentation/bursztein. [13]

Chandrashekar, A. and R. Granger (2012), "Derivation of a novel efficient supervised learning algorithm from cortical-subcortical loops", *Frontiers in Computational Neuroscience*, Vol. 5, http://dx.doi.org/10.3389/fncom.2011.00050. [31]

Dowe, D. and J. Hernandez-Orallo (2012), "IQ tests are not for machines, yet", *Intelligence*, Vol. 40, pp. 77-81, https://doi.org/10.1016/j.intell.2011.12.001. [17]

George, D. et al. (2017), "A generative vision model that trains with high data efficiency and breaks text-based CAPTCHAs", *Science*, http://dx.doi.org/10.1126/science.aag2612. [14]

Gershgorn, D. (2020), "GPT-3 is an amazing research tool. But OpenAI isn't sharing the code", *Medium,* 20 August, https://onezero.medium.com/gpt-3-is-an-amazing-research-tool-openai-isnt-sharing-the-code-d048ba39bbfd. [23]

Granger, R. (2020), "Toward the quantification of cognition", *arXiv*, Vol. 2008.05580, https://arxiv.org/abs/2008.05580. [19]

Haibe-Kains, B. et al. (2020), "Transparency and reproducibility in artificial intelligence", *Nature*, Vol. 546, pp. E-14–E-16, https://doi.org/10.1038/s41586-020-2766-y. [21]

Hernandez-Orallo, J., D. Dowe and M. Hernandez-Lloreda (2014), "Universal Psychometrics", *Cognitive Systems Research*, Vol. 27, pp. 50-74, https://doi.org/10.1016/j.cogsys.2013.06.001. [18]

Jones, N. (2018), "How to stop data centres from gobbling up the world's electricity", *Nature*, Vol. 561, pp. 163-166, https://doi.org/10.1038/d41586-018-06610-y. [1]

Kurakin, A., I. Goodfellow and S. Bengio (2016), "Adversarial examples in the physical world", *arXiv*, Vol. 1607.02533, https://arxiv.org/abs/1607.02533. [9]

LeCun, Y., L. Bottou and P. Haffner (1998), "Gradient-based learning applied to document recognition", *Proceedings of the IEEE*, Vol. 86/11, pp. 2278-2324, http://dx.doi.org/10.1109/5.726791. [11]

Legg, S. and M. Hutter (2007), "Universal intelligence: A definition of machine intelligence", *Minds and Machines*, Vol. 17/4, pp. 391-444, http://dx.doi.org/arXiv:0712.3329. [16]

Lillicrap, T. et al. (2020), "Backpropagation and the brain", *Nature Reviews Neuroscience*, Vol. 21, pp. 335-346, https://doi.org/10.1038/s41583-020-0277-3. [30]

Marblestone, A., G. Wayne and K. Kording (2016), "Toward an integration of deep learning and neuroscience", *Frontiers in Computational Neuroscience*, Vol. 10/94, https://doi.org/10.3389/fncom.2016.00094. [29]

Marcus, G. and E. Davis (2020), "Experiments testing GPT-3's ability at commonsense reasoning", Department of Computer Science, New York University, https://cs.nyu.edu/faculty/davise/papers/GPT3CompleteTests.html. [25]

Marcus, G. and E. Davis (2020), "GPT-3, Bloviator: OpenAI's language generator has no idea what it's talking about", *MIT Technology Review* 22 August, https://www.technologyreview.com/2020/08/22/1007539/gpt3-openai-language-generator-artificial-intelligence-ai-opinion/. [24]

Marcus, G. and E. Davis (2019), *Rebooting AI*, Penguin Random House, New York. [26]

McCarthy, J. et al. (2006), "A proposal for the Dartmouth summer research project on artificial Intelligence, August 31, 1955", *AI Magazine*, Vol. 27/4, p. 12, https://doi.org/10.1609/aimag.v27i4.1904. [27]

McKinney, S. et al. (2020), "Reply to: Transparency and reproducibility in artificial intelligence", *Nature*, Vol. 586, pp. E-17–E-18, https://doi.org/10.1038/s41586-020-2767-x. [22]

McKinney, S. et al. (2020), "International evaluation of an AI system for breast cancer screening", *Nature*, Vol. 577/7788, pp. 89-94, https://doi.org/10.1038/s41586-019-1799-6. [20]

Nguyen, A., J. Yosinski and J. Clune (2015), *Deep neural networks are easily fooled: High confidence predictions for unrecognizable images*. [10]

NHTSA (2018), "A framework for automated driving system testable cases and scenarios", *Report*, No. DOT HS 812 623, National Highway Traffic Safety Administration, Washington, DC, https://www.nhtsa.gov/sites/nhtsa.dot.gov/files/documents/13882-automateddrivingsystems_092618_v1a_tag.pdf. [3]

Pearce, F. (2018), "Energy hogs: Can world's huge data centers be made more efficient?", *Yale Environment*, Vol. 350, https://e360.yale.edu/features/energy-hogs-can-huge-data-centers-be-made-more-efficient. [2]

Rodriguez, A., J. Whitson and R. Granger (2004), "Derivation and analysis of basic computational operations of thalamocortical circuits", *Journal of Cognitive Neuroscience*, Vol. 16, pp. 856-877, http://dx.doi.org/10.1162/089892904970690. [28]

SAE (2018), "Taxonomy and definitions for terms related to driving automation systems for on-road motor vehicles", *Revised report*, No. J3016_201806, SAE International, Warrendale, PA, https://www.sae.org/standards/content/j3016_201806/. [4]

Serre, T. (2019), "Deep learning: The good, the bad, and the ugly", *Annual Review of Vision Science*, Vol. 5, pp. 399-426, https://doi.org/10.1146/annurev-vision-091718-014951. [15]

Szegedy, C. et al. (2014), "Intriguing properties of neural networks", *arXiv*, Vol. 1312.6199, http://dx.doi.org/arxiv.org/abs/1312.6199. [8]

Tian, Y. et al. (2018), "DeepTest: Automated testing of deep-neural-network-driven autonomous cars", *arXiv*, Vol. 1708.08559, http://dx.doi.org/arXiv:1708.08559. [5]

TKS Lab (2019), *Experimental Security Research of Tesla Autopilot*, Tencent Keen Security Lab, Shenzen, https://keenlab.tencent.com/en/whitepapers/Experimental_Security_Research_of_Tesla_Autopilot.pdf. [7]

Von Ahn, L. et al. (2003), *CAPTCHA: Using hard AI problems for security*, Springer, http://dx.doi.org/10.1007/3-540-39200-9_18. [12]

Notes

[1] Several recent efforts have tried to demonstrate an abstract mathematical relationship between some operations that brains may actually do and the back-propagated error correction of ANNs. Such convoluted arguments may serve in part as reminders of the radical differences between ANNs and brains. They emphasise the comparative lack of research on how the desired computational ends may be instead achieved via different algorithms that do not rely on this form of error correction (Rodriguez, Whitson and Granger, 2004[28]; Chandrashekar and Granger, 2012[31]; Marblestone, Wayne and Kording, 2016[29]; Lillicrap et al., 2020[30]).

[2] One sample failed story produced the following text: "At the party, I poured myself a glass of lemonade, but it turned out to be too sour, so I added a little sugar. I didn't see a spoon handy, so I stirred it with a cigarette. But that turned out to be a bad idea because it kept falling on the floor. That's when he decided to start the Cremation Association of North America, which has become a major cremation provider with 145 locations." This example is also discussed in Chapter 12.

14. On evaluating artificial intelligence systems: Competitions and benchmarks

Anthony G. Cohn, School of Computing, University of Leeds

This chapter discusses some approaches and methods used by the artificial intelligence (AI) community to measure and evaluate AI systems. It looks at the evolution of competitions, giving special attention to the Turing Test and the Winograd Schema Challenge. It also looks at the fascination of researchers for testing AI through games such as chess and Go. Several tests for measuring intelligence proposed for AI systems are examined, as well as the role of benchmark datasets in evaluating AI systems. The chapter ends with a discussion of the benefits and limitations of four approaches: custom dataset, benchmarks, competition and qualitative evaluation.

Introduction

This chapter discusses approaches and methods used by the artificial intelligence (AI) community to measure and evaluate AI systems. Ever since researchers started building AI systems, they have wanted to evaluate them. Some have sought to measure these systems against human benchmarks (such as playing human experts at chess or other games). Others have measured them against other AI systems. Some have done both.

Finding good benchmarks for evaluating systems, and conducting tests, is harder than it might seem. This is especially true because good methods apparently exist to evaluate human intelligence via standard tests and examinations.

Two major challenges for AI systems revolve around their "brittleness" and narrow scope (see also Chapter 2). AI systems that perform well have generally either been hand-engineered for particular problems, or trained on data relating to a particular task in a particular domain. This brittleness makes it hard to get good generalisation. Worse, AI systems are generally narrow in scope. They are able to tackle a limited set of problems and have limited knowledge of the world in general. Therefore, they will usually not even know when a problem lies beyond their competence.

Part of that challenge relates to the absence of common sense knowledge. One of the earliest challenges issued to the community was the need for AI systems to have "common sense" (McCarthy, 1959[1]). However, this remains one of the hardest aspects of human intelligence to build into an AI system.

That said, there are numerous examples of AI systems performing at, or even beyond human level in narrow, specialised domains and tasks. AlphaFold, for example, recently tackled protein structure prediction. It has been claimed that the latest version has solved a 50-year-old grand challenge problem in biology (Jumper et al., 2021[2]).

Tests for measuring intelligence

This section discusses tests for measuring intelligence proposed for AI systems. It then examines some of the competitions created to compare AI systems before looking at some benchmark datasets.

The Turing Test and other inducement prizes

There have been many tests proposed to evaluate AI systems. The Turing Test is probably the most famous (Turing, 1950[3]; Shieber, 2016[4]).[1] There have been various Turing Test competitions, including for the annual Loebner Prize.[2] The sometimes entertaining results have helped promulgate ideas about AI to the general public. However, the entrants have arguably not demonstrated any real important progress in AI.

Turing himself never proposed the test as a serious way of measuring AI systems or of measuring progress (Shieber, 2016[4]). The idea has been called "misguided and inappropriate" (Shieber, 1994[5]; Hayes and Ford, 1995[6]). Instead, Shieber (2016[4]) argues for new "inducement prize" contests – "award programs established to induce people to solve a problem of importance by directly rewarding the solver."

Inducement prizes have been around for centuries. Perhaps the most famous historical example is the Longitude Prize offered by the UK government in 1714. More recently, the IBM Watson AI XPRIZE[3] "challenges teams to demonstrate how humans can work with AI to tackle global challenges". The winner, scheduled to be announced in June 2021, was to win USD 5 million.

Five principles for inducement prizes

Shieber (2016[4]) suggests prizes should adhere to five principles (Table 14.1). Annual "bake-offs" favoured by some funding agencies between rival teams funded in a research programme tend to reduce diversity and do not reward ambitious risky approaches. As such, they have driven incremental progress towards specific targets. However, Shieber (2016[4]) argues that neither the "bake-offs" nor Turing Test prizes such as the Loebner Prize meet all five proposed principles.

Table 14.1. Five principles for inducement prizes

Principle	Rationale/description
Occasionality of occurrence	This ensures that awards are only given when warranted rather than automatically through an annual prize.
Flexibility of award	This aims to apply the spirit, rather than the letter, of the rules to determine a winner.
Transparency of result	The public should be able to inspect the results for both transparency and replicability.
Absoluteness of criteria	The award should satisfy absolute rather than relative criteria. In other words, it is not enough to be the best entrant in a competition to win an award.
Reasonableness of goal	This sets a constraint on the nature of the target(s); they should be beyond the current state of the art, but not impossibly so.

When the test involves a quantitative metric, it should have "headroom" (Shieber, 2016[4]). In other words, the level set should be a real indicator of a system that performs at human level. A speech recognition system, for example, may perform within a few percentage points of human-level performance. However, such a system may still be unable to understand speech in arbitrary contexts.

The question of "headroom" is also relevant to systems devised to address the Winograd Schema Challenge (WSC). The WSC aimed to ensure that statistics derived from mining large corpora could not be used to build successful systems rather than ensuring that the systems really understood the questions. At least for the current test set, statistics have been far more successful than anticipated (Kocijan et al., 2020[7]).

The Turing Test fails the reasonableness test since, in its full form, it is clearly too challenging for the current state of the art (Shieber, 2016[4]). He concludes a suitable form of a general AI inducement prize meeting all his principles is not possible in the short term.

The role of competitions in measuring artificial intelligence systems

This section discusses the main competitions that have come to play an increasingly important role in the AI community (Cohn, Dechter and and Lakemeyer, 2011[8]). Competitions of various kinds exist in nearly every subfield of AI. They range from theorem proving and SAT solvers… to trading agents, computational models of argumentation (Gaggl et al., 2018[9]), poker and other games… to object detection and recognition, among many others.[4]

Robot competitions are some of the few competitions that cover multiple aspects of AI and aim to evaluate integrated systems. Creating a competition that focuses on one specific aspect might seem to be a more achievable way of measuring progress in AI. However, in many cases, the sub-problem tackled is often called "AI complete" (Mallery, 1988[10]). In other words, the sub-problem is at least as hard as the AI problem in general. A solution to the "sub-problem", then, could be used to solve any AI challenge.

Box 14.1. The Winograd Schema Challenge

Levesque (2011[11]) first proposed an alternative to the Turing Test, which he called the Winograd Schema Challenge. A follow-up paper (Levesque, Davis and Morgenstern, 2012[12]) elaborates on the idea. The name derives from a pair of sentences given by Terry Winograd, a well-known early AI researcher:

1. The city councilmen refused the demonstrators a permit because they advocated violence.
2. The city councilmen refused the demonstrators a permit because they feared violence.

The two sentences only differ in one word, but the noun phrase referred to by the pronoun "they" changes. Winograd chose this sentence as a test for machine translation systems since, when translated to a gendered language such as French, "they" would be rendered as "elles" in the first sentence and "ils" in the second.

A key aspect of choice in such a pair of sentences is the requirement for general knowledge, or common sense knowledge, to identify the correct referent for the pronoun. It should not be possible to use selectional restrictions to solve the problem. In the sentence pair, "The men ate the burgers because they were [hungry/delicious]", for example, the AI only needs to know that men can eat things but burgers can't, and only food (not people) can be delicious. Similarly, using statistics of occurrence should not help determine the referent. For example, in the pair of sentences, "The motorcycle overtook the push-bike because it was going so fast/slow", the appropriate referent can be found through Google search statistics.

Competitions serve several purposes

Drive progress

Competitions drive progress in a community. The very announcement of a competition usually stimulates many of those working in the particular field to enter. Researchers enter both to evaluate their methods and techniques but also to gain kudos and potential career enhancement.

Measure the state of the art relatively objectively

Competitions provide a way of measuring the state of the art in a relatively objective way. The test is set externally rather than by the system's authors of what is now within the scope of the best systems and methods worldwide.

Build on failures

The failures of the systems in one competition generally drive improvements. This allows achieving a better performance in the following incarnation, thus setting short-term goals for improvements.

Build community

Competitions help the community come together and share expertise. While the events themselves are usually intensively competitive, there is usually a requirement for entrants to make their code available after the event. Workshops are organised to share what went well and what did not.

Criteria for successful competitions

A successful competition must set tasks that are neither too easy nor too hard but rather at the "cutting edge" so the best systems can at least partially succeed. As a field progresses, and successive instances of a competition are held in subsequent years, the tasks are typically made harder and more challenging.

This typically adds further realism so the challenges are closer to a real task that might be useful for humankind; the datasets might be larger; the robot environment less contrived, or stricter time limits applied.

In some cases, the initial competition was just too hard for the current state of the art. For example, a competition around the WSC at IJCAI-16 (Davis, Morgenstern and Ortiz, 2017[13]) contained two tasks. No system performed well enough in the first task (a simple pronoun resolution) to advance to the second round of solving a WSC.

Subsequently, neural language models such as BERT have achieved surprisingly good performance (at least to the challenge setters) (Devlin et al., 2019[14]). The best of these models achieved around 90% accuracy on the WSC273 dataset. Kocijan et al. (2020[7]) conclude that "new tests, more probing than the Winograd Schema Challenge, but still easy to administer and evaluate" are required.

Will progress via competitions that become incrementally harder every year eventually lead to human-level general intelligence and performance? Arguably, this is the way that humans learn, via the graded examinations in schools. However, whether this is the best approach to achieving AI is an open question.

Disadvantages of competitions

Competitions have undoubtedly led to much progress in AI, but they also have disadvantages. A new entrant to the field can sometimes generate new and innovative approaches. However, competitions more often drive incremental improvements to systems and methods.

The need for elaboration tolerance

Competitions can also blinker the community into solving problems that are easily assessable via a competition rather than pursuing fundamental and long-term research. Good competition design can address this issue. Ideally, the systems that researchers enter into competitions should have what has been called "elaboration tolerance" (McCarthy, 1959[1]).

A system has elaboration tolerance if it does not require substantive modification if the challenge itself was not substantively modified. More specifically, if a new version of the challenge is based on the original but differs in certain ways, the representation a system uses to solve the challenge only needs to be adapted or modified proportionately to the degree of the changes to the challenge.

Elaboration tolerance is a requisite of potential solutions to the challenge problems listed on the "Commonsense Reasoning" page.[5] Sloman (n.d.[15]) has suggested that elaboration tolerance is related to the ability of a system to "scale out" (rather than "scale up").

Whereas McCarthy's notion is purely about the *representation* a system uses, Sloman is concerned about the whole system. Sloman asks whether the system can be "combined with new mechanisms to perform a variety of tasks".

Hypothetically, a vision system could be used to label images in arbitrary sized corpora but cannot be used in a myriad of other contexts. These contexts might include producing descriptions of scenes or helping a robot in its activities as being able to scale up but not out. Both of these ideas are related to what has been called "brittleness" – the notion that a system can solve a particular class of problems well but fails on related but perhaps only subtly different problems.

Early artificial intelligence research with the games of chess and Go

AI has had a long fascination with game playing (Box 14.2). Indeed, some of the earliest AI researchers worked on building systems to play games, such as the simple tic-tac-toe/noughts and crosses, and checkers/draughts. Chess was a long-time challenge for AI, but in 1997 IBM's Deep Blue beat the reigning world champion, Gary Kasparov. A key part of the success was the sheer brute computational force of Deep Blue.[6]

Box 14.2. Why has AI had such a fascination with game playing?

Since the earliest days of AI, for many reasons, researchers have been trying to develop programs to play games at a human level. Playing games well is often taken as a sign of intelligence, in particular more "intellectual" games such as chess. Developing an AI game player can thus be taken as a mark of progress towards machine intelligence.

Many skills required to play games well are also required in real life. These include the ability to plan a sequence of actions to achieve some goal. It also includes how to reason under uncertainty (what moves the other player will make or a selection of cards in a game of poker). There is also the use of probability to optimise decision making and machine learning to improve the system's performance, among other aspects.

Games have the advantage that framework can be easily implemented (i.e. the basic rules of the game rather than the best strategy), and can be simulated in a virtual world. A machine can learn by playing against itself (e.g. the AlphaZero system learning Go).

Of course, many aspects of real life such as language and vision are not present in game playing, at least as normally tackled in the AI game-playing literature. Real life is also much less structured than game playing, which usually has strict rules about turn taking and what moves are legal. Games are usually a "closed world" – everything about the game (except how to play well) can usually be completely described in a short tutorial, including all the possible objects and moves.

The game of Go, which is considerably more complex than chess, was considered to be a long way from being played well by computers. However, in 2016, a program named AlphaGo from DeepMind, beat the 9-dan player, Lee Sedol, 4-1 in a five game match. The match was organised as a formal competition, with a prize of US$1M for the winner. AlphaGo involved much human-coded knowledge, but a subsequent version, AlphaZero, was able to learn with only the rules of the game, just by playing itself many times.

Togelius (2016[16]) has gathered advice on running competitions for the AI game-playing community. Much of this advice is pragmatic. For example, any software developed to support the competition should be platform-agnostic. Other advice mirrors Shieber's principles, particularly that everything should be open-source.

Video games and virtual worlds

Togelius (2016[17]) has also argued that video games make an excellent test bed and benchmark for AI (as well as providing technology for new kinds of video games). He acknowledges that robots, operating in the real world, are perhaps the most obvious test bed for AI. However, he also notes several disadvantages of this approach. These include expense, speed (i.e. experiments typically take non-trivial amounts of time to perform, making learning difficult) and physical complexity.

Thus, games in their virtual worlds seem attractive, and in particular video games rather than board games. Moreover, designing AI systems to solve particular games does not necessarily imply anything about their ability to solve games in general. Therefore, Togelius records this as a motivation for designing the General Video Game Playing Competition (GVGAI) (Box 14.3). In this competition, entrant programs have to play ten new games known only to the competition organisers. To support the GVGAI, the General Video Game AI Framework has been developed to specify new games (Perez-Liebana et al., 2019[18]).

Specifying a standard framework for problems is a recurrent theme across competitions. For example, all the many problems in the Thousands of Problems for Theorem Provers (TPTP) benchmark (see below) are specified in the same way. Meanwhile, the Planning Domain Definition Language[7] has been used in the automated planning community and competitions for many years now. Over the years, it has had several extensions to increase its expressivity.

Many games that have been tackled are about perfect information: the only unknown is what actions the other player(s) might take. But other games mirror real life better in that only partial information is available and actions may be random. For example, the American Contract Bridge League has organised an annual competition since 1996. It has successively stronger players but still not at human champion level. Great progress has been made in poker in recent years, culminating in the *Libratus* system for which Sandholm and Brown (2018[19]) were also awarded the prestigious Minsky Medal.[8] Subsequently, Brown and Sandholm (2019[20]) designed a system called "Pluribus" that beat multiple human players simultaneously.

In response to the worldwide popularity of the game *Angry Birds*, an annual competition, AIBIRDS, has been held since 2012 (Renz et al., 2019[21]). While still played in an artificial environment, the game has many attractions from the point of view of measuring AI. First, there is incomplete knowledge of the physical parameters of the game. Second, there is an infinite set of actions available to an agent at any time. Third, it combines planning, image interpretation, reasoning, hypothesis formation and learning.

Interestingly, Renz et al. (2019[21]) point out that, "learning-based approaches have been largely unsuccessful. Despite all the successes of deep learning in the past few years, no deep-learning based agent has yet entered the semi-final round of [the] competition." Each year, humans also compete against the machines, and have dominated so far.

Box 14.3. General Video AI Framework

The core of the framework is a Video Game Description Language (VGDL). This provides a way of describing 2D video games concisely in a few dozen lines of plain text and can model both single and multi-player games. The originator listed desirable features of such a language. It should be

> *clear, human-readable and unambiguous. Its vocabulary should be highly expressive from the beginning, yet still extensible to novel types of games. Finally, its representation structure should be easy to parse and facilitate automatically generated games, in such a way that default settings and sanity checks enable most random game description to be actually playable (Schaul, 2013[22]).*

Games descriptions have two main parts. Level descriptions specify the 2D layout of the screen using different symbols. The game description proper specifies what the symbols in the level description mean in terms of the VGDL ontology (e.g. monsters or goals). Other parts of the game description define a possibly hierarchical set of objects with reference to an ontology, the set of interactions that happen when objects collide (e.g. swords kill monsters) and a set of termination criteria that define how/when the game ends.

The role of benchmarks in evaluating artificial intelligence systems

Benchmarks have long played an important role in AI. A benchmark is a dataset which is proposed as, or has become, a dataset by which different solution techniques are evaluated.[9] Benchmarks provide a way of comparing different solution methods on a single dataset or set of datasets. Sometimes they have emerged from research, initially used by a single group before becoming more widely used. Sometimes they have been explicitly created to stimulate research or evaluate competing methods.

In competitions, such as those discussed in the previous section, a test dataset is usually required. It then serves as a benchmark for the competition itself and often for subsequent research. However, it must be made openly available. Sometimes the dataset is reserved to be reused in subsequent competitions as it can be hard to create) as in the WSC. Often, there is an explicit call for benchmarks (e.g. in the International Competition on Computational Models of Argumentation (Gaggl et al., 2018[9]).

Some of the major benchmarks are listed here.[10] The TPTP (Box 14.4) has been an ongoing library of benchmark problems. Many are no longer seriously challenging, but it is continuously being extended with new problems (Sutcliffe, 2017[23]). In natural language understanding, General Language Understanding Evaluation (GLUE) (gluebench.com) is a "collection of NLU tasks including question answering, sentiment analysis and textual entailment, and an associated online platform for model evaluation, comparison and analysis." One task is a text inference variant of the WSC dataset. Performance of systems trained jointly on all the tasks in GLUE perform better than those trained on each task separately (Wang et al., 2019[24]).

Box 14.4. The TPTP Automated Theorem Proving Library

The TPTP library contains literally thousands of problems for evaluating and testing automated theorem provers (ATPs). AI has addressed the challenge of building ATPs since its earliest days [e.g. the geometry theorem proving system in Gelernter (Gelernter, 1959[25])]. The current library ranges from simple test problems for people to use to test their programs to open problems; some test problems used to be challenging for ATPs but are now trivial. The 53 domains range from Mathematics (relation, Boolean, Kleene algebras, set theory, topology, etc.) to Biology, Medicine, Social Sciences Philosophy, Puzzles and various aspects of Computer Science and AI (knowledge representation, planning, common sense reasoning, etc.). There is a standard representation for all problems in the library, making it easy for an ATP to read in a problem.

The benchmarks have all been more or less hand-curated/generated, but this can be expensive and difficult. Moreover, annotations on benchmarks are not always correct. Northcutt et al. (2021[26]), for example, estimated a 3.4% error rate across ten datasets surveyed. They thus recommend that special test sets should be carefully constructed to be error-free.

There is also work in the automatic generation of benchmarks and in particular, test sets. For example, a second competition is associated with AIBIRDS: the level generation competition (Stephenson et al., 2019[27]). Generating new levels or competitions is challenging, even for humans. This second level competition aims to stimulate and evaluate automated level generators. They note, "Submitted generators were required to deal with many physical reasoning constraints caused by the realistic nature of the game's environment, in addition to ensuring that the created levels were fun, challenging, and solvable."

In their discussion of the 2017 competition, Stephenson et al. (2019[27]) raise several interesting questions about the evaluation criteria for judging entries. At present, humans judge entries on a) "how fun and enjoyable the level is to play"; b) "how creative the level design is"; and c) "how well balanced the level of difficulty is". The definition of "fun" was left to the judges to avoid biasing them, and was used to determine the overall winner; b) and c) determine secondary prize-winners. The definition of similarity was also at the discretion of the judges.

No consideration has been given to automated judging, suggesting this task is not amenable to AI techniques. There has been some research on determining how to evaluate "fun" or creativity [e.g. (Boden, 1996[28]; Sweetser and Wyeth, 2005[29])], but this problem is still unsolved. The need for judges here contrasts with other competitions. Some are designed to avoid the need for human judges, such as the WSC (Levesque, Davis and Morgenstern, 2012[12]).

Conclusion

Doing research to win competitions risks being too driven by the goals of the competition. Entrants may seek to make enough incremental progress to do well in the next iteration rather than being truly innovative and devoting time to solve the fundamental problems remaining for machines to reach human-level AI.

Nonetheless, competitions have been valuable. Similarly, the availability of the many benchmark datasets allow for comparisons between different approaches. Table 14.2 summarises the main ways in which AI systems can be evaluated, along with the main advantages and disadvantages.

AI has always been most successful in narrow, usually technical domains, isolated from common sense and world knowledge. Finding good ways to measure common sense, and to build it into AI systems remains one of the greatest challenges for AI.

While many tests evaluate single aspects of intelligence, a challenge is to develop a comprehensive test for intelligence that is not susceptible to "game playing" by entrants. Typically, as soon as criteria are published which a system will be evaluated against, developers and researchers may try to optimise their systems for these specific criteria, rather than trying to create a more general system.

McCarthy's notion of elaboration tolerance is relevant in the search for a more general system. It tries to test whether a system has solved a specific problem, or instead has enough knowledge and reasoning/computational ability to solver a broad class of problems. Multiple conflicting evaluation criteria would be one strategy to address this issue: it would be impossible to optimise against all criteria simultaneously. Of course, having hold-out, unseen tests also helps.

This is part of a more general issue that AI systems do not know what they don't know. A typical AI system to, say, diagnose liver pathologies from data, only "knows" about (some) pathologies and aspects of the training data. It has no idea about more general knowledge of the world, even perhaps relatively related areas such as pathologies in a different part of the body. Even a young child has some appreciation of the limitations of its knowledge, and can confidently say "I don't know" in answer to some input.

Over the history of AI, many technologies have been held out as the solution for machine intelligence – from rule-based systems and early neural nets to Bayesian Networks and Deep Neural Nets. Many have demonstrated successes. However, they have all, so far, been shown to have problems. Not one offers a single solution to enable general machine intelligence.

Machine intelligence may well require an amalgam of different techniques and approaches. Recent initiatives towards "neuro-symbolic" systems, for example, combine neural net technologies with symbolic reasoning. Each has its advantages. Some have high-level reasoning and ability to explain reasoning for symbolic systems. Others can learn from large amounts of data for neural systems. The challenge is how to build integrated systems that successfully combine multiple technologies.

Table 14.2. A comparison of different approaches to evaluating AI systems

Evaluation method	Advantages	Disadvantages
Custom dataset	Can be designed specifically to test hypotheses. Potential for it to become a benchmark in the future.	Datasets can be hard to produce/acquire. No comparator results available.
Benchmark	Previous comparator results available. No effort required to produce it.	May not display all the advantages of the new method. Suitable benchmarks not always available. Tends to encourage research towards improving benchmark performance rather than AI in general.
Competition	Evaluation datasets/problem formulation specifically designed for problem. Encourages multiple entrants. Often there is a workshop associated where competitors discuss their results and the consequences for future progress.	Effort required to organise. Can be challenging to pitch competition at suitable level with respect to state of the art – it must be enough of a challenge but not too much. May require sponsorship to run or to encourage entrants.
Qualitative evaluation (i.e. human inspection of system)	Can be tailored to specific dataset and problem. Not rigorous but useful for understanding strengths and weaknesses of an approach.	Requires human effort and can suffer from selection bias.

Note: The different rows are often used in combination: for example, a competition might involve a custom dataset, as well as benchmarks and perhaps a qualitative evaluation.

References

Anderson, M. (11 May 2017), "Twenty years on from Deep Blue vs Kasparov: How a chess match started the big data revolution", The Conversation blog, https://theconversation.com/twenty-years-on-from-deep-blue-vs-kasparov-how-a-chess-match-started-the-big-data-revolution-76882. [32]

Boden, M. (1996), "Creativity", in Boden, M. (ed.), *Artificial Intelllgence (Handbook of Perception and Cognition)*, Academic Press, Cambridge, MA. [28]

Brown, N. and T. Sandholm (2019), "Superhuman AI for multiplayer poker", *Science*, Vol. 365/6456, pp. 885-890, http://dx.doi.org/10.1126/science.aay2400. [20]

Brown, N. and T. Sandholm (2018), "Superhuman AI for heads-up no-limit poker: Libratus beats top professionals", *Science*, Vol. 359/6374, pp. 418-424, http://dx.doi.org/10.1126/science.aao1733. [19]

Cohn, A., R. Dechter and G. and Lakemeyer (2011), "Editorial: The competition section: A new paper category", *Artificial Intelligence*, Vol. 175/9-10, p. iii, https://doi.org/10.1016/S0004-3702(11)00060-9. [8]

Commonsense Reasoning (n.d.), "Problem Page", webpage, http://commonsensereasoning.org/problem_page.html (accessed on 1 January 2021). [33]

Davis, E., L. Morgenstern and C. Ortiz (2017), "The first Winograd Schema Challenge at IJCAI-16", *AI Magazine*, Vol. 38/3, pp. 97-98, https://doi.org/10.1609/aimag.v38i4.2734. [13]

Devlin, J. et al. (2019), "BERT: Pre-training of deep bidirectional transformers for language understanding", *arXiv*, http://dx.doi.org/arXiv:1810.04805v2. [14]

Fox, M. and D. Long (2002), *PDDL+: Modelling Continuous Time-dependent Effects.* [31]

Gaggl, S. et al. (2018), "Summary report of the second international competition on computational models of argumentation", *AI Magazine*, Vol. 39/4, p. 73, https://doi.org/10.1609/aimag.v39i4.2781. [9]

Gelernter, H. (1959), "Realization of a geometry theorem proving machine", *IFIP Congress*, pp. 273-281. [25]

Hall, M. (3 April 2019), "What makes a good benchmark dataset?", Views and News about Geoscience and Technology blog, https://agilescientific.com/blog/2019/4/3/what-makes-a-good-benchmark-dataset. [30]

Hayes, P. and K. Ford (1995), "Turing test considered harmful", in *Proceedings of the Fourteenth International Joint Conference on Artificial Intelligence*, Morgan Kaufmann Publishers, San Francisco. [6]

Jumper, J. et al. (2021), "Highly accurate protein structure prediction with AlphaFold", *Nature*, Vol. 596/7873, pp. 583-589, http://dx.doi.org/10.1038/s41586-021-03819-2. [2]

Kocijan, V. et al. (2020), "A review of Winograd Schema Challenge datasets and approaches", *arXiv*, http://dx.doi.org/abs/2004.13831. [7]

Levesque, H. (2011), "The Winograd Schema Challenge", *AAAI Spring Symposium: Logical Formalizations of Commonsense Reasoning 2011*. [11]

Levesque, H., E. Davis and L. Morgenstern (2012), *The Winograd Schema Challenge*, AAAI Press, Palo Alto, CA. [12]

Mallery, J. (1988), "Thinking about foreign policy: Finding an appropriate role for artificially intelligent computers", in *The 1988 Annual Meeting of the International Studies Association, St. Louis, MO*. [10]

McCarthy, J. (1959), "Programs with common sense", in *Proceedings of Teddington Conference on the Mechanization of Thought Processes, 1959*, Stanford University, Stanford, CA, http://jmc.stanford.edu/articles/mcc59/mcc59.pdf. [1]

Northcutt, C., A. Athalye and J. Mueller (2021), "Pervasive label errors in test sets destabilize machine learning benchmarks", *arXiv preprint*, http://dx.doi.org/arXiv:2103.14749. [26]

Perez-Liebana, D. et al. (2019), *General Video Game AI*, Morgan Kaufman, San Francisco. [18]

Renz, J. et al. (2019), "AI meets Angry Birds", *Nature Machine Intelligence*, Vol. 1/328, https://doi.org/10.1038/s42256-019-0072-x. [21]

Schaul, T. (2013), "A video game description language for model-based or interactive learning", *2013 IEEE Conference on Computational Intelligence in Games (CIG)*, pp. 1-8, http://dx.doi.org/10.1109/CIG.2013.6633610. [22]

Shieber, S. (2016), "Principles for designing an AI competition, or why the Turing test fails as an inducement prize", *AI Magazine*, Vol. 37/1, pp. 91-96, https://doi.org/10.1609/aimag.v37i1.2646. [4]

Shieber, S. (1994), "Lessons from a restricted Turing test", *Communications of the ACM*, Vol. 37/6, pp. 70-78, http://dx.doi.org/10.1145/175208.175217. [5]

Sloman, A. (n.d.), "John McCarthy – Some Reminiscences", webpage, https://www.cs.bham.ac.uk/research/projects/cogaff/sloman-jmc-aisb.html (accessed on 1 January 2021). [15]

Stephenson, M. et al. (2019), "The 2017 AIBIRDS level generation competition", *IEEE Transactions on Games*, Vol. 11/3, pp. 275-284, http://dx.doi.org/10.1109/TG.2018.2854896. [27]

Sutcliffe, G. (2017), "The TPTP problem library and associated infrastructure: from CNF to TH0, TPTP v6. 4.0", *Journal of Automated Reasoning*, Vol. 59, pp. 583-402, http://dx.doi.org/doi.org/10.1007/s10817-017-9407-7. [23]

Sweetser, P. and P. Wyeth (2005), "GameFlow: A model for evaluating player enjoyment in games", *Computers in Entertainment*, Vol. 3/3, p. 3, https://doi.org/10.1145/1077246.1077253. [29]

Togelius, J. (2016), "AI researchers, video games are tour friends!", in Merelo, J. et al. (eds.), *Computational Intelligence.International Joint Conference, IJCCI 2015 Lisbon, Portugal, 12-14 November 2015, Revised Selected Papers*, https://doi.org/10.1007/978-3-319-48506-5_1. [17]

Togelius, J. (2016), "How to run a successful game-based AI competition", in *IEEE Transactions on Computational Intelligence and AI in Games*, http://dx.doi.org/10.1109/TCIAIG.2014.2365470. [16]

Turing, A. (1950), "Computing machinery and Intelligence", *Mind*, Vol. LIX/236, pp. 433-460, https://doi.org/10.1093/mind/LIX.236.433. [3]

Wang, A. et al. (2019), "Glue: A multi-task benchmark and analysis platform for natural language understanding", *arXiv*, http://dx.doi.org/arXiv:1804.07461. [24]

Notes

[1] Turing called the test the "imitation game" (now the title of a film about Turing's life). He proposed it as a test of whether a machine could display intelligent behaviour indistinguishable to a human observer confronted with both the machine and another human; the observer could only communicate with the machine and the other human by typed natural language.

[2] The Loebner Prize, instituted in November 1991, is normally held annually. Competitors enter chatbots and human judges (since 2019 members of the public) judge which is the most human-like. It was instituted by Hugh Loebner and offers a prize of USD 100 000 for "the first program that judges cannot distinguish from a real human in a Turing test that includes deciphering and understanding text, visual and auditory input. Once this is achieved, the annual competition will end."

[4] These are often associated with AI conferences, but sometimes independently organised. Kaggle hosts many competitions: www.kaggle.com/competitions, many with prizes. The platform www.aicrowd.com/ also hosts challenges to enable "data science experts and enthusiasts to collaboratively solve real-world problems". Another platform hosting competitions is https://competitions.codalab.org/. All URLs accessed 24 March 2021.

[5] As an example of elaboration tolerance, consider one of the problems on Commonsense Reasoning (n.d.[33]): "A gardener who has valuable plants with long delicate stems protects them against the wind by staking them; that is, by plunging a stake into the ground near them and attaching the plants to the stake with string". Variants of the problem suggested included: "What would happen: If the stake is only placed upright on the ground, not stuck into the ground? If the string were attached only to the plant, not to the stake? To the stake, but not to the plant? If the plant is growing out of rock? Or in water? If, instead of string, you use a rubber band? Or a wire twist-tie? Or a light chain? Or a metal ring? Or a cobweb? If instead of tying the ends of the string, you twist them together? Or glue them?..."

[6] Some have speculated that a bug in the code contributed to Deep Blue's success. In this view, the bug provoked a random move for the computer. Kasparov apparently attributed this move to a "deeper strategy" and subsequently played too cautiously (Anderson, 11 May 2017[32]).

[7] Fox and Long (2002[31]) write: "The adoption of a common formalism for describing planning domains fosters far greater reuse of research and allows more direct comparison of systems and approaches, and therefore supports faster progress in the field. A common formalism is a compromise between expressive power (in which development is strongly driven by potential applications) and the progress of basic research (which encourages development from well-understood foundations). The role of a common formalism as a communication medium for exchange demands that it is provided with a clear semantics."

[8] The team behind AlphaGo won the inaugural Minsky Medal. The Marvin Minsky Medal is awarded by IJCAI for "Outstanding Achievements in AI".

[9] Hall (Hall, 3 April 2019[30]) proposes characteristics of a good machine learning benchmarks such as being openly available, well documented, labelled and with an accompanying demonstration.

[10] There are many others including those included in repositories such as www.kaggle.com/ https://paperswithcode.com/datasets.

15. Assessing artificial intelligence capabilities

Guillaume Avrin, artificial intelligence

As artificial intelligence (AI) becomes more mature, it is increasingly used in the world of work alongside human beings. This raises the question of the real value provided by AI, its limits and its complementarity with the skills of biological intelligence. Based on evaluations of AI systems by the Laboratoire national de métrologie et d'essais in France, this chapter proposes a high-level taxonomy of AI capabilities and generalises it to other AI tasks to draw a parallel with human capabilities. It also presents proven practices for evaluating AI systems, which could serve as a basis for developing a methodology for comparing AI and human intelligence. Finally, it recommends further actions to progress in identifying the strengths and weaknesses of AI vs. human intelligence. To that end, it considers the functions and mechanisms underlying capabilities, taking into account the specificities of non-convex AI behaviour in the definition of evaluation tools.

Introduction

Based on evaluations of AI systems by the Laboratoire national de métrologie et d'essais (LNE) in France, this chapter proposes a high-level taxonomy of AI capabilities. It then generalises this taxonomy to other AI tasks to draw a parallel with human capabilities. It also presents proven practices for evaluating AI systems, which could serve as a basis for developing a methodology for comparing AI and human intelligence. Finally, it recommends further actions to identify the strengths and weaknesses of AI vs. human intelligence. To that end, it considers the functions and mechanisms underlying capabilities, taking into account the specificities of non-convex AI behaviour in the definition of evaluation tools.

The chapter uses the terms "evaluation" and "evaluation campaign". An evaluation is a single test that aims to measure the characteristics (performance, explainability, etc.) of an intelligent system. Conversely, an evaluation campaign represents the process of evaluating products either vertically (by observing a range of products at a given time) and/or horizontally (by observing the evolution of the product over time).

Disciplinary field of artificial intelligence evaluation

This section provides a framework for LNE's "evaluation" activities in the AI field, while presenting the good practices acquired since this activity was set up in 2008.

The good practices at the heart of LNE evaluation campaigns are mainly the result of the search for a compromise between realism and reproducibility of experiments. It has led to the identification of features to be presented in campaigns below. Some of these features relate to the general organisation of the evaluation campaign, while others are more specialised on the evaluation process.

- Scientific

Evaluation campaigns preserve the demonstration aspect typically associated with them. However, they are based on the scientific criteria of assessment objectivity, performance measurement repeatability and experiment reproducibility. They also respect the requirements imposed by metrological rigour.

- Benchmark-based

The intelligent systems are evaluated through benchmarks. This means they perform well-specified tests in realistic environments or on databases. In addition, their performance is assessed by applying quantitative metrics.

- Modular

It is often not satisfactory to evaluate only the robot as a whole. Thus, the elements constituting the robot's architecture are broken down into functionalities (e.g. obstacle detection). These are then combined to perform more complex tasks (e.g. semantic navigation). The evaluation thus consists in Functionality Benchmarks (FBMs) and Task Benchmarks (TBMs). FBMs evaluate specific capabilities with a limited utility when used alone, while TBMs evaluate more complex activities (see below).

- Periodical

Evaluation campaigns should be organised as recurring events offering a similar evaluation framework each time (similar testbeds, similar testing datasets, same evaluation tools, etc.). This framework enables monitoring of the technological progress of the community of developers as a whole.

- Structured

Evaluation campaigns are structured to optimise effort and maximise impact. As such, they provide the scientific community with a stable set of benchmarking experiments. This, in turn, enables objective comparison of research results and can act as the seed for the definition of standards.

- Synergic

Evaluation campaigns should build on the well-established framework originally created by RoCKIn[1] and Quaero[2] projects and subsequently validated, perfected and extended by RockEU2[3], SciRoc[4], ROSE[5] and METRICS[6] projects.

- Open

Evaluation tools and annotated datasets should be publicly available. This will enable research and industry to develop and fine-tune their own algorithms, systems and products. Existing and prospective actors gain access both to difficult-to-obtain data with associated ground truth and to validated evaluation tools. Importantly, these by-products benefit the evaluator and promote the long-term sustainability of its evaluation campaigns. Users of the open data and tools will naturally be inclined to participate in the campaigns, thus creating a virtuous circle enabling their success.

Functionality benchmarks and task benchmarks

Evaluation campaigns include two groups of benchmarks (Amigoni et al., 2015[1]; Avrin, Barbosa and Delaborde, 2020[2]).

Functionality benchmarks (FBMs)

A functionality is conventionally identified as a self-contained unit of capability, which is too low level to be useful on its own to reach a goal (e.g. self-localisation, crucial to most applications but aimless on its own). A single component or a set of components can provide a functionality, and usually involves both hardware and software.

An FBM is a benchmark that investigates the performance of a robot component when executing a given functionality. A functionality is as independent as possible of the other functionalities of the system. In this way, functionality can be controlled as the sole dependent variable in the evaluation.

Task benchmarks (TBMs)

A task is an activity of a robot system that, when performed, accomplishes a goal considered useful on its own. A task always requires multiple functionalities to be performed. Finding and fetching an object, for example, involves functionalities such as self-localisation, mapping, navigation, obstacle avoidance, perception, object classification/identification and grasping. A TBM is a benchmark that investigates the performance of a robot system when executing a given task. TBMs are designed by focusing on the goal of the task, without constraining the means by which such goal is reached.

Evaluating the overall performance of a robot system while performing a task is interesting for assessing the global behaviour of the application. However, it does not allow evaluation of the contribution of each component. Nor does it put in evidence which components are limiting system performance.

On the other side, the good performance of each element in a set of components does not necessarily mean that a robot built with such components will perform well. System-level integration has, in fact, a deep influence on this, which component-level benchmarking does not investigate.

For these reasons, combining a TBM with FBMs focused on the key functionalities required by the task provides a deeper analysis of a robot system and better supports scientific and technical progress. The objective is to address the evaluation needs of end-users, integrators and equipment manufacturers.

Fairness of evaluation campaigns

This section looks at how to ensure an optimally fair treatment of the campaign's participants. The notion of fairness is addressed in light of metrological considerations.

Simultaneity

The evaluator shall ensure a simultaneity of the evaluation, as required by the following considerations:

- **The difficulty to model the influence of environmental factors on the system's performance:** any outdoor experiment will never be completely repeatable. Clouds change in the sky; waves and tides modify visibility underwater, etc. This lack of repeatability, in addition to its influence on the metrological rigour of the evaluation, has an impact on the fairness of the evaluation between participating systems. It is not conceivable that one participant will have to operate in pouring rain, while another will suffer from maximum sunshine. In this regard, the evaluator shall define thresholds and limits in several parameters that are considered to influence performance of the devices. Outside of this acceptability range, the evaluator shall define remedial strategies to have intelligent systems compete in reasonably similar conditions.
- **The "*a priori* ignorance" imperative**: evaluated systems have a learning capability and consequently, should not have *a priori* knowledge of the testing environment (testbeds and testing datasets) used for the evaluation in order to avoid measurement bias and overfitting. This remark remains valid for systems that do not have learning skills since developers can influence the design of their systems if they have *a priori* information about testbeds and data.
- **The "*a posteriori* publication" imperative**: to ensure reproducibility of the evaluation experiments, testing environments used must be publicly described (and accessible if they are datasets) when the measurements and results are published.

This notion of "simultaneity" can sometimes be spread across the one or two days of the evaluation campaigns. The tolerance level about what may be considered "simultaneous" must, of course, be discussed on a case-by-case basis.

Impartiality

The evaluation must be carried out by a "trusted third party". This evaluator must have metrology expertise applied to the evaluated systems in order to develop an evaluation protocol common to all participants. In addition, it must guarantee there are no conflicts of interest between the campaigns' evaluator and participants.

Precise evaluation plan

Each evaluation must rely on an evaluation plan, a document that details the features of the following:

- one or more evaluation tasks that focus on a device or software performing a specification
- characteristics that need to be measured or estimated (performance, quality, safety, explainability, etc.)
- metrics (i.e. a formula that allows production of scores, such as accuracy, precision, recall, F-measure)
- test data or test environments (datasets or testbeds)
- evaluation tools (software for data collection, visualisation, comparison).

Evaluation task

The first step in organising an evaluation campaign is to specify and prioritise a set of evaluation tasks (FBMs and TBMs). They are deduced from the identification of scientific and technological barriers. The principal (campaign funder), who expresses the "business" need, defines the tasks rather than the evaluator (LNE and its potential partners). On the other hand, when a potential use case is identified, the evaluator must carry out the following checks:

- List solutions corresponding to the use case, with an estimate (when the information is accessible) of the performance limitations associated with their characteristics or conditions of use (costs, knowledge to be implemented for deployment, operation in highly constrained environments or for an extremely specific field, etc.).
- List the types of data required for the development and operation of such solutions, and their availability (considering regulatory or ethical limitations, the cost of collection, etc.).

During this stage, the evaluator checks the feasibility of the campaign using the following criteria:

- possibility of objectifying the evaluation criteria
- difficulty of collecting and transmitting test data to the participants of the challenge (confidentiality of data inherent to use cases, availability of data, etc.), or making test environments available
- comparability of solutions for the use case (systems that can potentially take in extremely varied types of data may lead to significant adaptation of evaluation protocols, or even incomparability).

Evaluation method

The evaluation paradigm generally consists in comparing reference and hypothesis data. Reference data are the ground truth annotated by human experts or provided by measuring instruments in the test facility. Conversely, hypothesis data are the behaviour or output produced automatically by the intelligent system. This comparison allows the estimation of the performance, the reliability and other characteristics such as efficiency of robots. The evaluation can concern the entire system (during TBM) or the main technological components taken independently (during FBM), as shown in Figure 15.1.

Figure 15.1. Evaluation method

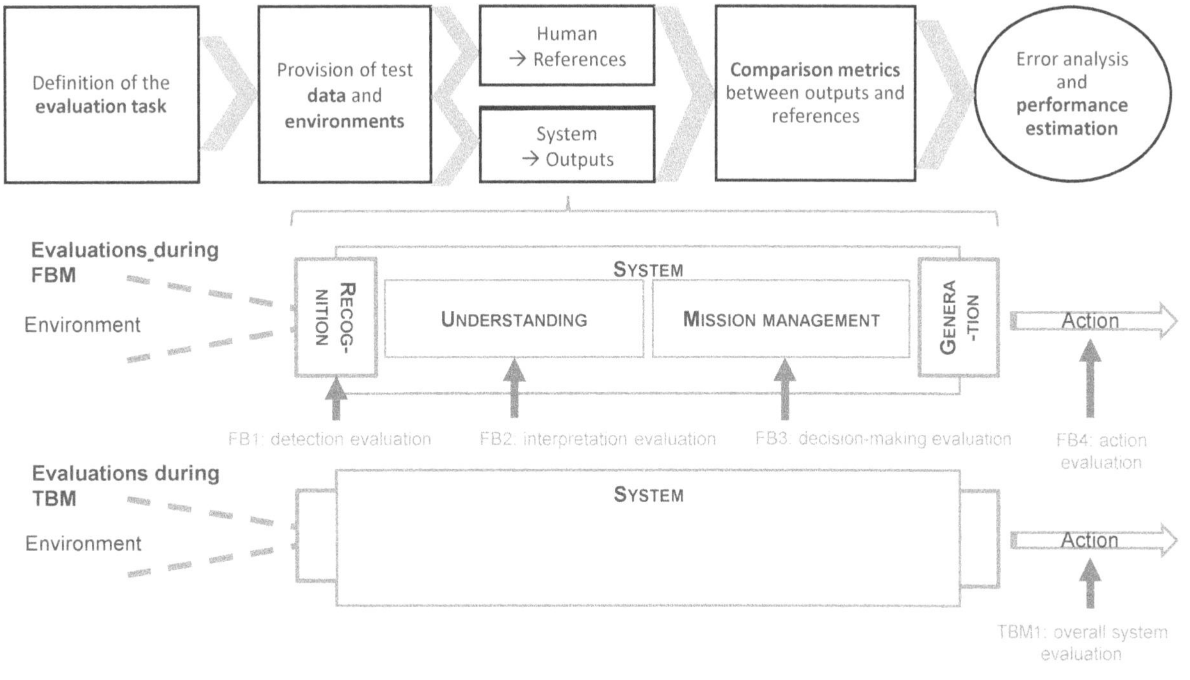

Some evaluation campaigns last several years and include several evaluations. The repeated evaluations allow the principal to assess the effectiveness of the funding granted for the organisation of the evaluation campaign. For example, this could estimate the performance of potential technological solutions that address its use case. For developers, repeated evaluations allow them to update the technological components of the intelligent system according to the quantitative results obtained.

The dry-run evaluation guarantees the smooth implementation of the campaign. It allows the evaluator to ensure its evaluation plan is both realistic with respect to the capabilities of the systems, and fair among the different technologies used by participants. Thus, the dry run can experiment with several test environments and metrics to define the best evaluation protocol that will be fixed during the official evaluation campaigns. Several official evaluation campaigns follow the dry run. These aim at objectively measuring the progress of participating robots in real field conditions. To this end, the evaluation plan is meant to be adapted throughout the campaign to accompany the evolutions of the participants' technological solutions. The steps of an evaluation campaign are presented in Figure 15.2.

Figure 15.2. Steps of a campaign involving several evaluations

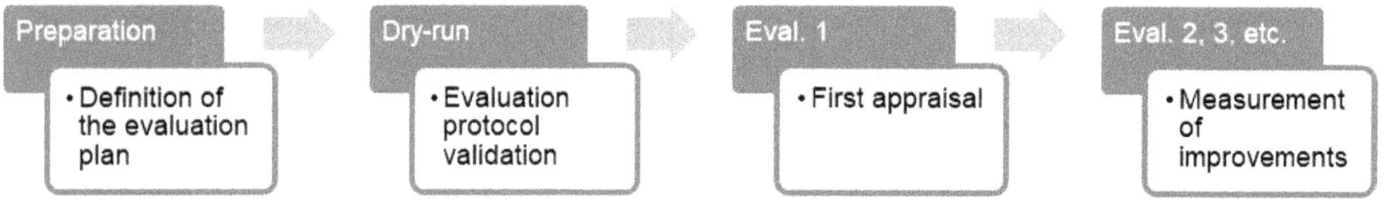

Comparison metrics

The capability measurements must be quantitative and provided by a formula ("metric") that indicates the distance between the reference and the hypothesis, or measures capacity directly. The distance between the reference and the hypothesis could be measured by the distance between a real or ideal trajectory in a navigation task, the number of false positives and false negatives in an image recognition task, the binary success of a task, etc. A direct measurement could be time to completion, distance covered, etc.

Test data or test environments (datasets or testbeds)

Evaluations can be based on physical (in real or laboratory conditions on testbeds) and/or virtual testing environments (simulators and testing datasets). Pros and cons of the two types of environments are presented below.

Table 15.1. Strengths and weaknesses of test environments

	Physical testbeds	Simulators	Datasets
Exhaustiveness of tested scenarios	*	***	**
Realism of tested scenarios	***	*	**
Generation of new data	***	**	*
Dynamic and closed-loop testing	***	**	*
Experimentations reproducibility and measure repeatability	*	***	***
Cost of each test	*	***	**
Data obsolescence	***	**	*

Note: The number of asterisks indicates the rank of the test environment type: *** corresponds to the best solution and * to the worst.

Taxonomy of evaluations carried out by Laboratoire national de métrologie et d'essais

Clustering Laboratoire national de métrologie et d'essais's artificial intelligence evaluations

LNE has carried out more than 950 evaluations of AI systems since 2008. These include areas such as language processing (translation, transcription, speaker recognition, etc.), image processing (person recognition, object recognition, etc.) and robotics (autonomous vehicles, service robots, agricultural robots, intelligent medical devices, industrial robots, etc.). Examples of evaluations in the context of research and development projects are presented in Table 15.2.

The evaluation tasks have been grouped into the capabilities of recognition, understanding, mission management and generation. These are an extension of the "sense-think-act" paradigm and an adaptation of the AI cycle "Perception – Learning – Knowledge representation – Reasoning – Planning – Execution" (Beetz et al., 2007).

These capabilities are also consistent with the NIST 4D/RCS reference architecture for autonomous vehicles (Albus, 2002[3]); with the "SPACE" breakdown into functions (Sense, Perceive, Attend, Apprehend, Comprehend, Effect action) that underpin intelligent behaviour (Hughes and Hughes, 2019[4]) and finally, with most cognitive architectures (Kotseruba and Tsotsos, 2016[5]; Ye, Wang and Wang, 2018[6]). These can be illustrated by the dialogue systems, for which such a division is common (see Figure 15.3) (Leuski and Traum, 2008[7]).

Figure 15.3. Division of dialogue systems capabilities

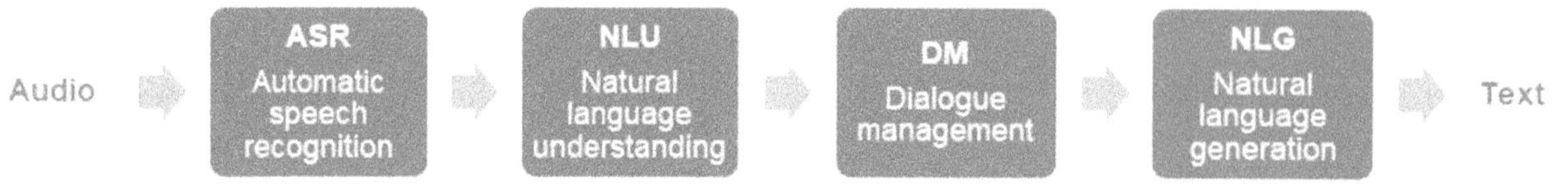

Comparative evaluation of different architectures

LNE also assesses another capability of AI, which cuts across the different capabilities listed above: the system's capability to learn and update its parameters throughout its lifecycle.

The differences between the cognitive architectures cited above do not result so much from divergent points of view on potential capabilities. Rather, they reflect two other factors. First, they were developed in different contexts (different perimeters and objectives of the associated research projects). Second, they have different hypotheses regarding the neural processes underlying these functionalities (symbolic, connectionist or hybrid architectures, centralised or decentralised processing, etc.). Summaries of the main cognitive architectures[7] and the main capabilities[8] covered by these architectures are available in the literature.

The comparative evaluation of these different architectures is a vast subject of research. The evaluation criteria considered include the generality of the architecture. This measures the types of tasks and environments that can be handled by systems developed according to this architecture. This measurement, in turn, is assessed in terms of versatility and taskability.

Versatility is defined as the number of ways in which the system designed according to the architecture can solve the same task, using different capabilities. Meanwhile, taskability is the number of different tasks that can be performed by the system receiving external commands. This generality feature of an architecture is directly related to the general intelligence of the resulting systems (Langley, Laird and Rogers, 2009[8]) and therefore directly relevant to this study.

There is a wide variety of terms within these cognitive architectures to describe their capabilities. The first column of Table 15.2 proposes a first equivalence between these terms.

These cognitive architectures are consistent with each other. They are not only interested in reproducing the external behaviour of biological intelligences but also in modelling the internal properties of their cognitive systems. They propose a blueprint for cognitive agents depicting the arrangement of functional units. This facilitates implementation of their principles in mechatronic systems [see "*architecture-as-methodology*" in Jiménez et al. (2021[9])].

These cognitive architectures also provide a formalism for presenting human capabilities (and dealing with the intrinsic complexity of cognitive systems) that can be reproduced in artificial systems.[9] In this way, they represent a bio-inspired and integrated taxonomy of human and artificial capabilities and they facilitate the comparison of these capabilities when they are implemented in humans and in machines.

Exclusivity and exhaustiveness

As these cognitive architectures are used to assemble technological components of intelligent mechatronic systems, each capability can be associated to an exclusive component or group of components. Mutual exclusivity between these capabilities is thus guaranteed. For obvious cost reasons, engineers using a cognitive architecture to design their intelligent systems would have no interest in building in functional redundancy between the different components (which must be distinguished from the redundancy intended to meet the safety requirements of critical systems).

This modular architecture therefore makes it possible to isolate the technological components that underpin the different capabilities (i.e. the functional units) and to carry out input-output evaluations on each component to evaluate each capability independently.

Various studies have investigated the exhaustiveness of the capabilities covered by these architectures. However, there does not yet seem to be a consensus regarding the most comprehensive architectures. Some prefer CLARION and AIS (Kotseruba and Tsotsos, 2016[5]), while others prefer OpenCogPrime (Ye, Wang and Wang, 2018[6]). Exhaustiveness can be measured by the "generality", i.e. the number of tasks and environments in which a system built according to this architecture can be used.

Table 15.2. LNE's evaluation tasks and metrics

Automatic information processing systems			
Task	**Capability**	**Metrics**	**Project**
Speaker verification for criminalistics application	Recognition	Equal error rate (EER), Detection cost function (DCF), Detection error trade-off (DET), Probabilistic linear discriminant analysis, etc. (Ajili et al., 2016[10])	FABIOLE (2013-16) VOXCRIM (2017-21)
Automatic speech recognition	Recognition	Word error rate (WER), Automatic transcription evaluation for named entities (ATENE), Word Information Loss (WIL), Relative information loss (RIL), IN, Near (Ben Jannet et al., 2015[11])	VERA (2013-15)
Speaker diarisation	Recognition	Diarisation Error Rate (DER) (Prokopalo et al., 2020[12])	ALLIES (2017-20)
Speaker diarisation across time	Recognition Learning	Average DER across audio file, weighted by duration of the file	
Lifelong learning diarisation	Recognition Learning	The DER is computed on the final version of the hypothesis for each document penalised by the cost of interacting with the user in the loop	
Translation, translation across time, lifelong translation	Recognition Understanding Generation Learning	Bilingual evaluation understudy (BLEU) adaptations similar to those of the speaker diarisation tasks	
Recognition of patients' vital signs (breathing, heart rate, etc.).	Recognition	Estimated global error rate (EGER), Precision, Recall, F-measure	AIR (2020-22)
Transcription from TV feeds	Recognition Generation	Word error rate (WER), (Galibert et al., 2014[13])	ETAPE (2010-12)
Named entity recognition (detection, classification, decomposition)	Recognition	Entity Tree Error Rate (ETER), Slot error rate (SER), Error per response (ERR), EDT value, Local entity detection and recognition (LEDR) (Ben Jannet et al., 2014[14])	QUAERO (2008-14)
Question-answering systems	Recognition Understanding Mission management Generation	QA distance measure, (Bernard et al., 2010[15])	
People recognition in multimodal conditions	Recognition	EGER (Kahn et al., 2012)	REPERE (2012-14)
Translation of newspapers articles and broadcast news transcriptions that come from various radio and television programmes	Recognition Understanding Generation	Translation Error Rate (TER), BLEU, Human-mediated translation edit rate (HTER)	TRAD (2012-14)
Area segmentation	Recognition	ZoneMap, Pset, DetEval, Jaccard, (Brunessaux et al., 2014[16])	MAURDOR (2012-14)
Identification of the writing type (handwritten, printed, unspecified)	Recognition	Accuracy	
Language identification	Recognition	Accuracy	
Text-to-text transcription and optical character	Recognition	WER, Character error rate	

recognition		(CER)	
Extraction of logical structure (logical connections between semantic areas)	Recognition Understanding	Precision, Recall, F-measure (Oparin, Kahn and Galibert, 2014[17])	
Synthesis of video and text information	Recognition Understanding Mission management Generation	WER, SER, Precision, Recall, F-measure	IMM (2013-16)
Automatic speech recognition performance prediction	Recognition Understanding	Mean Absolute Error (MAE) and Kendall (Elloumi et al., 2018[18])	Autre – 2018
Satellite image classification	Recognition	EGER, ZoneMap, Jaccard	Confidential (2019-20)
Recognition of aircraft movement patterns from radar data	Recognition	EGER, Precision, Recall, F-measure	Confidential (2019-20)
Transcription by smartphone intelligent personal assistant	Recognition	WER, Accuracy	Confidential (2019)
QA by smartphone intelligent personal assistant	Recognition Understanding Mission management Generation	Accuracy	
	Robotic systems		
Task	**Capability**	**Metrics**	**Project**
Crops and weeds recognition	Recognition	EGER, Precision, Rappel, F-measure (Avrin et al., 2019[19]); (Avrin et al., 2020[20])	Challenge ROSE (2018-21)
Mechanical/electrical weeding action	Generation	Accuracy	
Full agricultural weeding robot evaluation	Recognition Understanding Mission management Generation	Accuracy	
Advanced driver assistance (ADAS)	Recognition Understanding Mission management Generation	Time to collision, time exposed to time to collision, time to brake, time to steer, time to react	SVA/3SA (2015-22)
Climbing up 10 cm high stairs without handrail, climbing up 15 cm high stairs with handrail, walking over stepping stones, walking on a beam, walking on a flat ground, walking on a slope, walking over obstacles	Recognition Understanding Mission management Generation	Walked distance, success rate, max tracking error, duration of the experiment, etc. (Stasse et al., 2018[21])	Robocom++ (2017-20)
Human detection for logistics robots	Recognition	EGER, Precision, Recall, F-measure	Blaxtair Safe (2019-20)
Estimate the stopping distance (conventional or emergency) under load and maximum speed	Generation	Linear distance measurement	ECAI (2019-20)

As shown in Table 15.2 the same task may involve one or more capabilities depending on the context. For example, an information retrieval task may rely only on the mission manager if the information is stored in memory. It may require recognition and understanding if it involves searching for information in text. A medical diagnosis may be based solely on a capability for recognition, or may also involve a phase of reasoning. A medical prescription will involve the "mission manager" component.

Generalisation to other artificial intelligence tasks

The capabilities presented in the previous section are defined in more detail in Table 15.3 and generalised to other typical AI tasks. Table 15.4 illustrates the presence of these capabilities in AI systems. Table 15.5 provides an example of how to implement the evaluation process to assess these capabilities for a specific AI system.

Table 15.3. AI capabilities generalisation

AI capabilities (and equivalent words)	Examples of AI tasks	Example of AI output
Recognition: perception/acquisition of sensory information (vision, hearing, etc.)	Speech recognition, optical character recognition, tokenisation, named entity recognition, lemmatisation, parsing, pose estimation, face verification, scene segmentation, person reidentification, image classification, etc.	"object: glass", "position: falling" and "object: human arm", "position: stretched"
Understanding: contextualisation, interpretation, comprehension, conceptualisation, assimilation (relating to system state, storage, etc.)	Knowledge representation, 2D/3D mapping, information extraction, image captioning, etc.	"the human tries to catch the falling glass"
Mission manager: decision making, cogitation, cerebration, reasoning, inferring, arbitration (judgement), etc.	Prediction, planning, optimisation, selection between different options, self-check, etc.	Identification of the best trajectory to catch the glass safely before it touches the ground
Generation: action	Navigation, speech synthesis, locomotion, manipulation (grasping, etc.), content generation (image, etc.), etc.	Generation of the movement of the robotic arm and the gripping effector to catch the glass
Learning: adaptation, knowledge storing	Parameters update (supervised, unsupervised, reinforcement learning, etc.), operation algorithm change.	If "broken glass": failure, update the trajectory generation parameters, otherwise do nothing.

Table 15.4. Examples of AI capabilities for different tasks

Recognition	Understanding	Mission management	Generation
Autonomous car			
Traffic-sign recognition, obstacle recognition, etc.	Velocity synthesis, image plan mapping, relationship identification, etc.	Motion planning, risk assessment, etc.	Vehicle control, braking, steering, driver alert, etc.
Text summarisation			
Sentence segmentation, word segmentation, feature extraction	Feature frequency, similarity computation, sentences comparison and scoring	Sentences selection and assembly	Summary generation
Recommendation systems			
Analysis of rating	Analysis of behaviour, contextualisation based on location, time, user profile, etc.	Comparison to other user preferences	Recommendation of objects.

Table 15.5. Example of the evaluation steps for autonomous weeding robots (from ROSE and METRICS projects)

Step	Detail
Formalisation of the need	• What is the objective: autonomous weeding of the intra-row of agricultural plots. • Which crops should be considered in priority: corn and beans. • What are the weeds to be considered in priority: lamb's quarter, matricaria, ryegrass and wild mustard.
Feasibility analysis	• Mapping of weed control robots on the market. • Identification of the main capabilities useful for the task (weed detection, weeding decision making, weeding action). • Estimation of the costs associated with the evaluation of these different capabilities: cheap weed images to produce, expensive test farm to set up, etc.
Formalisation of the evaluation tasks	• Recognition: segmentation of weeds and crops on images. • Generation: navigation, weeding action. • Etc.
Formalisation of the evaluation criteria and metrics	• Segmentation metrics: estimated global error rate (EGER), Jaccard index, Zonemap, etc. • Generation metrics: biomass estimation, counting of weeds removed. • Etc.

Relevance of the proposed capability taxonomy

Relevance to artificial intelligence

This section reviews the mutually exclusive and collectively exhaustive capabilities (MECE character) of the different taxonomies to assess their relevance.

Exclusivity

The taxonomy proposed in the previous section is inspired (although simplified) by cognitive architectures. These are designed to assemble different functional units (each representing its own capability) to form an information processing pipeline. As each unit has its own function, these cognitive architectures are designed to ensure the mutually exclusive nature of the capabilities. In this way, they avoid any redundancy that would be detrimental in terms of the manufacturing cost of the AI system. However, with the rise of end-to-end learning (Shibata, 2017[22]), the boundary between these different functions is blurring as design moves from this traditional "pipeline".

Exhaustiveness

The proposed taxonomy seems to cover the capabilities of the main cognitive architectures, although with a high level of abstraction (Kotseruba and Tsotsos, 2016[5]; Ye, Wang and Wang, 2018[6]; Hughes and Hughes, 2019[4]). High-level capabilities could be further broken down into tasks, while retaining their MECE nature. Table 15.6 provides an example of the decomposition of a high-level capability, which is modality- and application-independent, into modality-dependent tasks and application-dependent sub-tasks. This division can be continued until specific tasks are reached (such as the manufacturing tasks proposed in Huckaby and Christensen (2012[23]): place, transport, retract, slide, insert, pick up, align, etc.).

Table 15.6. Example of breaking down the recognition capability into sub-tasks

Capability (modality- and application-independent)	Modality-dependant task	Modality- and application-dependant sub-task
Recognition	Image recognition	Optical character recognition
		Face recognition
		Pose estimation
		Etc.
	Language recognition	Tokenisation
		Lemmatisation
		Named entity recognition
		Etc.
	Etc.	Etc.

The breakdown of capabilities proposed for the taxonomy is also relevant given that substantial progress on a task in one capability advances AI performance on other associated tasks (see Table 15.2 for examples of tasks for each capability). This is, in particular, the consequence of the democratisation of the use of pre-trained algorithms and inductive transfer (Moon, Kim and Wang, 2014[24]).

This observation is even more striking for a particular modality related to a given capability [e.g. visual recognition (Razavian et al., 2014[25])or speech recognition (Howard and Ruder, 2018[26]; Peters et al., 2018[27]; Devlin et al., 2019[28])].

This is the case in part because the tasks for a given capability usually involve the same types of algorithms. For example, recognition tasks typically use classification, clustering or mapping algorithms. Conversely, mission management tasks will use more optimisation or regression algorithms. These correspondences between types of tasks to be automated and types of algorithms used for automation are discussed further below.

These dependencies between progress on tasks associated with the same capability are much more evident between high-level tasks and their sub-tasks. In particular, some work highlights the critical implications that progress in certain sub-tasks can have for AI as a whole (Cremer and Whittlestone, 2020[29]).

Relevance to humans

If the proposed taxonomy seems relevant to AI, another question arises: will it allow an effective comparison between human and AI capabilities? The answer requires two considerations.

First, this taxonomy is related to cognitive architectures. As such, they already provide an integrated view of human and artificial capabilities, with particular caution regarding jingle-jangle fallacies mentioned in Primi et al. (2016[30]). Indeed, such a taxonomy should be independent of the underlying methods and equipment used Shneier et al. (2015[31]).

Second, the idea of decomposing high-level capabilities into a pipeline of lower-level capabilities also seems relevant for the analysis of human capabilities. Tolan et al. (2020[32]) highlight this type of dependence between high-level capabilities and lower-level skills. This pipeline decomposition is also consistent with the levels of autonomy proposed in Huang et al. (2007[33]) to characterise the assistance of the machine to the human and vice versa.

The decomposition choices, of which a first example is provided in Table 15.6, are in turn complex to perform. A consensus seems to be found in the idea of starting the taxonomy with high-level capabilities that are non-specialised (Hernández-Orallo, 2017[34]). Neubert et al. (2015[35]) called these capabilities with a higher level of abstraction "Core domain skills", "Transversal skills" and "Basic cognitive skills", while O*NET[10] refers to them as "cross-occupational activities". Chapter 7 explores these skills in more detail.

The question of correspondence with the taxonomies of human capabilities also arises (Hernández-Orallo, 2017[34]; Hernández-Orallo, 2017[36]; Tolan et al., 2020[32]). An association is proposed in Table 15.7.

Table 15.7. Correspondence between AI and human capabilities

Human abilities	Capabilities
Memory processes	Mission management Learning
Sensorimotor interaction	Recognition Understanding Mission management Generation
Visual processing	Recognition
Auditory processing	Recognition
Attention and search	Recognition
Planning and sequential decision making and acting	Mission management Generation
Comprehension and compositional expression	Understanding Mission management Generation
Communication	Mission management Generation
Emotion and self-control	Mission management Generation
Navigation	Mission management
Conceptualisation, learning and abstraction	Understanding Learning
Quantitative and logical reasoning	Mission manager
Mind modelling and social interaction	Understanding Mission manager Generation
Metacognition and confidence assessment	Mission manager

Source: Hernández-Orallo, (2017[36]).

The human capabilities shown in Table 15.7 are mainly inspired by psychometrics, comparative psychology and cognitive science. They correspond to combinations of different capabilities proposed for AI, although the proposed taxonomy has a high level of abstraction. As a consequence, transcribing these human capabilities into an AI system would require different functional units. These capabilities would be called "composite". In AI, composite capabilities are complex to evaluate. The modular organisation of capabilities within cognitive architectures instead allows each technological component to be evaluated independently, through input-output evaluations, as discussed in Section 3.

Relevance of evaluation methods to compare human and artificial capabilities

Relevance of artificial intelligence tests

This chapter presents an approach used by LNE to evaluate AI systems based on the implementation of benchmarks (i.e. standard tests). The test-based approach is also commonly used to assess human capabilities. School exams and neuropsychological evaluations (perceptual, motor, attentional tasks, etc.) rely on tests. Moreover, the *a priori ignorance, a posteriori publication and impartiality* requirements are equally important for such human dedicated tests. Even the adaptive/adversarial testing approaches used for AI have their equivalent for human testing. Adaptive testing is found in GRE, as well as in oral tests such as the one used by German dual vocational education and training (see Chapter 9).

Since the test-based approach is already used to evaluate both biological and artificial capabilities, it would be interesting to compare these competences. In most of the LNE data-based evaluations, humans perform the reference annotations against which the outputs of the intelligent system under evaluation are compared (see sub-section "Precise evaluation plan"). In practice, several humans annotate each piece of data in the test database[11] to carry out inter- and intra-annotations agreement analyses (Mathet et al., 2012[37]) and to verify the ground truth associated with the test data. Therefore, most evaluations of AI systems include, from the beginning, a comparison with humans.

Tests designed for AI are also interesting because they are modular (cf. sub-section "Disciplinary field of AI evaluation"). As well, the evaluation tasks (task benchmarks and functionality benchmarks) follow the division of human capabilities into functional units proposed by cognitive architectures (cf. sub-section "Clustering LNE's AI evaluations"). Thus, they are optimal to compare human and artificial capabilities.

For these reasons, tests specifically designed for AI systems could occupy a prominent place in the OECD's *Artificial Intelligence and the Future of Skills* project.

Relevance of human tests

Many tests designed for humans seem unsuitable for AI.

First, tests are generally conducted with environments whose size (questionnaire, duration of driving licence exams, etc.) is not adapted to the specifics of AI behaviour. Indeed, AI behaviour is largely non-convex and non-linear. It is not possible to evaluate its performance at a few points and deduce by interpolation and extrapolation its performance on the whole operating domain. Thus, testing environments are set up to maximise the exhaustiveness of the test scenarios covered (e.g. virtual testing). On the contrary, humans have much less chaotic behaviour. This is why a driving exam of less than 60 minutes, or a written test with about 20 questions, is sufficient to test a human's performance.

Second, they sometimes focus on tasks (e.g. IQ tests) that can be easily overfitted by AI. Conversely, the risk of human overfitting of tasks designed to evaluate AIs seems much lower.

Third, LNE has never evaluated some human capabilities presented in Table 15.7 in AI. Perhaps the task was not immediately relevant to the machine kingdom (e.g. it has no "self-control"). Or perhaps it was not evaluated as part of a specifically dedicated task, even it was a sub-component of a more complex task being evaluated (e.g. memory processes, quantitative and logical reasoning). As another possibility, no client may have ever asked LNE to assess this capability (e.g. "Emotion", "Mind modelling and social interaction").

This third finding is informative for the OECD study because it may indicate one of two things:

- AI is too immature to perform this task. Therefore, there is no system on the market that can perform it and useless to organise an evaluation campaign for it.
- Economic stakeholders have not yet deemed the assessment of this capacity as useful.

The latter does not necessarily mean the automation of this capability has no market value. Indeed, most often only the "critical" systems incorporating AI (which present a risk to goods and/or people) are assessed by trusted third parties such as the LNE, in line with European regulations.[12]

Finally, human tests are designed to assess abilities, some of which have a name that may be questionable for AI. A somewhat simplistic understanding of the "memory" capability in Table 15.7, for example, could suggest this task is not relevant for AI, since AI never forgets. On the contrary, if this task concerns the ability to store, recognise and re-use knowledge in general, then it seems a critical step not yet reached in AI development (Cremer and Whittlestone, 2020[29]).

Similarly, many tasks automated by AI, such as optimising movements on a farm plot to weed a maximum of weeds in a minimum of time call for "quantitative and logical reasoning" skills (Avrin et al., 2020[20]; Avrin et al., 2019[19]). However, it is not clear whether this task is more consistent with this capability than with "planning and sequential decision making and acting" or even "navigation".

Relevance of test methods specific to the intelligences being tested

With respect to the non-convexity of AI behaviour and the convexity of human behaviour, and given the risks of overfitting, evaluation tools must generally be defined according to the intelligence to be evaluated. Two elements generally define the testing tools (measuring instrument, test dataset, etc.) to be used in an evaluation. First, there are the expected functionalities (image recognition, scene understanding, etc.) of the evaluated intelligent system. Second, there are the technological solutions underpinning these functionalities, be they algorithms (CNN, SVM, etc.) or biological neural networks.

Another taxonomy relating to the type of technical solution (algorithms, biological neural architectures, etc.) used to achieve the functionality could therefore be established. This "mechanisms taxonomy" would be used to define the test protocol used (sampling and number of tests/questions, etc.) to evaluate the skills listed in the "capabilities taxonomy" and offered by the intelligent system under study.

This does not mean that some systematic correspondences between the "capabilities taxonomy" and the "mechanisms taxonomy" cannot be found. For example, recognition tasks are often automated by deep learning algorithms. In addition, comprehension tasks often rely on knowledge graphs and mission management tasks on reasoners.

This conclusion, moreover, is quite logical with regard to certain specificities of AI and human intelligence:

- Other elements than capabilities can influence human performance, such as traits, interests and values (De Fruyt, Wille and John, 2015[38]). The socio-emotional characteristics of human performance must be considered when designing the test. This is not the case for AI.
- AI can be duplicated and simulations run in parallel to test a large number of test scenarios; it is not possible to do the same for humans.

Relevance of task assessments

Although the assessment of AI and human intelligence capabilities are the focus of the study, task-based assessments may still be useful given the two points below:

- **There is no single combination of capabilities to perform a given task**. Each type of agent will try to rely on its best capabilities: AI systems will rely on their remembering and retrieving skills, their unbounded working memory, their speed of calculation, their perfect attention span; humans will rely on their unrivalled manipulation skills, common sense reasoning, frugal learning skills, etc.
- **The end-to-end learning approach of AI can render obsolete/impossible the evaluation of certain capabilities** (e.g. it is not possible to evaluate the performance of an end-to-end dialogue system in named entity recognition).

Relevant commonalities between all test methods

The test-based evaluation approach is common to both AI and human intelligence. It seems to be a crucial avenue to compare them. The Animal-AI testbed is, for example, dedicated to the evaluation of non-specific capabilities in both animals and AIs. How could standard test modules, such as ASTM E2919-14 for "Pose measurement", be designed for AI in many different applications in manufacturing, construction, medicine and aerospace, to evaluate human performance?

In addition, the test-based approach has other attributes that can inspire the expert judgement-based method of this study:

- The assessments should be **modular** (in agreement with the taxonomical approach of the OECD project), as already discussed above.
- The **impartiality** of evaluations should be ensured: an expert could underestimate or overestimate the capabilities of AI systems due to a conflict of interest.

Recommendations

This chapter capitalises on LNE's experience in evaluating AI systems to address two main questions:

- Which taxonomy should be used to compare AI and human intelligence capabilities?
- What evaluation tools and methods should be used to compare these capabilities?

It proposed a first taxonomy, simple but relevant to both biological and artificial intelligences. It then made recommendations regarding assessments to compare these intelligences. To make progress in answering the two questions above, and to pursue the impulse launched by the OECD in a particularly constructive, methodical, concerted and transparent spirit, the following actions would be useful:

- **Classify human and AI capabilities in terms of functions and mechanisms**

Intelligent systems (human or machine) perform very different functions (e.g. face recognition and bipedal walking, medical diagnosis and navigation of an unmanned aerial vehicle) using information processing mechanisms that rely on the same elementary principles. Conversely, within the same category of functions, different mechanisms can be used (rational or intuitive channels for humans, neural networks or expert systems for AI). For AI, grouping by evaluation metrics, types of automated tasks (classification, segmentation, etc.) and types of algorithms used (CNN, SVM, etc.) are examples of interesting avenues.

- **Organise evaluation tools around this double classification (function and mechanism)**

The general architecture and the hardware devices of the test benches to be set up (input/output channels, feedback, real time, etc.) are closely related to the mission of the system to be evaluated. Conversely, protocols to be followed (sampling and annotation of the operating domain, number of tests, etc.) will be determined mainly by the cognitive or computer mechanisms involved. In a maths competition, for example, a grading scale and a reader are mobilised; in a singing or figure skating competition, a jury is set up; in a sitting trial, both a professional legal judge and a popular jury are involved.

- **Formalise the influence of the non-convexity and intra-task variability of behaviour on the evaluation tools to be implemented**

AI generally has a non-convex behaviour with significant intra-task performance variability, while humans have a convex and stable behaviour. The behaviour convexity has a direct impact on the evaluation methods. It constitutes a gap between AI and human testing approaches that makes any assimilation difficult at this stage, in either direction. The evaluator of an intelligent machine has no choice but to go through the operating domain in all its corners. It must be tested at each of its operating points with a sampling step that is immediately related to the extremely unstable, non-linear character of its reactions.

The evaluator of a human being will be much less precise. The evaluator will be satisfied with probing the acquisition of a know-how by putting the person in "typical" situations that solicit the various components of the competence (e.g. the driving licence exam vs. the long test campaigns of the autonomous vehicle). The evaluator thus hypothesises that the person has regulation capabilities and mental resources more general and common to the ordinary human being that will make him/her able to face any intermediate situation.

The machine does not have them yet. This is probably because of its specialisation and its relative simplicity. However, it is also undoubtedly because of the technologies and processes used, which are not, or not sufficiently, superimposable on the natural cognitive mechanisms, composite and articulated, inherited from evolution.

These major differentials – the instability of intelligent systems – are of course to be nuanced precisely according to these technologies and applications. This is the main criterion on which to base improvements of the proposed taxonomy for comparison and cross-fertilisation between the two disciplines.

- **Deepen the discussion concerning the inter-task and intra-capability repercussions of the progress made in AI, to identify the root of AI capabilities and, by analogy, that of the human being**
- **Develop a broadly shared set of resources, methodologies and evaluation metrics that will enable these analyses to be conducted and AI/human progress to be tracked**

The strengths and weaknesses of human intelligence compared to AI by a technical and comparative rapprochement in terms of taxonomic and methodological unity of appreciation should be identified as soon as possible. This should accompany the progress in AI and cognitive sciences and, in particular, pilot what contributes to identify their "greatest common divisors".

AI seems to be the source of changes that are extremely favourable to the destiny of humanity, such as a radical emancipation from work. Therefore, this evolution should be supported by seeking to control the risks rather than pushing it back or slowing it down. Otherwise, humans will end up enduring AI without having prepared for it sufficiently.

References

Ajili, M. et al. (2016), "FABIOLE, a speech database for forensic speaker comparison", in *Proceedings of the 10th International Conference on Language Resources and Evaluation, LREC 16*, European Language Resources Association, Luxembourg, https://www.aclweb.org/anthology/L16-1115.pdf. [10]

Albus, J. (2002), "4D/RCS: a reference model architecture for intelligent unmanned ground vehicles.", in *Proceedings Volume 4715, Unmanned Ground Vehicle Technology IV*, Aerosense 2002, Orlando, FL, https://doi.org/10.1117/12.474462. [3]

Amigoni, F. et al. (2015), "Competitions for benchmarking: Task and functionality scoring complete performance assessment", *IEEE Robotics & Automation Magazine*, Vol. 22/3, pp. 53-61, https://doi.org/10.1109/MRA.2015.2448871. [1]

Avrin, G., V. Barbosa and A. Delaborde (2020), "AI evaluation campaigns during robotics competitions: The METRICS paradigm", presented at the First International Workshop on Evaluating Progress in Artificial Intelligence - EPAI 2020, Santiago de Compostela, Spain, https://dmip.webs.upv.es/EPAI2020/papers/EPAI_2020_paper_5.pdf. [2]

Avrin, G. et al. (2020), "Design and validation of testing facilities for weeding robots as part of ROSE Challenge", presented at the First International Workshop Evaluating Progress in Artificial Intelligence of the European Conference on Artificial Intelligence, Santiago de Compostela, Spain, https://dmip.webs.upv.es/EPAI2020/papers/EPAI_2020_paper_6.pdf. [20]

Avrin, G. et al. (2019), "Boosting agricultural scientific research and innovation through challenges: The ROSE Challenge example", 3rd RDV Techniques AXEMA, SIMA, https://www.seaperch.org/challenge. [19]

Ben Jannet, M. et al. (2014), "ETER: A new metric for the evaluation of hierarchical named entity recognition", in *Proceedings of the Ninth International Conference on Language Resources and Evaluation, LREC 2014*, European Language Resources Association, Reykjavik, https://www.aclweb.org/anthology/volumes/L14-1/. [14]

Ben Jannet, M. et al. (2015), "How to evaluate ASR output for named entity recognition?", in *Proceedings of the Annual Conference of the International Speech Communication Association, INTERSPEECH, 2015-January*, International Speech Communication Association, Baixas, France. [11]

Bernard, G. et al. (2010), "A question-answer distance measure to investigate QA system progress", in *Proceedings of the Seventh International Conference on Language Resources and Evaluation, LREC 2010*, European Language Resources Association, Malta, https://www.aclweb.org/anthology/volumes/L10-1/. [15]

Brunessaux, S. et al. (2014), "The Maurdor Project: Improving automatic processing of digital documents", *11th IAPR International Workshop on Document Analysis Systems*, pp. 394-354, http://dx.doi.org/10.1109/DAS.2014.58. [16]

Cremer, C. and J. Whittlestone (2020), "Canaries in technology mines: Warning signs of transformative progress in AI", *International Journal of Interactive Multimedia and Artificial Intelligence*, Vol. 6, pp. 100-109, http://doi.org/10.9781/ijimai.2021.02.011. [29]

De Fruyt, F., B. Wille and O. John (2015), "Employability in the 21st century: Complex (interactive) problem solving and other essential skills", *Industrial and Organizational Psychology-Perspectives on Science and Practice*, Vol. 8/2, pp. 276-U189, http://dx.doi.org/10.1017/iop.2015.33. [38]

Devlin, J. et al. (2019), "BERT: Pre-training of deep bidirectional transformers for language understanding", presented at NAACL HLT 2019 – 2019 Conference of the North American Chapter of the Association for Computational Linguistics. [28]

Elloumi, Z. et al. (2018), "Analyzing learned representations of a deep ASR performance prediction model", in *Proceedings of the 2018 EMNLP Workshop Blackbox NLP: Analyzing and Interpreting Neural Networks for NLP*, Association for Computational Linguistics, Brussels, https://www.aclweb.org/anthology/W18-5402/. [18]

Galibert, O. et al. (2014), "The ETAPE speech processing evaluation", in *Proceedings of the Ninth International Conference on Language Resources and Evaluation, LREC 2014*, European Language Resources Association, Reykjavik, https://www.aclweb.org/anthology/volumes/L14-1/. [13]

Hernández-Orallo, J. (2017), "Evaluation in artificial intelligence: From task-oriented to ability-oriented measurement", *Artificial Intelligence Review*, Vol. 48/3, pp. 398-447. [34]

Hernández-Orallo, J. (2017), *The Measure of All Minds: Evaluating Natural and Artificial Intelligence*, Cambridge University Press, New York. [36]

Howard, J. and S. Ruder (2018), "Universal language model fine-tuning for text classification", *arXiv*, pp. 328-339, http://nlp.fast.ai/ulmfit. [26]

Huang, H. et al. (2007), "Characterizing unmanned system autonomy: Contextual autonomous capability and level of autonomy analyses", *Proceedings Volume 6561, Unmanned Systems Technology IX*, https://doi.org/10.1117/12.719894. [33]

Huckaby, J. and H. Christensen (2012), "A taxonomic framework for task modeling and knowledge transfer in manufacturing robotics", *AAAI Workshop – Technical Report*, No. WS-12-06, Association for the Advancement of Artificial Intelligence, Paolo Alta, CA, http://dx.doi.org/www.aaai.org. [23]

Hughes, C. and T. Hughes (2019), "What metrics should we use to measure commercial AI?", *AI Matters*, Vol. 5/2, pp. 41-45, https://doi.org/10.1145/3340470.3340479. [4]

Jiménez, J. et al. (2021), "Methodological aspects for cognitive architectures construction: A study and proposal", *Artificial Intelligence Review*, Vol. 54/32133-2192, https://doi.org/10.1007/s10462-020-09901. [9]

Kotseruba, I. and J. Tsotsos (2016), "A review of 40 years of cognitive architecture research: Core cognitive abilities and practical applications", *arXiv*, Vol. 08602, http://dx.doi.org/arXiv:1610.08602. [5]

Langley, P. (2006), "Cognitive architectures and general intelligent systems", *AI Magazine*, Vol. 27/2, p. 33, https://doi.org/10.1609/aimag.v27i2.1878. [39]

Langley, P., J. Laird and S. Rogers (2009), "Cognitive architectures: Research issues and challenges", *Cognitive Systems Research*, Vol. 10/2, pp. 141-160, https://doi.org/10.1016/j.cogsys.2006.07.004. [8]

Leuski, A. and D. Traum (2008), "A statistical approach for text processing in virtual humans", https://www.researchgate.net/publication/228597921_A_statistical_approach_for_text_processing_in_virtual_humans. [7]

Mathet, Y. et al. (2012), "Manual corpus annotation: Giving meaning to the evaluation metrics", in *Proceedings of COLING 2012: Posters*, The COLING 2012 Organizing Committee, Mumbai, https://www.aclweb.org/anthology/C12-2079. [37]

Moon, S., S. Kim and H. Wang (2014), "Multimodal transfer deep learning with applications in audio-visual recognition", *arXiv*, Vol. 3121, http://arxiv.org/abs/1412.3121. [24]

Neubert, J. et al. (2015), "The assessment of 21st century skills in industrial and organizational psychology: Complex and collaborative problem solving", *Industrial and Organizational Psychology*, Vol. 8/2, pp. 238-268, https://doi.org/10.1017/iop.2015.14. [35]

Oparin, I., J. Kahn and O. Galibert (2014), "First Maurdor 2013 evaluation campaign in scanned document image processing", *2014 IEEE International Conference on Acoustics, Speech and Signal Processing*, pp. 5090-5094, https://doi.org/10.1109/ICASSP.201. [17]

Peters, M. et al. (2018), "Deep contextualized word representation", *arXiv*, Vol. 04365, http://allennlp.org/elmo. [27]

Primi, R. et al. (2016), "Mapping questionnaires: What do they measure?", *Estudos de Psicologia (Campinas)*, Vol. 36/e180138., https://doi.org/10.1590/1982-0275201936e180138. [30]

Prokopalo, Y. et al. (2020), "Evaluation of lifelong learning systems", in *Proceedings of the 12th Language Resources and Evaluation Conference*, European Language Resources Association, Luxembourg, https://hal.archives-ouvertes.fr/hal-02496971. [12]

Razavian, A. et al. (2014), "CNN features off-the-shelf: An astounding baseline for recognition", *arXiv*, Vol. 6382, http://dx.doi.org/arXiv:1403.6382. [25]

Shibata, K. (2017), "Functions that emerge through end-to-end reinforcement learning –The direction for artificial general intelligence", *arXiv*, Vol. 0239, http://dx.doi.org/arXiv:1703.02239. [22]

Shneier, M. et al. (2015), "Measuring and representing the performance of manufacturing assembly robots", *NIST Interagency/Internal Report (NISTIR)*, No. 8090, National Institute of Standards and Technology, Gaithersburg, MD, https://doi.org/10.6028/NIST.IR.8090. [31]

Stasse, O. et al. (2018), "Benchmarking the HRP-2 humanoid robot during locomotion", *Frontiers Robotics AI, 5(NOV)* 8 November, https://doi.org/10.3389/frobt.2018.00122. [21]

Tolan, S. et al. (2020), "Measuring the occupational impact of AI: Tasks, cognitive abilities and AI benchmarks", *Journal of Artificial Intelligence Research*, Vol. 71, https://doi.org/10.1613/jair.1.12647. [32]

Ye, P., T. Wang and F. Wang (2018), "A survey of cognitive architectures in the past 20 years", *IEEE Transactions on Cybernetics*, Vol. 48/12, pp. 3280-3290, https://doi.org/10.1109/TCYB.2018.2857704. [6]

Notes

1 http://rockinrobotchallenge.eu/

2 www.quaero.org/

3 www.eu-robotics.net/eurobotics/about/projects/rockeu2.html

4 https://sciroc.org/

5 http://challenge-rose.fr/

6 https://metricsproject.eu/

7 https://bicasociety.org/cogarch/

8 https://web.archive.org/web/20100315140823/http://ai.eecs.umich.edu/cogarch0/common/capa.html

9 The usefulness of having cognitive architectures to produce general artificial intelligence is presented in Langley (2006[39]).

10 www.onetonline.org/

11 Learning data is also often subject to human annotation, which can be related to the concept of *Fauxtomation*.

12 https://ec.europa.eu/growth/single-market/goods/building-blocks/conformity-assessment_en

16. Assessing Natural Language Processing

Yvette Graham, Trinity College Dublin

This chapter details evaluation techniques in Natural Language Processing, a challenging sub-discipline of artificial intelligence (AI). It highlights proven methods to provide both fair and replicable results for evaluation of system performance, as well as methods of longitudinal evaluation and comparison with human performance. It recaps pitfalls to avoid in applying techniques to new areas. In addition to direct measurement and comparison of system and human performance for individual tasks, the chapter reflects on the degree of shared human-machine task, scalability and potential for malicious application. Finally, it discusses the applicability of human intelligence tests to AI systems and summarises considerations for devising a general framework for assessing AI and robotics.

Introduction

Artificial intelligence (AI) and robotics aim to automate tasks that would otherwise require human intelligence and/or physical ability to complete. The Future Skills Expert Meeting aimed to devise and compile skills and tests suitable for assessing AI and robotics. The best place to look for appropriate tests is within each sub-discipline itself. Within each sub-discipline, tests for experimental procedures provide evidence that a proposed method or approach for improving system performance works or is worthwhile.

This chapter looks at how Natural Language Processing, for example, has developed such tests. Analysis of emerging methods shows that valid and reliable measurements have already been developed for some areas. This include methods of comparison with human performance, as well as longitudinal evaluation of systems. It also includes several examples where evaluation has resulted in inaccurate conclusions to demonstrate the challenge of getting it right.

Taking inspiration from successful methods of Natural Language Processing evaluation, the chapter identifies considerations for constructing a general-purpose framework for measuring system performance in AI and robotics. Such a framework would allow measurement of improvements to systems over time and comparison of system and human performance.

Furthermore, it identifies the three most important factors beyond individual system performance for predicting or measuring the impact of a given AI technology on society and the future workforce. First, a framework should quantify the human effort saved by an AI system by automating an individual task. This would consider if the approach were unsupervised or supervised (and any human effort involved in labelling data). Second, it should quantify the human effort saved by a given AI system when applied at scale. This would consider the feasibility of substantially growing the deployment of a given technology. Third, it should quantify the human effort in monitoring and retraining the technology. In addition, it should identify the potential impact of the technology on society: do most people think it is safe and desirable?

In all of the above, the framework should provide the context in which the results of a given test or measurement apply. The degree to which results apply outside the testing context cannot be assumed. AI technologies, when applied to a new domain, will inevitably require varying retraining and produce distinct results.

Finally, the chapter considers the importance of the AI system's ability to operate independent of humans. Rarely will an AI system fully replace human workers. Rather, AI technologies are much more likely to substantially enhance the way in which humans complete tasks. In some cases, they will vastly increase both the efficiency and scale by which task completion will be possible.

To that end, the chapter outlines a potential method of measuring this dimension of AI system performance, as opposed to pitting human against machine in tests. This attempts to measure the amount of human effort saved in a hybrid human-machine setting.

Existing methods

The vast majority of AI research aims to automate the completion of individual tasks and tests. Methods are subsequently designed to evaluate the system with respect to specific individual tasks. In Natural Language Processing, for example, the research area is generally divided into the following sub-areas with individual tasks evaluated with appropriate methodology for that task:[1]

- Dialogue and interactive systems: development of systems capable of engaging with humans through natural language.

- Discourse and pragmatics: the study of language in its context of use, with pragmatics focusing on the effects of context on meaning, and discourse analysis focusing on written and spoken language and social context.
- Natural language generation: automated creation of natural language text or speech from natural language and/or abstract representations.
- Information extraction: automated retrieval of information from text or speech.
- Information retrieval: obtaining information or resources relevant to user information need.
- Text mining: automatic derivation of high-quality information from text or speech data.
- Language grounding to vision and robotics: techniques for linking language to objects, actions, sounds, images and so on.
- Cognitive modelling: computer science that deals with simulating human problem solving and mental processing in a computerised model.
- Psycholinguistics: the study of the mental aspects of language and speech.
- Machine translation: automated translation of text or speech from one natural language to another.
- Phonology: the study of the patterns of sounds in a language and across languages.
- Morphology and word segmentation: the study of words, how words are formed and the relationship of words to other words in a single language.
- Question answering: automatic retrieval or generation of answers to questions posed in natural language text or speech.
- Semantics: Lexical, Sentence level and Textual Inference: study of the meaning of natural language text or speech and what can be logically inferred from it.
- Sentiment analysis, stylistic analysis and argument mining: automatic classification of the sentiment, style or argument encoded in natural language text or speech.
- Speech and multimodality: automatic processing of spoken and multimodal (image, video, etc.) data.
- Summarisation: automatic extraction of information from natural language text or speech rendered as more concise text or speech.
- Syntax: analysis of the underlying grammatical structure of natural language text or speech (tagging, chunking and parsing).

Machine translation leads the way

In Natural Language Processing, machine translation has led the way in terms of rigorous evaluation methodology. Its evaluation methods that aim to provide a realistic reflection of system performance were first established and later adapted to other Natural Language Processing areas.[2] In machine translation, benchmark-shared tasks in which multiple research teams compete with each other are evaluated. A win at this task identifies the team as a world leader in this research area (Barrault et al., 2020[1]).

Teams are provided with a set of test documents, freshly sourced on line and unseen by participants in the task to translate automatically with systems. In blind tests, large numbers of human assessors provide quality judgements of translations produced by each system. Finally, human evaluation quality ratings are combined into a meaningful statistic/overall score for a given system. The best performing system(s) is identified, considering statistical significance when small differences occur between results for systems.[3]

Figure 16.1. A German test sentence translated by machines and humans

		Rating
Source :	*Im Ziel warf er sein Paddel vor Freude weg und reckte beide Arme siegessicher in die Hohe - wohlwissend, " dass es mindestens fur eine Medaille reichen würde.*	
Machine :	At the finish, he threw away his paddle for joy and raised both arms in victory - knowing that it would be enough for at least one medal.	23.4
Human :	He threw his paddle with joy at the finishing line and, confident of victory, threw both arms in the air - safe in the knowledge that his efforts would secure him a medal.	67.5

Source: Graham et al. (2020[2]).

Tests to evaluate machine translation systems are designed to provide a realistic and fair ranking of systems.

First, to help provide realistic results, test data used in translation are freshly sourced and unseen by systems. This is akin to a realistic use case where the input text to be translated will certainly neither be known nor predictable to systems. Figure 16.1 provides example test data freshly sourced from an online news article and used in a past machine translation competition.

Second, human assessment is employed as opposed to an automatic metric of some description. Automatic metrics, even popular ones such as BLEU (Papineni et al., 2002[3]), are known to disagree with human assessment of translations to varying degrees and in different ways. Therefore, human assessment of translation quality has been established within machine translation as the most valid form of ground truth in tests (Callison-Burch, Osborne and Koehn, 2006[4]).

Third, and also of high importance for valid measurement, is the employment of suitable statistics that can accurately reflect the performance of systems. For example, a meaningful intuitive statistics for which established methods of statistical significance testing exist are superior to ad hoc measures. Examples include the mean or median for central tendency or Pearson correlation for association.

As a final consideration, since human evaluation of systems at scale takes substantial time and resources, many tests employ crowdsourced human assessments of system performance. This makes the validity of measurements highly dependent on strict quality checks. These test the reliability of human assessors for whom little to no verifiable information is known.

Direct assessment vs. crowdsourcing

The measure of choice for machine translation was coined by Conference on Machine Translation (Barrault et al., 2020[1]) as "direct assessment" (Graham and Liu, 2016[5]). It employs human assessors at scale with strict quality checks. Ratings of translation quality are collected on a 0-100 rating scale.

These ratings result in score distributions for systems for which accurate statistical tests can be applied. Such tests can avoid conclusions of differences in average ratings that are likely to occur simply by chance. Replication of experiments showed an almost perfect correlation with past results (Graham, Awad and Smeaton, 2018[6]).

Direct assessment has been used in machine translation for longitudinal evaluations that showed an average 10% improvement in system performance for machine translation of European languages over five years (Graham et al., 2014[7]). Furthermore, direct assessment made possible for the first time accurate comparison of human and machine translation system performance at the Conference on Machine Translation's news translation task (Barrault et al., 2019[8]).

Results of tests showed that machine translation systems can outperform a human translator when sufficient training data are available. More recent results provide evidence that even professional human translators can vary substantially in performance in tests. Consequently, a win over an individual human translator in a competition does not imply that a given system outperforms human translation or human translators in general (Barrault et al., 2020[1]).

Furthermore, direct assessment has been applied to additional AI tasks, such as video captioning (Awad et al., 2019[9]) and multilingual surface realisation (Mille et al., 2019[10]). Both of these Natural Language Generation research areas previously suffered from lack of reliable evaluation methodologies and low agreement in human assessment.

Avoiding evaluation pitfalls

Valid and reliable measures of system performance have not always been available within Natural Language Processing. This section provides examples of inaccurate or even misleading evaluation measures that were a necessary part of the process of producing more reliable measures. Several pitfalls, noted below, can be avoided when adapting evaluation techniques to new areas of AI:

- inappropriate application of statistical significance testing[4]
- application of statistics that allow machine learning algorithms to game the measure employed to evaluate systems[5]
- reporting results on selective subsets of data that show system/metric in most favourable light[6]
- lack of rigour in test settings that allow unfair advantage of systems due to unrealistic test data.[7]

Human-system hybrids and performance tests

In many AI applications, technologies will aid humans rather than replace them. As a result, pitting the performance of each against the other, only in isolation and working as entirely independent agents, could oversimplify reality. It would misjudge how technologies will be used in practice. Therefore, in addition to pitting human against machine in blind tests, performance of human/machine hybrids should ideally be measured where one or more human workers is *aided by* as opposed to being *replaced by* a given AI technology.

Relative participation of humans and machines

A hybrid test setting is more complex since it introduces the additional dimension of the relative participation of human and machine within completion of a given task. Participation can potentially range anywhere from *almost entirely automated* (with minimal human input) to *almost entirely manual* (with minimal AI).

Such a dimension brings the evaluation into a more realistic and therefore better setting where results can have a stronger impact. This added complication does raise questions. How should the degree of hybridisation be measured? Where would emerging AI and robotics technologies be placed along such a scale?

Creating a realistic, valid and reliable scale

A measure should make the resulting scale realistic, valid and reliable. The scale should provide a real-world reflection of how AI technologies rank against each other in terms of human participation. Valid measurements would accurately reflect the degree of hybridisation. The scale would also be highly reliable so that subsequent measurements would produce the same conclusions.

In terms of hybridisation, one method would be to measure the amount of human effort saved by a given AI technology. This could be estimated, for example, by giving sets of human workers tasks to complete with and without the aid of the AI technology. Tests could include measurements of, for example, average times saved for completion of a single task with the AI system. This could result in a time-saving scale for single-task completion for AI technologies (from no or little savings to vast savings).

Measuring the social value of AI technology

Measurement must also consider the scale at which society needs such tasks and the level of scaling of tasks possible. A task completed by AI could produce vast time savings but society or the workforce may not value this extra time. Indeed, a task that saves less time but is needed by many people or businesses or people worldwide will have more impact. Similarly, a technology with vast time-saving potential but is not scalable will also not have a high impact on employment or the workplace; measurements should try to reflect this.

An additional dimension is the potential social benefit of the AI technology. For example, a robot that can safely detect and deactivate bombs, aid a human perform life-saving surgery or prevent manipulation of the democratic process would rank highly on a scale of societal importance.

Measuring potential risks

Finally, an additional scale should ideally measure the risks of malicious applications of a given AI technology. Emerging Natural Language Processing technologies, for example, can generate fake news articles thought to be indistinguishable from human-authored news articles (Zellers et al., 2019[11]).

Such technologies might be widely deployable, scalable and operate almost independently of human input. However, identification of their malicious potential will increase the likelihood of developing preventative measures within society. This could aid against abuses such as manipulation of democratic processes. Numerous other examples of malicious AI exist such as deep fakes, drones and use of AI in the military.

In summary, tests can be devised to measure the human effort saved by assisting human workers with a given AI technology. There are four main considerations:

- time saved by hybridisation corresponding to a straightforward measurement of reduction in task completion time for a given task with and without the aid of the specific AI technology
- scalability, the need within society or the workforce for deployment of the technology at scale and whether scaling is possible
- importance of the task to individuals within society or society as a whole
- potential malicious impact of the AI technology and if this will require significant resources to deter or reduce risks on society or individuals within society.

Recommendations

As several of the earlier chapters have noted, computer scientists will argue that intelligence tests designed for humans are not suitable for measuring the performance of systems. This is primarily because human intelligence tests make assumptions about the basic abilities and skills of the test candidate; these are generally true of humans but not of AI. The main shortcomings with respect to natural language understanding are:

- Human intelligence tests require basic human abilities that AI systems do not have. For example, understanding questions in natural language and relating the meaning to real-world objects, and understanding how real-world objects might fit together, be manipulated and so on.

- Most areas of AI research do not attempt to obtain a general understanding of natural language. For example, they do not attempt to make the system complete tasks as a human would. Rather, they select a specific human intelligence task and work towards producing a system that can complete that single task.
- The vast majority of successful AI systems are evaluated in terms of completion of such tasks in isolation of other tasks. Humans are able to integrate separate basic skills, such as natural language understanding and object recognition. However, little AI research attempts to integrate multiple distinct task completion skills. For this reason, as well as other basic assumptions of human intelligence test results, these skills would be inappropriate for testing AI systems.
- An AI system trained to perform well on a specific human intelligence test is not a proof of intelligence in the way it would be of a human candidate. Even a minor change to the test format would likely lead to system failure due to its lack of general natural language understanding. Additionally, an AI system might perform well even on multiple human intelligence tests. However, without a function beyond these tests, it would be a system with few practical applications.

Highly sophisticated AI systems may one day possess basic human abilities such as general natural language understanding, general knowledge and understanding of the real world. They may learn to integrate such skills when faced with a new task. In this scenario, human intelligence tests would then be relevant.

However, measurement of system performance would need to mitigate against "gaming" the test. Just as humans can memorise answers to questions, a system can simply tune to the tests instead of demonstrating skill integration and general natural language understanding.

Although human intelligence tests are not appropriate for testing AI systems, other tests for humans could apply. Possibilities include vocational and educational tests, or indeed neuropsychological and developmental psychological tests that focus on the low-level skills possessed by most humans but not AI.

However, such tests are also likely to be misleading when directly applied to AI systems. Similar to human intelligence tests, such tests were devised with human candidates in mind. As a result, testing procedures include many assumptions about intelligence based on human intelligence alone. These assumptions, such as memory limitations and skills transfer, simply do not apply to AI.

Whether human or AI-specific tests are adapted or new tests are developed to evaluate AI systems, these should consider the following guidelines:

- **Avoid direct adoption of tests designed to test humans**

Assumptions about human candidates do not hold for AI systems.

- **Examine testing scenarios employed in each area**

These scenarios should include evaluation procedures in research papers, particularly for those employed in benchmark tasks.

- **Use human ratings of performance for evaluation**

This approach must ensure human assessors are blind to whether a system or other human is performing the task as opposed to metrics.

- **Employ realistic, unseen test data**

Freshly sourced data will help AI systems avoid tuning to the test data.

- **Include multiple humans in tests**

Humans should receive the same input data and/or environment as systems to represent human variance in performance, as well as to compare systems realistically.

- **Measure reliability of test results**

Reliability can be measured by repeating experiments with different human assessors or by measuring agreement of human assessors using inter-annotator and intra-annotator agreement measures, such as Kappa coefficient.

- **Repeat tests at regular intervals with new data and track improvements over time**
- **Report meaningful statistics and account for variance and statistical significance**
- **Measure performance of aid for a human rather than replacement**

Include a hybridisation scale that considers the performance of a given AI technology when aiding as opposed to entirely replacing a human.

- **Quantify the human effort involved to create key training data**

Include quantification of the human effort involved in creating training data for supervised AI technologies, monitoring and retraining systems.

- **Include scalability**

Technologies that can be deployed at scale are likely to have a higher impact on society.

- **Include both the potential and risk for society**

Include both the potential for positive impact and negative impact (e.g. malicious application of the AI technology).

References

Awad, G. et al. (2019), "Trecvid 2019: An evaluation campaign to benchmark video activity detection, video captioning and matching, and video search & retrieval", *arXiv*, Vol. 2009.09984v1, http://dx.doi.org/arXiv:2009.09984v1. [9]

Barrault, L. et al. (2020), "Findings of the 2020 conference on machine translation", in *Proceedings of the Fifth Conference on Machine Translation*, Association for Computational Linguistics (online), https://aclanthology.org/2020.wmt-1.1. [1]

Barrault, L. et al. (2019), "Findings of the 2019 conference on machine translation", in *Proceedings of the Fourth Conference on Machine Translation*, Association for Computational Linguistics, Florence, http://dx.doi.org/10.18653/v1/W19-5301. [8]

Callison-Burch, C., M. Osborne and P. Koehn (2006), "Re-evaluatiing the role of Bleu in machine translation research", *Proceedings of the 11th Conference of the European Chapter*, Association for Computational Linguistics, Trento, https://aclanthology.org/E06-1032. [4]

Graham, Y. (2015), "Improving evaluation of machine translation quality estimation", *Proceedings of the 53rd Annual Meeting of the Association for Computational Linguistics and the 7th International Joint Conference on Natural Language Processing*, Association of Computational Linguistics, Beijing, http://dx.doi.org/10.3115/v1/P15-1174. [15]

Graham, Y. (2015), "Re-evaluating automatic summarization with bleu and 192 shades of rouge", *Proceedings of the 2015 Conference on Empirical Methods in Natural Language Processing*, Association for Computational Linguistics, Lisbon, http://dx.doi.org/10.18653/v1/D15-1013. [16]

Graham, Y., G. Awad and A. Smeaton (2018), "Evaluation of automatic video captioning using direct assessment", *PLOS ONE*, Vol. 13/9, pp. 1-20, https://doi.org/10.1371/journal.pone.0202789. [6]

Graham, Y. et al. (2014), “Is machine translation getting better over time?”, *Proceedings of the 14th Conference of the European Chapter*, Association for Computational Linguistics, Gothenburg, http://dx.doi.org/10.3115/v1/E14-1047. [7]

Graham, Y. et al. (2020), “Assessing Human-Parity in Machine Translation on the Segment Level”, *Findings of the Association for Computational Linguistics: EMNLP 2020*, http://dx.doi.org/10.18653/v1/2020.findings-emnlp.375. [2]

Graham, Y., B. Haddow and P. Koehn (2020), “Statistical power and translationese in machine translation evaluation”, *Proceedings of the 2020 Conference on Empirical Methods in Natural Language Processing*, Association of Computational Linguistics (online), http://dx.doi.org/10.18653/v1/2020.emnlp-main.6. [17]

Graham, Y. and Q. Liu (2016), “Achieving accurate conclusions in evaluation of automatic machine translation metrics”, in *Proceedings of the 15th Annual Conference of the North American Chapter*, Association for Computational Linguistics: Human Language Technologies, San Diego, http://dx.doi.org/10.18653/v1/N16-1001. [5]

Graham, Y., N. Mathu and T. Baldwin (2014), “Randomized significance tests in machine translation”, *Proceedings of the ACL 2014 Ninth Workshop on Statistical Machine Translation*, Association for Computational Linguistics, Baltimore, http://dx.doi.org/10.3115/v1/W14-3333. [13]

Koehn, P. and C. Monz (2006), “Manual and automatic evaluation of machine translation between European languages”, in *Proceedings on the Workshop on Statistical Machine Translation*, Association for Computational Linguistics, New York, https://aclanthology.org/W06-3114. [12]

Ma, Q. et al. (2017), “Further investigation into reference bias in monolingual evaluation of machine translation”, in *Proceedings of the 2017 Conference on Empirical Methods in Natural Language Processing*, Association of Computational Linguistics, Copgenhagen, http://dx.doi.org/10.18653/v1/D17-1262. [14]

Mille, S. et al. (2019), “The second multilingual surface realisation shared task (SR’19): Overview and evaluation results”, in *Proceedings of the 2nd Workshop on Multilingual Surface Realisation (MSR 2019)*, Association of Computational Linguistics, Hong Kong, http://dx.doi.org/10.18653/v1/D19-6301. [10]

Papineni, K. et al. (2002), “BLEU: A method for automatic evaluation of machine translation”, in *Proceedings of the 40th Annual Meeting*, Association for Computational Linguistics, Philadelphia, https://doi.org/10.3115/1073083.1073135. [3]

Zellers, R. et al. (2019), “Defending against neural fake news”, in Wallach, H. et al. (eds.), *Advances in Neural Information Processing System*, Curran Associates, Inc., New York. [11]

Notes

[1] List adapted from http://https://2020.emnlp.org and intended only as a high-level summary of topics of interest in NLP as opposed to an exhaustive list. There is often overlap between different areas but no hierarchical relationship between tasks.

[2] A main venue for development of rigorous evaluation techniques was the Conference on Machine Translation (WMT) (Koehn and Monz, 2006[12]).

[3] It is not unusual for more than one system to be tied for first position in the competition.

[4] Competing statistical significance tests applied in machine translation evaluation were widely believed to provide substantially distinct conclusions. These were based on results of an oft-cited publication that claimed approximate randomisation to be superior to bootstrap resampling. Analysis in repeat experiments showed apparent differences in test results were simply due to an unwittingly bad comparison of one-sided and two-sided test results in experiments (Graham, Mathu and Baldwin, 2014[13]). Other analysis claimed that direct assessment introduced substantial bias in evaluation results. Yet this method had been employed since 2017 at the Conference for Machine Translation to produce official results. On further inspection and analysis of experiment data, claims of bias were revealed as unfounded due to the application of inappropriate analysis techniques (Ma et al., 2017[14]).

[5] An active area of Natural Language Processing is quality estimation. This involves application of machine learning algorithms to predict the quality of system-generated language, commonly applied to machine translation output. Within this area in benchmark tasks, systems were tested using measures such as mean absolute error. These yielded results in favour of systems that produced overly conservative quality estimates. Analysis of results showed an unfair advantage for systems that accurately predicted the mode of the test set score distribution and produced a conservative quality estimate located close to this mode. Subsequently, a more suitable measure was recommended that avoided the bias of previous results using a unit free measure, the Pearson correlation (Graham, 2015[15]). In evaluation of machine translation metrics, inappropriate statistical significance tests were applied to evaluation of metrics. This resulted in high proportions of over-estimates of statistically significant differences in performance, prior to identification and correction of this problem (Graham and Liu, 2016[5]).

[6] Inappropriate measures of Natural Language Processing were also widely applied in automatic summarisation. In this sub-discipline, an automatic metric was produced and widely adopted based on an apparent higher correlation with human assessment than BLEU. However, closer inspection of experiment data showed results were only presented for metric scores calculated on a subset of the relevant data. Consequently, they did not hold true for the full set of human annotations (Graham, 2015[16]).

[7] In machine translation research, an early Chinese to English machine translation system achieved performance on-par with a human translator. These results were identified as inaccurate for several reasons. First, there was a lack of context provided to human assessors in the evaluation. Second, it included reverse-created test data. Third, there was a lack of statistical power analysis to ensure sufficient sample size when concluding ties. Correction of such experiments is described in detail in Graham, Haddow and Koehn (2020[17]). They show that, contrary to initial claims, the human translator did not outperform the system when evaluated in the most appropriate way.

17. Common sense skills: Artificial intelligence and the workplace

Lucy Cheke, Leverhulme Centre for the Future of Intelligence, Cambridge

Marta Halina, Leverhulme Centre for the Future of Intelligence, Cambridge

Matthew Crosby, Leverhulme Centre for the Future of Intelligence, Cambridge

This chapter looks at the basic "common sense" skills needed by artificial intelligence (AI) to perform in the workplace. It begins by focusing on the challenge for AI of navigating in complex and unpredictable environments. It explores the common sense skills underpinning these apparently sophisticated skills, including spatial memory, object representation and causal reasoning. The chapter continues with an exploration of social challenges of the workplace, including the need to predict or infer meaning from ambiguous behaviour and the "common sense" skills underpinning these. Finally, it looks at the challenges, opportunities and next steps related to testing AI with tasks from psychology.

Introduction: Workplace perspectives on artificial intelligence

Artificial intelligence (AI) systems can be trained to exceed human performance on certain tasks such as playing games, recognising images, controlling large cooling systems or painting cars. However, these systems cannot act outside of the task for which they have been trained. Moreover, they often fail under even minor deviations from their expected inputs (Shevlin et al., 2019[1]; Shanahan et al., 2020[2]).

From a workplace perspective, this means that AI and robotics can only be used for highly specific, well-defined tasks or where the environment can be strictly controlled. If AIs are to move beyond these narrow tasks, a number of skills are required.

In contrast, humans and other animals are generalists, able to perform a wide variety of both complex and seemingly simple tasks. Humans have a great capacity for specialisation. However, they are also highly capable of retraining and adapting to novel situations without losing previous knowledge. At the chess world championship, the grand masters are expected to be able to make a coffee, find the bathroom and operate the lights in their hotel rooms, all while remaining chess experts.

The maturation of these common sense abilities often marks the biggest developmental leaps in children: a step-change in behavioural flexibility and independence occurs, for example, at around 12 months, with the ability to search for an object (be that food, toy or parent) that has moved out of sight (Piaget, 1977[3]).

Non-human animals are also adept at common sense reasoning (Tomasello and Call, 1997[4]). A cat that sees a mouse enter a tunnel can flexibly deploy a range of behaviours – waiting at the tunnel mouth, running to the other end of the tunnel, digging or reaching in a paw. These are all based on the understanding that while the mouse can no longer be seen, it still exists and is potentially accessible. Both of these examples involve the cognitive capacity known as "object permanence": an assumed skill in human adults, an area of study in child and animal psychology, and yet to be achieved in AI.

The need for common sense skills

AI and robotics can, of course, develop far beyond current capabilities without developing common sense reasoning. However, the presence or absence of common sense will fundamentally change the types of positions AI can take in the workplace of the future. Without common sense, AI will arguably remain a highly specialised tool limited in application (Shanahan et al., 2020[2]). It will either require specifically designed workspaces, potentially costly human supervision or be bounded by inflexibility and overfitting.

Without common sense, attempts to broaden AI usage will result in applications for which it is fundamentally ill-suited, leading to potentially disastrous outcomes. In the United Kingdom, recent problems surrounding the algorithm used to assign A-level results to students illustrate the issue. Overfitting and other problems led to bias in the predicted grades assigned to students by the algorithm. Kelly (2021[5]) calls this probably the greatest policy failure in modern times for the examination system in England.

If AI can develop some common sense skills, it might be able to expand into a new category of roles within the workplace.

In some cases, these common sense skills will be needed for an artificial agent to perform any kind of role. This is likely the case for roles that require diverse and context-sensitive interactions with the physical and social environment, such as social care. Care-taking, in the form of both child and elder care, involves physical activities (feeding and cleaning) and social ones (play and friendship). Both are highly reliant on flexibility and common sense reasoning (Lin, Abney and Jenkins, 2017[6]).

In other cases, common sense skills would be required for the benefit of automation to outweigh the cost and effort required for supervision and restructuring the environment. Traditional autonomous-vehicle designs, for example, have led to accidents in cases where a vehicle encounters a situation not included in its training data (Shafaei et al., 2018[7]).

To avoid such accidents, autonomous vehicles require continuous human supervision while on the road. Companies like iSee, however, are moving away from such designs to "autonomy powered by humanistic common sense". It aims to build cars that can flexibly respond to new situations without human supervision (www.isee.ai/about-us).

Both the knowledge and appropriate tools are currently lacking to assess the capacity of AI to take on some or all of these common sense skills. At present, tests of AI capacity are neither sufficiently cognitively defined nor sufficiently general to answer this question. The next two sections review spatial and social cognition respectively, exploring the implications of these two key categories of common sense for the future of work. The final section then proposes a way to assess AI progress towards solving these challenges.

Spatial navigation and treatment of objects

Spatial navigation and the appropriate treatment of objects is a challenge for robotics. Robots are either limited in application or rely on environments that have been specifically adapted to ensure highly consistent inputs. Any input that deviates from expectation may result in damage or disruption. Amazon, for example, designs its warehouses to allow the Kiva robots to transport entire racks on predefined paths, while picking and packaging products still requires human common sense and dexterity (Wurman, D'Andrea and Mountz, 2008[8]). Outside of certain factories and warehouses with a large scope for investment, many workplaces cannot easily be adapted to address such strict limitations.

Issues faced by developers of robotic vacuums and lawn mowers exemplify the challenges faced by bringing robotic assistance into a wider range of workplaces. These robots must navigate a messy real-world environment with unpredictable inputs and obstacles: is that grass, the edge of a flowerbed or an object on the lawn?

To some extent, the environment can be controlled and accounted for. The edge of a lawn, for example, could be delineated with a guide wire. Similarly, the ceiling can be scanned for room size or the robot can be re-routed after a detected collision. However, in practice, many users spend more time managing the robot than they save by owning it.

A taxonomy of common sense space and object skills

Non-human animals can navigate complex environments with varying levels of sophistication, with even the simplest creatures showing some capacity. Animals use spatial skills but also memory to deal with obstacles, objects and affordances. The following taxonomy is a subset of skills identified by psychologists as key features of physical cognition in human and non-human animals. It focuses on skills relevant for basic physical tasks common in human workspaces, such as navigating a space and interacting with objects.

1. Spatial memory and navigation:
   - **Path integration**: using self-motion cues (like limb velocity and movement) to navigate. This ability allows navigation without relying on external landmarks (Etienne and Jeffery, 2004[9]).
   - **Cognitive maps**: building an internal representation of external space, routes and landmark arrangements (Kitchin, 1994[10]). Cognitive maps can rely on path integration, as well as other sources of information (like external landmarks).
   - **What-where-when episodic-like memory**: remembering the location of a specific object at a specific time (Clayton et al., 2001[11]).

- **Episodic memory**: remembering entire events (including spatial context, perceptual information and internal processes) (Tulving, 1983[12]; Cheke and Clayton, 2015[13]). Episodic memory includes information about the "what", "where" and "when" of an event (as in episodic-like memory). However, it also involves mentally reconstructing the experience of an event and other features.

2. Object representations:
   - **Object-level representations**: Representing visual input as objects (that may be movable, able to block other objects, act as containers etc.), rather than patterns of light (Gregory, 1997[14]).
   - **Object permanence**: knowing something is still there, even if you can no longer detect it (Baillargeon and DeVos, 1991[15]). Object permanence requires representing visual input as objects.
   - **Affordance-level representations**: relating perceptual information into affordance information to, for example, predict whether a large object will fit through a narrow aperture (Scarantino, 2003[16]). An "affordance" is a property of an object that makes clear how it can be used. Affordance-level representations often depend on representing visual input as objects (e.g. representing something as an "apple" may evoke the affordance "eat-able").
3. Causal reasoning
   - **Spatial and object inference:** the ability to infer the location or properties of an object through inference, based on prior knowledge and context. For example, eliminative reasoning can conclude that out of a large and small container, only the former can contain a large object (Shanahan et al., 2020[2]).
   - **Folk physics**: the capacity to predict outcomes based on some understanding of the physical mechanisms involved (e.g. predicting a spherical object on a slope will roll downwards) (Povinelli, 2003[17]). The capacity for folk physics will typically rely on a capacity for spatial and object inference.

AI's space and object skills in the workplace

Even the simplest tasks may require these skills if the inputs themselves cannot be strictly controlled and defined. Various standard software programs involve "objects" (e.g. text boxes or pictures). These objects often vary in whether they can be manipulated, and whether they can obstruct or occlude other objects, etc.

An AI without common sense space and object skills may be useful in processing documents but only if these take predictable forms and are accompanied by sufficient metadata. Such an AI would be unable to cope with variations in input type (such as new software). Even in perfect conditions, it would make regular "obvious" errors that would necessitate extensive oversight.

In contrast, an AI with an ability to represent objects and their affordances – i.e. able to continue to reason about them when they are no longer perceptible (object permanence) and infer new information based on this reasoning – would be able to process a large range of files from even unfamiliar sources. Furthermore, if an agent is also able to develop a folk physics of the software environment (e.g. image software that would contain multiple "layers" with "transparency"), when encountering an entirely unfamiliar object or software, it could make inferences based on its knowledge of similar objects/software. While this may sound sophisticated, it is the minimum required for an agent skilled in PowerPoint, for example, to learn to use Google Slides without explicit retraining.

Robots' space and object skills in the workplace

Robots in controlled environments

For robots in physical workplaces, common sense space and object skills are even more crucial. Robotics is currently limited to strictly controlled environments. Even within workplaces (such as factories and warehouses), that can provide a relatively high level of consistency, space and object common sense would vastly expand the types and range of tasks available to robotics.

Amazon illustrates the limitations of robots in the workplace. Even if all packages were labelled with (say) barcodes representing their content, a robot would still need several skills to locate a given package reliably. First, it would need to represent objects ("this visual input is a *package*"). Second, it would need to represent the affordances of the object ("the package is *x shape and x weight* and this translates into manipulating it in *x manner*"). Third, it would often require object permanence and inference ("the barcode may be behind or under the visible surface of the package, which may mean rotating the package to find it"). For basic navigation of the warehouse without predefined routes, the robot would need to use path integration, as well as the capacity to deal with obstacles.

Robots in uncontrolled environments

In many cases, where robot assistance would be most useful – such as medical and elderly care – are within uncontrolled environments (Lin, Abney and Jenkins, 2017[6]). However, these environments, by definition, create challenges. At its most fundamental level, assistance in the home will require the capacity to navigate a home environment, and to share it with someone who may behave in unpredictable ways.

A robot carer without space and object common sense would function only within a consistently clear-floored home. It could not interact with objects that might take unpredictable forms or be in unpredictable locations. It could also not perform any bodily care (assisting with dressing, washing or toileting) without risk of causing pain and injury. Without at least episodic-like memory, it would be unable to locate items that can vary in their location *even if it has seen where they were last placed.*

Conversely, a robot carer in possession of space and object common sense has multiple tools at its disposal. When confronted with an obstacle, a robot carer could use a cognitive map to choose an alternative route. It could also use affordance-level representations to relocate the obstacle and clear its own path. If an important item (e.g. a hearing aid or pacifier) falls out of sight, it could use object permanence to retrieve it. If delivering a drink, it could use folk physics regarding support and flatness to identify a surface that can safely hold the receptacle. Episodic memory would be crucial for a robot carer to monitor cognitive health of their charge. Speech repetition, for example, could be a warning sign for memory decline.

In a wide range of workplace contexts, space and object common sense skills make a fundamental difference to the types of roles an agent can play. Some of these skills (e.g. path integration) are already common features in AI. Others (e.g. folk physics) are clearly some way off. For most skills in between, tools are not yet available to assess them in AI, and thus it is not clear how close they are to development (Crosby, 2020[18]; Shanahan et al., 2020[2]).

Pragmatics and social interaction

Recent advances in linguistic AI have led to an explosion of interactive applications. Chatbots, for example, are now routinely used as a first port of call for online customer enquiries and tech support. GTP-3 is a language model with 175 billion parameters that can perform at close to human level in many few-shot learning tasks.

However, GPT-3 performs relatively poorly on common sense reasoning tests that involve inferring meaning from non-explicit reference or background knowledge (Floridi and Chiriatti, 2020[19]). Within linguistics, this kind of reasoning using language is known as pragmatics. However, the skill of predicting or inferring meaning from ambiguous behaviour is an issue beyond linguistics and common to all social interaction.

In the physical realm, a considerable challenge for AI has been to detect, identify and interpret inputs associated with objects and their affordances. Much the same challenge is magnified when it comes to human behavioural cues. For example, gaze direction indicates the subject of an individual's attention (and therefore their behaviour and verbal reference). It provides the means for responding appropriately to a host of otherwise ambiguous behaviours. However, gaze indicates a direction but not a specific target. The target must be inferred from the context and behaviour of the individual, in combination with prior knowledge. These forms of common sense inferences are a particular challenge for current AI.

A taxonomy of common sense social and communicative skills

Animal and developmental cognition research distinguishes between two categories of social skills. Social learning and communication is the exchange of information with others. Meanwhile, social cognition is understanding and predicting the behaviour and mental states of others. The following taxonomy represents capacities widely accepted by researchers in this area (Hoppitt and Laland, 2013[20]; Shettleworth, 2013[21]).

1. Social learning and communication
   - **Local or stimulus enhancement:** increased attention to areas or objects manipulated by others (Hoppitt and Laland, 2013[20]). Stimulus enhancement may be involved in other forms of learning (e.g. observational conditioning). For example, an observer may respond more to a stimulus as a result of witnessing another agent interact with that stimulus.
   - **Observational conditioning:** learning action-outcome relationships through watching others. For example, a rhesus monkey with no prior exposure to snakes will not behave fearfully in response to a snake. However, if the monkey observes another individual responding fearfully to snakes, it will begin responding this way as well (Cook et al., 1985[22]).
   - **Emulation:** targeting a goal or outcome after observation without imitating exact behaviour (Tomasello, 1990[23]). An adult carrying a stack of books might use her elbow to switch on a light; a child emulating this adult would know they could use alternative means (e.g. their hand) to achieve this same goal of switching on the light.
   - **Imitation:** reproduction of observed behaviour. In the previous example, a child learning through imitation would use their elbow to switch on the light if they had observed the adult doing so (Hoppitt and Laland, 2013[20]). Whether it is more appropriate to emulate or imitate will depend on the situation. In this example, the exact behaviour was the product of a limitation on the adult (the stack of books) that the child did not suffer. It was therefore more efficient to emulate the target of the action than to imitate the exact action itself.
   - **Non-verbal communication:** production and comprehension of non-linguistic communicative signals, such as gestures and facial expressions (Kendon, 2004[24]). Some argue that non-verbal communication paves the way for language development in humans (Iverson and Goldin-Meadow, 2005[25]).
   - **Linguistic communication:** production and comprehension of language for communicative purposes (Tomasello, 2009[26]).
2. Social cognition
   - **Co-operation:** co-ordinating behaviour with another individual to achieve a shared goal (Henrich and Muthukrishna, 2021[27]). Some researchers suggest that co-operative activities,

like helping and sharing, depend on understanding the goals of others or a minimal theory of mind (Tomasello and Vaish, 2013[28]).

- **Behaviour reading:** inferring likely behaviour from context and behaviour or facial expressions (Perner and Ruffman, 2005[29]).
- **Minimal theory of mind:** inferring the goal (or outcome to which a behaviour is directed) of another agent from context and behaviour or facial expressions, without reference to mental states (Butterfill and Apperly, 2013[30]). This ability goes beyond behaviour reading (in involving goal attribution) but does not require a full-blown theory of mind.
- **Theory of mind:** inferring and reasoning about internal epistemic and motivational states, such as intentions, desires and knowledge (Wellman, 1992[31]). Agents with this capacity can predict and explain behaviour by attributing a range of mental states, rather than just reasoning about behaviour or goal-directed behaviour.
- **Empathy:** inferring and reasoning about the emotional states of others (Decety and Lamm, 2006[32]).

Social skills in the workplace

All AIs interface, either directly or indirectly, with people. The form of this interaction is dictated by the capacity of the AI to deal with variability and unpredictability in its social input. This, in turn, dictates the roles available to AI in the workplace. This social input may take multiple forms, most commonly: information communicated via language, behaviour and social context.

Predicting human behaviour is a major challenge for AI in multiple contexts. Yet this will be central to enable AIs to expand into new roles in the workforce, even if those roles do not involve an explicitly social component. For robots, human workers represent a unique form of (highly fragile) obstacle with complex and unpredictable trajectories.

Predicting trajectories is particularly challenging for autonomous vehicles. Unless AI controls all road users, autonomous vehicles must predict and account for human behaviour to avoid collision. An AI without social common sense may learn a range of predictors of collision-risk. For example, it can learn that a child on the pavement may lead to a child on the road. However, it will struggle with anything that falls outside of its specific training or requires inference. For example, it may not understand that a ball on the road may lead to a child on the road, with potentially fatal consequences.

Much more explicitly social roles for AI and robots may be widely useful (as can be seen with chatbots). However, these roles would require more refined social skills beyond predicting simple behaviours (such as trajectory). A robot carer, for example, may not require full-blown empathy. However, it must identify needs that may be communicated both verbally and nonverbally. It must also deal with issues as they arise such as distress, confusion and injury.

A robot carer would also need to be able to monitor health and cognitive function. This would require learning enough information about a specific individual to detect changes in personality or behaviour (such as aggression, delirium or unresponsiveness).

A robot carer would also benefit from being able to distinguish between intentional (lying down) and unintentional (falling down) behaviour. This fundamental element of social common sense in practice requires elements of theory of mind, making it one of the more complex social cognition skills. However, it is required ubiquitously in the workplace: papers placed on the desk should be processed, while papers placed on a shelf that *fall* onto the desk should not.

Requirements for training and supervision will significantly influence the potential for proliferation of AI within the workforce. If expert oversight is required for every new task, then such agents may be useful if produced and programmed or trained at scale. However, they will not be appropriate for niche or specified applications. The ability of agents to "learn on the job" – to learn by watching and perhaps practising with oversight – vastly expands potential applications.

AIs are currently adept at some forms of social learning such as observational conditioning. They can be trained on predictive relationships, for example, by watching YouTube videos. However, other forms are much more challenging. Emulation – arguably the most effective form of task learning – requires the goal or affordances of a situation to be extracted from observation.

Testing artificial intelligence with tasks from psychology: Challenges, opportunities and next steps

The above review of a taxonomy of common sense skills draws on research in developmental and comparative psychology. It focused on those capabilities that represent the breadth of current work in human and non-human animal research (Tomasello and Call, 1997[4]; Shettleworth, 1998[33]; Hoppitt and Laland, 2013[20]) and which are also immediately relevant to the development of AI in the workplace (Shanahan et al., 2020[2]). These capabilities are ubiquitous in humans and often common throughout the animal kingdom, but currently pose a challenge for AI.

A research agenda for artificial intelligence in the workplace

Lack of these common sense skills in AI limits the tasks that can be performed. It means many types of jobs will be safe from full automation until significant progress is made. To understand what jobs these are, and whether that progress is likely, requires basic research in two areas. First, researchers need to identify which common sense skills are necessary for specific roles within the workplace, and to what extent. When most humans have a capability, it is an assumed skill within an employee, but this would not be the case with an AI. Second, they need to understand current AI capacity, and measure progress, in common sense skills.

This chapter has taken a first step towards addressing the first area of research. Concerning the second area, a number of well-established tests within animal and developmental cognition research assess these skills. However, there are no "off-the-shelf" cognitive tests appropriate for testing AI. Many tests designed for children are diagnostic, rather than evaluative, and others rely heavily on species-specific tendencies or biases.

Conversely, a number of benchmarking assessments for AI have been the basis for cognitive claims. However, these assessments are rarely cognitively defined, and suffer from a construct validity problem. In particular, it is easy for a programmer (either consciously or subconsciously) to design and train an AI to pass a particular *test,* rather than to possess a particular *skill* (Hernández-Orallo, 2017[34]; Shevlin and Halina, 2019[35]).

Animal-AI Testbed

One solution to the validity problem is to assess AI in the same way that psychologists assess animals and children, such as through the Animal-AI Testbed (Crosby, Beyret and Halina, 2019[36]; Crosby, 2020[18]; Crosby et al., 2020[37]). Here, agents can be trained in a 3D environment and learn about the spaces, objects, agents and rules it contains.

During training, agents can access a simulated arena with simple objects of various sizes, shapes and textures. The objects behave as they would in the real world (subject to forces of gravity, friction, etc.) due to simulated physics. The arena also contains positive and negative rewards, which an agent can attempt to acquire or avoid, respectively.

This training arena is designed to reproduce the type of environment an animal might encounter before they are tested in a typical animal-cognition task. Crucially, in both cases, an agent is not trained on the test itself. It thus cannot learn to simply pass this test through hundreds or thousands of trials. Instead, it must rely on general common sense skills like those reviewed above.

The most recent version of the Animal-AI Testbed consists of 300 tasks. Each task includes a set of objects, including a positive reward and a time limit. An agent succeeds in the task if it retrieves sufficient reward (i.e. reward above some threshold) within the time limit. The tasks are either drawn directly from or inspired by tests used in developmental and comparative psychology.

Each task aims to probe an agent's ability to engage in common sense reasoning. For example, several tasks test an agent's spatial memory and navigation abilities: these include mazes of various difficulties, from simple mazes in the shape of the letter "Y" to more difficult radial mazes with many arms. Another category of task tests an agent's capacity for object permanence by moving rewards out of sight.

Finally, other tasks probe an agent's capacity for folk physics by presenting agents with problems that require a tool. To solve these problems, the agent must have some understanding of the physical mechanisms involved and how the tool can produce a causal effect that will lead to a reward.

Using a wide range of cognitively defined tests, it is therefore possible to provide a robust and valid assessment of AI common sense capacities (Crosby et al., 2020[37]).

Emerging tests for social and communicative skills

Using the Animal-AI Testbed, it has been demonstrated that state-of-the-art artificial agents display path integration and the beginnings of cognitive maps and spatial and object inference (Crosby et al., 2020[38]; Voudouris et al., 2021[39]). However, episodic memory, object-level representations, object permanence, affordance-level representations and folk physics require further research. Solving these tasks in Animal-AI will be the first step towards solving them in real-world settings and will be a marker for the possible landscape of AI integration into the future workplace.

The Animal-AI Testbed tests for common sense space and object skills rather than social and communicative skills. However, the latter is a promising area for future research. In psychology, there is a rich literature of tasks for testing social and communicative capacities in human and non-human animals (see references in taxonomy above).

Researchers have begun to apply these tasks to AI systems. For example, AI and robotics researchers have tested artificial systems on theory of mind tasks (Winfield, 2018[40]; Rabinowitz et al., 2018[41]). Further research in this area would benefit from distinguishing abilities like behaviour reading and theory of mind, given these capacities may have differential effects on performance (Shevlin and Halina, 2019[35]).

Enhancing human performance with artificial intelligence

Moving beyond autonomous artificial agents, another use of AI in the workplace involves enhancing or augmenting human performance. Such enhancement technologies include tools like speech assistants, navigation systems and other decision-support systems (Hernández-Orallo and Vold, 2019[42]; Shevlin et al., 2019[1]; Sutton et al., 2020[43]).

These tools are likely to present unique challenges to implementing AI in the workplace. Human-AI collaboration would also benefit from AI with common sense skills. For example, common sense space and object skills would facilitate guidance and navigational systems that can respond to the challenges and affordances of the environment rather than relying on pre-programmed routes and rules.

Common sense social abilities may be even more important in the context of AI-human collaboration: an agent that can detect and respond to needs, interpret verbal and non-verbal communication, and engage in joint problem solving will be a far more effective assistant in whatever capacity it is employed. Thus, common sense skills are important for not only enabling AI to operate autonomously or semi-autonomously in the workplace but also for enhancing human performance.

Recommendations

- **Use animal and child development tests to test common sense skills**

While AIs excel at highly specified skills, they struggle with basic skills humans take for granted. Without such fundamental skills, they cannot navigate the world in a flexible and adaptive manner. It is challenging to test these skills in AI precisely because they are not generally assessed in human adults. Instead, tests developed for animals and young children provide the most appropriate means to assess AI systems in these areas.

- **Adapt these tests for AI**

However, such tests cannot be simply used, off the shelf, to assess AI. They must be adapted into contexts appropriate for AI and administered to ensure assessment of the skill and not expertise at the test. The Animal-AI Testbed provides an example of how this can be done.

- **Extend tests to social and communicative common sense**

The Animal-AI Testbed should be extended to tests for social and communicative common sense. These will be important for the integration of AI into workplaces that involve human interaction.

References

Baillargeon, R. and J. DeVos (1991), "Object permanence in young Infants: Further evidence", *Child Development*, Vol. 62, pp. 1227-1246, http://dx.doi.org/10.1111/j.1467-8624.1991.tb01602.x. [15]

Butterfill, S. and I. Apperly (2013), "How to construct a minimal theory of mind", *Mind & Language*, Vol. 28, pp. 606-637, http://dx.doi.org/10.1111/mila.12036. [30]

Cheke, L. and N. Clayton (2015), "The six blind men and the elephant: Are episodic memory tasks tests of different things or different tests of the same thing?", *Journal of Experimental Child Psychology*, Vol. 137, pp. 164-171, http://dx.doi.org/10.1016/j.jecp.2015.03.006. [13]

Clayton, N. et al. (2001), "Elements of episodic-like memory in animals", *Philosophical Transactions of the Royal Society of London. Series B: Biological Sciences*, Vol. 356, pp. 1483-1491, http://dx.doi.org/10.1098/rstb.2001.0947. [11]

Cook, M. et al. (1985), "Observational conditioning of snake fear in unrelated rhesus monkeys", *Journal of Abnormal Psychology*, Vol. 94/4, pp. 591-610, http://dx.doi.org/10.1037//0021-843x.94.4.591. [22]

Crosby, M. (2020), "Building thinking machines by solving animal cognition tasks", *Minds and Machines*, Vol. 30, pp. 589-615, http://dx.doi.org/10.1007/s11023-020-09535-6. [18]

Crosby, M., B. Beyret and M. Halina (2019), "The Animal-AI Olympics", *Nature Machine Intelligence*, Vol. 1, pp. 257-257, http://dx.doi.org/10.1038/s42256-019-0050-3. [36]

Crosby, M. et al. (2020), *The Animal-AI Testbed and Competition*, https://proceedings.mlr.press/v123/crosby20a.html. [38]

Decety, J. and C. Lamm (2006), “Human empathy through the lens of social neuroscience”, *The Scientific World Journal*, Vol. 6, pp. 1146-1163, http://dx.doi.org/10.1100/tsw.2006.221. [32]

Escalante, H. and R. Hadsell (eds.) (2020), “The animal-AI testbed and competition”, *Proceedings of the NeurIPS 2019 Competition and Demonstration Track*, No. 123, PMLR, http://proceedings.mlr.press/v123/crosby20a.html. [37]

Etienne, A. and K. Jeffery (2004), “Path integration in mammals”, *Hippocampus*, Vol. 14, pp. 180-192, http://dx.doi.org/10.1002/hipo.10173. [9]

Floridi, L. and M. Chiriatti (2020), “GPT-3: Its nature, scope, limits, and consequences”, *Minds and Machines*, Vol. 30, pp. 681-694, http://dx.doi.org/10.1007/s11023-020-09548-1. [19]

Gregory, R. (1997), *Eye and Brain: The Psychology of Seeing – Fifth Edition*, Princeton University Press, http://dx.doi.org/10.2307/j.ctvc77h66. [14]

Henrich, J. and M. Muthukrishna (2021), “The origins and psychology of human cooperation”, *Annual Review of Psychology*, PMID: 33006924, pp. 207-240, http://dx.doi.org/10.1146/annurev-psych-081920-042106. [27]

Hernández-Orallo, J. (2017), “Evaluation in artificial intelligence: From task-oriented to ability-oriented measurement”, *Artificial Intelligence Review*, Vol. 48, pp. 397-447, http://dx.doi.org/10.1007/s10462-016-9505-7. [34]

Hernández-Orallo, J. and K. Vold (2019), “AI extenders: The ethical and societal implications of humans cognitively extended by AI”, *Proceedings of the 2019 AAAI/ACM Conference on AI, Ethics, and Society*, Association for Computing Machinery, http://dx.doi.org/10.1145/3306618.3314238. [42]

Hoppitt, W. and K. Laland (2013), *Social Learning: An Introduction to Mechanisms, Methods, and Models*, Princeton University Press, http://dx.doi.org/10.1515/9781400846504. [20]

Iverson, J. and S. Goldin-Meadow (2005), “Gesture paves the way for language development”, *Psychological Science*, PMID: 15869695, pp. 367-371, http://dx.doi.org/10.1111/j.0956-7976.2005.01542.x. [25]

Kelly, A. (2021), “A tale of two algorithms: The appeal and repeal of calculated grades systems in England and Ireland in 2020”, *British Educational Research Journal*, Vol. n/a, http://dx.doi.org/10.1002/berj.3705. [5]

Kendon, A. (2004), *Gesture: Visible Action as Utterance*, Cambridge University Press, http://dx.doi.org/10.1017/CBO9780511807572. [24]

Kitchin, R. (1994), “Cognitive maps: What are they and why study them?”, *Journal of Environmental Psychology*, Vol. 14, pp. 1-19, http://dx.doi.org/10.1016/S0272-4944(05)80194-X. [10]

Lin, P., K. Abney and R. Jenkins (eds.) (2017), *Robot Ethics 2.0: From Autonomous Cars to Artificial Intelligence*, Oxford University Press, http://dx.doi.org/10.1093/oso/9780190652951.001.0001. [6]

Perner, J. and T. Ruffman (2005), "Infants' insight into the mind: How deep?", *Science*, Vol. 308/5719, pp. 214-216, http://dx.doi.org/10.1126/science.1111656. [29]

Piaget, J. (1977), *The Development of Thought: Equilibration of Cognitive Structures (Trans A. Rosin)*, Viking. [3]

Povinelli, D. (2003), *Folk Physics for Apes: The Chimpanzee's Theory of How the World Works*, Oxford University Press, http://dx.doi.org/10.1093/acprof:oso/9780198572190.001.0001. [17]

Rabinowitz, N. et al. (2018), "Machine theory of mind", *Proceedings of the 35th International Conference on Machine Learning*, No. 80, Dy, J. and A. Krause (eds.), PMLR, http://proceedings.mlr.press/v80/rabinowitz18a.html. [41]

Scarantino, A. (2003), "Affordances explained", *Philosophy of Science*, Vol. 70, pp. 949-961, http://dx.doi.org/10.1086/377380. [16]

Shafaei, S. et al. (2018), "Uncertainty in machine learning: A safety perspective on autonomous driving", Conference: First International Workshop on Artificial Intelligence Safety Engineering, Västerås, Sweden. [7]

Shanahan, M. et al. (2020), "Artificial intelligence and the common sense of animals", *Trends in Cognitive Sciences*, Vol. 24, pp. 862-872, http://dx.doi.org/10.1016/j.tics.2020.09.002. [2]

Shettleworth, S. (2013), *Fundamentals of Comparative Cognition*, Oxford University Press, https://global.oup.com/academic/product/fundamentals-of-comparative-cognition-9780195343106. [21]

Shettleworth, S. (1998), *Cognition, Evolution, and Behavior*, Oxford University Press, https://global.oup.com/academic/product/cognition-evolution-and-behavior-9780195319842. [33]

Shevlin, H. and M. Halina (2019), "Apply rich psychological terms in AI with care", *Nature Machine Intelligence*, Vol. 1, pp. 165-167, http://dx.doi.org/10.1038/s42256-019-0039-y. [35]

Shevlin, H. et al. (2019), "The limits of machine intelligence", *EMBO Reports*, Vol. 20/10, p. e49177, http://dx.doi.org/10.15252/embr.201949177. [1]

Sutton, R. et al. (2020), "An overview of clinical decision support systems: Benefits, risks, and strategies for success", *npj Digital Medicine*, Vol. 3, p. 17, http://dx.doi.org/10.1038/s41746-020-0221-y. [43]

Tomasello, M. (2009), *Constructing a Language*, Harvard University Press, https://www.hup.harvard.edu/catalog.php?isbn=9780674017641. [26]

Tomasello, M. (1990), "'Language' and intelligence in monkeys and apes", in *Cultural Transmission in the Tool Use and Communicatory Signaling of Chimpanzees*, Parker, S. and K Gibson (eds.), Cambridge University Press, http://dx.doi.org/10.1017/cbo9780511665486.012. [23]

Tomasello, M. and J. Call (1997), *Primate Cognition*, Oxford University Press, https://global.oup.com/academic/product/primate-cognition-9780195106244. [4]

Tomasello, M. and A. Vaish (2013), "Origins of human cooperation and morality", *Annual Review of Psychology*, PMID: 22804772, pp. 231-255, http://dx.doi.org/10.1146/annurev-psych-113011-143812. [28]

Tulving, E. (1983), *Elements of Episodic Memory*, Oxford University Press. [12]

Voudouris, K. et al. (2021), "Direct Human-AI Comparison in the Animal-AI Environment", https://doi.org/10.31234/osf.io/me3xy. [39]

Wellman, H. (1992), *The Child's Theory of Mind*, MIT Press, https://mitpress.mit.edu/books/childs-theory-mind. [31]

Winfield, A. (2018), "Experiments in artificial theory of mind: From safety to story-telling", *Frontiers in Robotics and AI*, Vol. 5, p. 75, http://dx.doi.org/10.3389/frobt.2018.00075. [40]

Wurman, P., R. D'Andrea and M. Mountz (2008), "Coordinating hundreds of cooperative, autonomous vehicles in warehouses", *AI Magazine*, Vol. 29, pp. 9–19. [8]

Part IV. Reflections and a pragmatic way forward

18. Tasks and tests for assessing artificial intelligence and robotics in comparison with humans

Art Graesser, Psychology and the Institute for Intelligent Systems, University of Memphis

This chapter reflects on major issues recounted in the previous chapters, raising three key questions relevant to the *Artificial Intelligence and the Future of Skills* project. What is the value in identifying ideal models when comparing humans with artificial intelligence (AI) and robotic systems? How might systematic mapping occur between skill taxonomies, tasks, tests and functional AI components? How can major differences be handled in targeted skills, different occupations and changes in the world? Some suggestions are offered on next steps in addressing these questions.

Introduction

The OECD *Artificial Intelligence and the Future of Skills* (AIFS) project attempts to understand the educational implications of artificial intelligence (AI) and robotics. The goal is to create an ongoing programme to assess the capabilities of AI and robotics, and to compare them with human capabilities. The 5-6 October 2020 meeting featured presentations from 13 experts who spanned diverse fields, including educational assessment, AI, robots, cognitive science and workforce training. Additional experts from various fields (and countries) also contributed. This chapter raises some questions and suggestions, and offers other reflections on major issues covered at that meeting and recounted in the preceding chapters.

Identifying knowledge, skills and abilities

One central issue is to identify the set of knowledge, skills and abilities (KSAs) to assess. Psychology has proposed comprehensive taxonomies with psychometric *tests*. These include the three-level Carroll-Horn-Cattell model presented by Kyllonen (see Chapter 3). This has a long history of validation in humans and quantitatively tuned factor analyses.

There are abilities and skills identified in industrial-organisational psychology and business that involve *tasks* specific to particular occupations. This allows adults to be trained and certified to practice in the occupation. For example, Dorsey and Oppler (see Chapter 10) described the O*NET (Occupational Information Network) in the US Department of Labor. It identifies KSAs for occupation categories (e.g. manufacturing, health care). Rüschoff (see Chapter 9) presented the vocational education and training framework in Germany. It has an intense two-day assessment that has practical, written and oral components, including answering questions to justify actions.

Greiff and Dörendahl (see Chapter 7) pushed the envelope beyond basic cognitive skills and domain-specific skills into the realm of *transversal* skills that have increasing importance in the 21st century. These comprise problem solving, collaboration, creative thinking and global competency. Wooley (see Chapter 6) echoed the importance of social intelligence and collaboration. Conversely, De Fruyt (see Chapter 5) emphasised social skills and emotion regulation skills.

The AI/robotics contingency did not offer taxonomies of KSAs, as pointed out by Hernández-Orallo (see Chapter 11). Instead, it focuses on *functional components* of intelligent mechanisms, such as knowledge representation, reasoning, planning, learning, perception, navigation and natural language processing. They evaluate how well the various computational models in AI/robotics compare with humans on tasks that focus on these functional components.

Forbus and Davis (see Chapter 2 and Chapter 12, respectively) pointed out which components are easier for computers to achieve (such as remembering and accessing facts) and which are easier for humans (such as common sense reasoning). Avrin (see Chapter 15) discusses systematic evaluations of over 900 AI systems on recognition capabilities, learning, understanding, generation and mission navigation.

Nearly all of these evaluations of systems in AI/robotics have been on practical tasks. Such tasks, such as autonomous cars and text summarisation, have objective criteria of success. Moreover, the tasks typically focus on those performed by adults in the workforce. However, Cheke (see Chapter 17) covered low-level skills of animals whereas Chokron (see Chapter 4) focused on cognitive and social skills of children. These presentations address the second central issue of the expert meeting, namely identifying differences in what can be accomplished by humans versus AI/robotic systems.

With this context in mind, the chapter raises three questions, with associated reflections and suggestions. These aim to shed light on the primary goal and two central issues of the AIFS project.

What is the value in identifying ideal models when comparing humans and artificial intelligence/robotic systems?

Towards a formal model of performance to assess and compare humans and AI robots

One could imagine an ideal specification (i.e. formal model) of performance on tasks, tests and functional components. Such a model could serve as a standard to assess and compare humans versus AI/robots. That would go a long way in providing a fair comparison on the capabilities of the two systems.

A perfect ideal model is perhaps illusory, but there can be approximations. For example, accomplished human experts can specify ideal responses to tasks and tests. These could either solve a problem or meet a level of mastery in achieving particular tasks. Such a specification has both a content analysis and a threshold analysis.

Content analysis

The content specification would declare the particular behaviours and products that correspond to a successful accomplishment of a task. This approach is adopted by designers of intelligent tutoring systems (Koedinger, Corbett and Perfetti, 2012[1]; Graesser, Hu and Sottilare, 2018[2]). These systems identify *knowledge components* required to master a subject matter or skill (e.g. algebra, physics). They also prepare a *Q-matrix* that specifies the knowledge components associated with each particular problem, task, or item along with behavioural manifestations of each knowledge component mastery. A complete and accurate solution would be needed, but it might also consider intermediate levels of achievement.

Threshold analysis

While the content analysis is applied to each individual item on a test or step in a task, the threshold analysis is applied to an aggregate score from the entire test/task. The threshold analysis identifies points on a continuum of scores that predict practical external criterial outcomes (which is infrequently conducted). This contrasts with exclusively psychometric indices or breakpoints in the distribution of scores (which is routinely conducted). Analyses can assess how well a population of humans or AI/robot systems meet the various thresholds of scores.

How well would a human vs. an artificial intelligence/robotics system perform?

Each adult has decades of experiences to fortify them in a task. Information about this past is either non-existent or minimally specified through demographic data or assorted tests. AI/robotics systems are unlikely to have such data available. However, there are AI systems that learn with experience. The *Never-Ending Language Learner* (NELL), for example, runs 24 hours per day learning to read the web and grow a knowledge base of beliefs (Mitchell et al., 2018[3]).

There are several possible approaches to understanding how AI/robotics system can be put on an even playing field with a human. The first approach assumes AI systems and humans have a different array of assets and resources, and thus are rarely on an even playing field. However, they can still be compared on tasks, which the AIFS project is planning. A second approach puts the system through a practice set of benchmark tasks for a month. It then grades performance on a test set, as in the case of the NIST methodology. A third approach is to conduct an AI/simulation over a long stretch of time or epochs of experiences. The performance produced in such tasks is then observed, as in NELL (Mitchell et al., 2018[3]).

A computational or information-theoretic analysis could specify a problem space, combinatorial landscape or another type of formal, quantitative model that identifies hypothetical alternatives and bone fide solutions. For example, the iconic "travelling salesman problem" attempted to find a route between 40 cities that minimised the distance in travel time. The problem proved to be so hard that it would have required over 1 000 years on the fastest computer that existed 20-30 years ago. Who knows how the travelling salesman problem fares 30 years later? However, computational analyses like these can be posed for a fair comparison between humans and AI/robotics.

Ideal models with both computational and human constraints

There are ideal models that incorporate both computational and human constraints. For example, cognitive scientists often perform tasks analyses on particular problems or problem sets that decompose the solution plans and concrete steps in executing solutions (Anderson, 2009[4]; Laird, 2012[5]). Researchers can compute the probability and time of (sub)task completion, as well as the assorted solution strategies.

A good example of this approach is the models of lower-level perceptual-motor tasks. These include the Goals, Operators, Methods and Selection Rules (GOMS) model (Card, Moran and Newell, 1983[6]) and CogTool (John, 2013[7]). A researcher first specifies a set of tasks and the model generates expected task completion times and other aspects of performance.

GOMS and CogTool are based on an ideal rational model (much like Anderson's ACT-R and Laird's SOAR) and psychological components (such as perception-cognition-action cycles, production rules) and psychological laws. Fitt's law, for example, computes the time to move a part of a body to a target. Hick's law specifies that the time taken for a decision is a logarithmic function of the number of alternatives. The power law of practice specifies an exponentially decreasing function of task completion time as a function of number of attempts to complete a task.

GOMS and CogTool are remarkably accurate in predicting performance in some tasks (Graesser et al., 2018[8]) but not others that require higher-order reasoning. Consequently, GOMS is best used to complement rather than replace expert judgements of task difficulty. Nevertheless, for lower-level tasks involving well-practised procedures, researchers would have a foundation for comparing humans and robots. Similar approaches could be proposed for problem solving, reasoning, collaboration and other transversal skills (Sinatra et al., 2021[9]).

How might systematic mapping occur between skill taxonomies, tasks, tests and functional artificial intelligence components?

Participants at the expert meeting presented several skill taxonomies, tasks, tests and functional AI components. A few of these are presented below.

- Carroll's (1993) 3-Stratum model presented by Kyllonen (Chapter 3)

As discussed in Chapter 3, Stratum 3 is a general intelligence factor, whereas Stratum 2 has eight constructs manifested in factor analyses: fluid intelligence, crystallised intelligence, general memory/learning, broad visual perception, broad auditory perception, broad retrieval ability, broad cognitive speediness and processing speed. Stratum 2 depends on the measures collected in Stratum 1, which consists of dozens of precisely operationalised tests from the field of psychological assessment (Carroll, 1993[10]).

- Greiff and Dörendahl's (Chapter 7) taxonomy (which includes transversal skills)

There is a distinction between transversal skills (problem solving, collaboration, creativity, digital competence, global competence), core domain skills (mathematical, reading and science literacies) and basic cognitive skills (general mental ability, fluid reasoning, comprehension knowledge, working memory, and others discussed in Chapter 3).

- Cheke, Halina and Crosby's taxonomy (which emphasises lower-level cognitive skills in both animals and humans)

Object and space skills include spatial memory and navigation, object representations and causal reasoning. Social and communicative skills include social learning and communication and social cognition, with several behavioural tests or tasks to operationalise these major categories.

- Hernández-Orallo's functional components in AI/robotics

The functional components in the list were knowledge representation, reasoning, planning, learning, perception, navigation and natural language processing. However, there may be others as AI/robotics evolves.

There were other taxonomies and distinctions discussed in the expert meeting. These included emotion regulation, empathy, trust and other dimensions of human experience (smell was an intriguing example). Perhaps the detection of misinformation would be particularly relevant in the age of social media (Rapp and Braasch, 2014[11]). In theory, there are subcomponents of misinformation detection, such as identifying the expertise of the source of information, comparisons to information in other documents, status of the media outlet, and sophistication of the language or information delivery.

Many skill categories and distinctions are potential candidates. Consequently, there are challenges in identifying which skills to include. One could adapt at least four approaches to meeting the challenges.

- The *comprehensive approach* would include any skill included by two or more stakeholders in the project, noting the lack of impact of unusual singletons.
- The *consensus approach* would include those skills that a sufficient number of stakeholders would endorse.
- The *intersection approach* would include those skills that can be measured in both humans and AI/robots in action.
- The *theoretical approach* would adopt one singular model for all to adhere to.

Approaches to selecting functional components

The *intersection approach* is compelling because the measures are available. This means it would be pragmatically strategic to implement them. However, this approach would need to consider comprehensiveness and the theoretical landscape. Perhaps the most pragmatic solution is to identify a small number of skills/tasks that represent different areas of the theoretical landscape.

A *consensus approach* would require a mapping between the different categories in different taxonomies, as exemplified above. Such a mapping could facilitate understanding of the landscape of skills, but major difficulties will emerge. For example, reasoning in one taxonomy will mean something different in another taxonomy. To address these concerns, stakeholders would need to negotiate a common ground of conceptual meanings of constructs, operational definitions of measures and other considerations. These would need to be familiar to those in the world of assessment, as well as science more generally.

Differences in semantics will occur between different research communities and stakeholders. For example, mundane plausible reasoning in human taxonomies is different from the formal reasoning in propositional calculus, AI's theorem proving and even the inference rules in the CYC computer system that represents world knowledge (Lenat et al., 2010[12]). Humans do well on *modus ponens* (If X, then Y; X; therefore Y). They consistently fail on *modus tollens* (If X, then Y; not Y; therefore, not X). Further, they often embrace the abductive reasoning that has an illegitimate formal foundation (Rips, 1994[13]). Similarly, statistical reasoning is different in formal systems vs. humans. Humans are prone to have, for example, base-rate and hindsight biases (Kahneman, Slovic and Tversky, 1982[14]). Such facts, of course, have relevance to what humans versus AI/robot systems can accomplish. That is apparent and also interesting.

In sum, reasoning is different in the various taxonomies. These differences reflect the goals of the projects, differences in fields and history. Negotiations among relevant stakeholders will be necessary to converge on a common ground. The achievement of a deep mapping of categories between taxonomies is complex and perhaps unlikely at the fine-grain levels but routinely successful at a course-grain level.

There are caveats, of course. The achievement of a loose mapping of categories is easier but possibly misleading because of non-trivial differences that end up getting missed. A small number of broad categories risks glossing over major differences in specific tasks/tests selected to represent the broad categories. Mapping between taxonomies is thus beset with serious challenges, but the history of assessment offers encouragement that the goals can be pragmatically achieved.

Mapping taxonomies and tasks, tests and functional artificial intelligence components: the Q-matrix

It is essential to generate mappings between particular taxonomies and the specific tasks, tests and functional AI components. In some circles, these are called a *Q-matrix*. Each item (e.g. question to answer, alternative to select, action to perform) in an evaluation scenario is assigned a code of attributes being assessed by an item (i.e. knowledge component, knowledge, skill, strategy, ability).

There can be primary, secondary and tertiary codes in these expert annotations. Stakeholders from different professional communities can annotate the items in a candidate scenario on the taxonomy categories of importance. The analysts in each stakeholder community could adopt whatever standards and criteria they wish to adopt, as long as other stakeholders can understand them.

What can be accomplished with the Q-matrices at hand from various stakeholders on candidate items in scenarios? The different stakeholders can evaluate and give feedback on whether the scenarios and items have a sufficient representation of the important taxonomic categories in their community.

Just as countries give such feedback in OECD international assessments (e.g. the Programme for International Student Assessment), stakeholders from relevant communities can give their feedback. Approval of scenarios and tasks depends on constraint satisfaction and negotiation. Relevant stakeholders need input on most phases of the assessment – from selecting relevant scenarios and tasks to developing illuminating items with the associated constructs they manifest. Items in this context may be actions in addition to verbal contributions and decisions in conventional assessments.

How can major differences be handled in targeted skills, different occupations and changes in the world?

The tasks, tests and functional components under focus are different. Occupations have different expectations. Subject matters are different among the occupations. The world also changes in trajectories that differ among countries, languages and cultures. How can these differences be accommodated in AI systems?

Consider separate implementations for each occupation, skill and time slice

As one simple answer, AI will need separate implementations for each occupation, skills and time slices being considered. This can be accomplished surprisingly quickly if certain conditions are met:

- a sufficient corpus of data for training and testing with machine learning
- a sufficient crew of knowledge engineers for annotation of data (needed for supervised machine learning) and development of scripts, rules or other modules with authoring tools.

This would require funding. However, it is a matter of availability of resources, engineering and investments as opposed to a devasting bottleneck. Whether general AI principles and mechanisms can be gleaned from such activities is an open question.

Compare timepoints

It is, of course, important to address bias in many of these questions, as well as changes that occur over time. As articulated by Greiff and Dörendahl in Chapter 7, there is a shift in the need for transversal skills. Therefore, problem solving, collaboration, reasoning and creativity in the world will have a higher impact on predicting the successful workforce profile than will memory and routine perceptual-motor skills. The workforce data clearly reveal this shift (Autor, Levy and Murnane, 2003[15]; Elliott, 2017[16]).

It could be argued that AI/robotics systems have not made significant headway in self-regulated activities and many of the transversal skills. This puts them at a disadvantage in these 21st century KSAs in contrast to their clear superiority in retrieving facts. Nevertheless, these points are non-problematic. For now, the goal is to identify what skills can be accomplished by humans vs. AI/robotics systems.

Some comparisons between timepoints might help project the workforce of the future and assess generalisation of claims. In one approach, a collection of scenarios and tasks is representative of the past, vs. the present vs. the future in the ultimate assessment. That is, a subset of the scenarios would represent the world of ten years ago; another subset the present; and another subset the uncertain science fiction of the future.

The three time points would crudely track trends over time on the measures collected from individuals at different age partitions and occupations. To increase the precision of temporal trends, time can be divided into finer slices. It would move from the past through present so that linear and non-linear trends can be detected and projected according to different quantitative models. However, projections would be considered with caution. Revolutionary disruptive historical changes periodically occur, such as war, pandemics like COVID-19 and the escalation of technology.

The selection of assessment scenarios and items will need to accept how existing tests, tasks and functional AI components have a distinctive history that may resist compromise. Perhaps the assessment materials that end up being selected/created will be a blend of the different traditions. In this way, they may have a chance to pacify multiple stakeholders. Perhaps the selected assessment scenarios will be fortified by Q-matrices that have tentacles to most or all of the stakeholders. Whatever scenarios end up selected, systematic comparisons will be needed between humans and AI/robotic systems.

Recommendations

- **Use ideal models as a neutral standard**

Some ideal models could serve as a neutral standard in comparisons of AI/robotics systems and humans. Ideal models are likely to stimulate exciting research on the data that end up being collected. However, they could potentially influence the selection of scenarios/tasks/tests.

- **Adopt an intersection approach to select tasks/tests for the comparisons**

The intersection approach uses tasks/skills/components that have been investigated both in psychology and AI, and covers different regions in the theoretical landscape. These decisions will require negotiations among stakeholders, as has been routinely accomplished for decades in the world of assessment.

- **Select scenarios, tasks and skills that present past, present and future**

The tasks performed in the work and daily lives of adults are known to vary over decades. Therefore, it would be prudent to select scenarios, tasks and skills that are representative of the past, present and future. This would permit detection of trends over time for participants in different age groups, occupations and demographic characteristics. However, projections must be tempered with caution to the extent there are disruptive historical events such as COVID-19.

References

Anderson, J. (2009), *How Can the Human Mind Occur in the Physical Universe?*, Oxford University Press. [4]

Autor, D., F. Levy and R. Murnane (2003), "The skill content of recent technological change: An empirical exploration", *The Quarterly Journal of Economics*, Vol. 118/4, pp. 1279-1333, https://doi.org/10.1162/003355303322552801. [15]

Card, S., T. Moran and A. Newell (1983), *The Psychology of Human-computer Interaction*, Lawrence Erlbaum Associates, Hillsdale, NJ. [6]

Carroll, J. (1993), *Human Cognitive Abilities: A Survey of Factor-analytic Studies*, Cambridge University Press, New York. [10]

Elliott, S. (2017), *Computers and the Future of Skill Demand*, Educational Research and Innovation, OECD Publishing, Paris, https://dx.doi.org/10.1787/9789264284395-en. [16]

Graesser, A., X. Hu and R. Sottilare (2018), "Intelligent tutoring systems", in Fischer, F. et al. (eds.), *International Handbook of the Learning Sciences*, Routledge, New York. [2]

Graesser, A. et al. (2018), "Via: Using GOMS to improve authorware for a virtual internship environment", in Roscoe, R., S. Craig and I. Douglas (eds.), *End-user Considerations in Educational Technology Design*, IGI Global. [8]

John, B. (2013), *Cogtool (Version 1.2.2) (Software)*, https://github.com/cogtool/cogtool/releases/tag/1.2.2. [7]

Kahneman, D., P. Slovic and A. Tversky (1982), *Judgment Under Uncertainty: Heuristics and Biases*, Cambridge University Press, New York. [14]

Koedinger, K., A. Corbett and C. Perfetti (2012), "The Knowledge-Learning-Instruction (KLI) framework: Bridging the science-practice chasm to enhance robust student learning", *Cognitive Science*, Vol. 36/757-798, http://dx.doi.org/10.1111/j.1551-6709.2012.01245.x. [1]

Laird, J. (2012), *The SOAR Cognitive Architecture*, MIT Press, Cambridge, MA. [5]

Lenat, D. et al. (2010), "Harnessing Cyc to answer clinical researchers' ad hoc queries", *AI Magazine*, Vol. 31/3, pp. 13-32, http://dx.doi.org/10.1609/aimag.v31i3.2299. [12]

Mitchell, T. et al. (2018), "Never-ending learning", *Communications of the ACM*, Vol. 61/5, pp. 103-155, http://dx.doi.org/10.1145/3191513. [3]

Rapp, D. and J. Braasch (eds.) (2014), *Processing Inaccurate Information: Theoretical and Applied Perspectives from Cognitive Science and the Educational Sciences*, MIT Press, Cambridge, MA. [11]

Rips, L. (1994), *The Psychology of Proof: Deduction in Human Thinking*, MIT Press, Cambridge, MA. [13]

Sinatra, A. et al. (2021), *Design Recommendations for Intelligent Tutoring Systems: Competency-based Scenario Design*, Army Research Laboratory, Orlando, FL. [9]

19. Questions to guide assessment of artificial intelligence systems against human performance

Eva L. Baker, the University of California, Los Angeles

Harold F. O'Neil Jr., University of Southern California's Rossier School of Education

This chapter takes a step back from the project, reviewing practical issues around the assessment of artificial intelligence (AI) that must guide the next phase of research. It provides guidance for setting up a general analytical framework for assessing AI capabilities with regard to human skills. Questions are centred around three main issues. First, the chapter looks at how the various parameters of measurement depend on the objectives of the assessment. Second, it examines selection of tests for comparing AI and human skills. Third, it discusses selection and training of raters. The chapter concludes with a summary of issues considered in planning the study.

Introduction

Alongside the continuing explosion of artificial intelligence (AI) in research and applications is an accompanying need to understand its capabilities in a variety of settings. There is ample evidence that AI systems can solve particular problems far better and certainly faster than humans attempting comparable tasks. There is also a long list of AI accomplishments on sets of tasks humans could not directly attempt because of size and complexity of data. Nonetheless, there is a need to characterise and perhaps benchmark the capability of these systems.

This chapter discusses general precepts for consideration with respect to the planning study. It focuses on the issues surrounding the use or development of taxonomies and relevant measures to allow a new level of understanding of current and future capacity of AI systems. To that end, it poses questions and provides guidance related to the characterisation of AI systems, issues in organisation and task selection for assessment, and the rating process.

Why measure or characterise artificial intelligence systems?

This OECD project expects that measures of AI performance can be developed to compare to human performance based on a set of existing (or potentially newly constructed) measures. The approach tests the assumption that clear correspondences are possible for performance by AI systems and people. The project must clarify its needs before the best strategy and operational objectives can be proposed. Three large considerations are discussed below to stimulate reflection.

- Compare AI to human intelligence or to human workplace or academic outcomes?

Comparing AI to human intelligence versus comparing it to workplace or academic performances have implications for the tests and types of skills or tasks used for its assessment.

Based on the October 2020 workshop, the project is focused on the ability of AI to reproduce human performance in any conceivable human work or activity. To compare AI and human workplace performances, bounded domains of skills and content are much more useful than measurement approaches that depend on general constructs, e.g. math ability. However, to ensure that all human activities are covered, reflecting on the selected set of skills and content periodically is recommended. This helps map activities that were left out of the first selection (for example because they have emerged since the first assessment).

- Compare outcomes alone or combine outcomes and process?

AI systems use powerful computational approaches to solve problems or consolidate information. However, they can vary in terms of their strategy. Strategies include bottom-up exploratory approaches in machine learning, such as unsupervised learning, and a combination of rule-based and bottom-up data processing, such as expert systems.

The assessment may seek to understand at a conceptual level how AI systems accomplish their tasks rather than simply their success in solving problems or reaching conclusions. In this case, AI and human process performance could also be compared. Indeed, as all AI systems profit from their massive computer power to solve specific tasks, they may be subject to measurement error when a problem involves non-linear relationships, while humans may be able to adjust to such complexities. Hence, a focus on both process and outcomes will have implications for task selection for the study.

The OECD is primarily interested in outcomes, which is a suitable approach if AI produces the same outcomes with a different process. However, process is likely to be important for generalising the results of a test to a broader range of capabilities than are directly tested. Some generalisations may be valid for humans but not AI and vice versa. For example, common sense inferences can usually be assumed in humans, while they must be tested explicitly in AI. That stems in part from process differences in developing expertise in AI and humans – AI expertise does not rest on a base of common sense the way it does for humans. Human-AI process differences should be kept in mind since they probably have implications for what needs to be tested on the AI side. Overall, some analyses or explanation of how the system arrived at its outcome must be provided for both acceptance and fairness issues.

- Document changes or improvements in AI competence over time or describe a single temporal point?

If the project intends to map the progress of AI systems across one or more sets of performance variables, then task selection should anticipate increases in AI performance over time. Thus, the tasks selected initially should represent areas in which AI currently does not do well. This would avoid a ceiling effect for later measurement occasions. If the intention is for one time only, then a range of "difficulty" or complexity of tasks could be included to detail how the systems operate today.

Issues in organisation and task selection

Selection of organisational systems and the corresponding tasks are major concerns for the study. The questions below amplify or posit additional prerequisite concerns about the best ways to determine taxonomies or organisational structures. Should the project select existing ones or develop new ones? The questions also examine assessments to be considered by the project.

Context of assessment tasks

- How important are the contexts of use of the assessment tasks? Is the project considering tasks drawn primarily from an academic disciplinary base or tasks that include context, such as general and specific knowledge?

If the interest is in disciplinary tasks, such as tasks of mathematical concepts or interpretations of the meaning of texts, then the project might assemble data from an age- or experience-based range of subjects. It would then ask the project rating team to make their estimates of AI systems using categories that capture span of respondent differences. For example, raters might be asked to estimate how well upper elementary students, junior high school algebra pupils or college math students could solve a problem. This distribution of performers would provide a database to guide the raters in making their judgements. However, disciplinary assessments often include implicit contextual information. This could have unintended effects on performance, especially if background characteristics of respondents are unknown or not included in the study.

Instead of disciplinary-focused tasks or items, tasks could be selected from areas in workforce skills. While there is contextual and background knowledge here, much of it is job related, which the systems could learn. This strategy is recommended if the project wants to influence how AI is perceived, used or evaluated in workplace context. It involves the selection of tasks (both comparable and unique) from two or more sectors, e.g. finance, hospitality. This method would indicate the robustness of a system across sectors and types of embedded domain knowledge. It could also give useful information for similar tasks, as well as about the functionality of the system on tasks wholly unique to a particular sector (think of a Venn diagram showing shared and unique knowledge and skills).

Prescriptive vs descriptive functions

- Is prescription or description intended as the principal function of the taxonomy or other organising system with respect to the experimental tasks? Is the taxonomy to guide selection of tasks, i.e. to function prescriptively? Or, will the organising system reflect, in a verbal or symbolic way, the nature of the tasks to describe and represent them as they exist?

If tasks were selected, perhaps from domains assessed by the German vocational/technical system – in two or more sectors – what similarities and differences in assessment design structure would be found? How might such bottom-up analyses among sectors relate to the O*NET structure? Depending on the sphere of influence selected as the major target of the project, a hybrid approach could be adopted. This approach would use formulations that guided the creation of tasks, i.e. German sector tasks or O*NET descriptors. It would then modify organisational structures and descriptions depending upon whether the purpose was an existence-proof project or the beginning of a sequence of connected studies about AI progress.

The OECD has been thinking of the taxonomies functioning prescriptively to guide task selection for two reasons. First, taxonomies can serve as a reminder of capabilities or workplace activities that need to be assessed. Second, they can point to capabilities or workplace activities that are closely related and should be assessed together. A descriptive approach would make sense if specific work contexts are of interest, each with well-defined authentic tasks that would provide the natural focus of assessment. In that approach, the OECD could see the need to make sense of the various tasks that arise in those contexts. However, in principle, the full range of current and possible work contexts must be addressed. Therefore, a prescriptive framework will more likely be needed. This framework would consist of taxonomies of capabilities and workplace activities to use in specifying what needs to be assessed.

Characterising artificial intelligence systems

- How can AI systems be characterised?

Following the selection of tasks and tests, experts will be selected to rate how computer systems would fare with the problems or tasks. This can be done in multiple ways.

First, through procedures that permit the direct processing of tests or tasks by existing AI systems themselves.

Second, by assessing the capabilities of specific AI systems. This approach may yield more reliable judgements but involves some decisions:

  - Which system attributes will be included? How will they be represented or described? Will experts be expected to understand a subset of common AI systems?
  - Should AI systems that have an explanatory function be selected or rather those that do not (e.g. a white box or black box)?

Third, by assessing the capabilities of current technology in general. In this case, experts rate the feasibility of well-defined tasks based on their knowledge of state-of-the-art AI approaches.

There are many ways to combine and describe data for this project. For example, there are several options regarding how judgements are made, how ratings are tabulated and summarised. Rating can take place in various forms such as through interface development options.

Rating process

This section suggests refinements to the human rating process. These approaches are derived from the extensive experiences available from ratings of open-ended achievement tests, as well as from new ideas stimulated by the workshop. All decisions regarding rating have cost implications, to be discussed in a subsequent section. Cost implications interact with decisions to have one status study or a sequence of studies over time assessing AI system progress.

Raters

Raters need to understand the nuance, as well as the operations, of their study tasks. They must have both adequate training and time to accomplish their judgements. Finally, the processes they used must be documented.

Number and selection of raters

- Who will be the actual raters of the assessment materials? Who will determine if the selected tasks are representative of the desired domain or sub-domain? How will the tasks be identified and selected?

Published tasks may or may not be thoroughly vetted. Unless there is robust evidence on this process, a group of experts should examine and judge their quality and suitability for the study.

A second group of raters, the AI experts, will then judge how well AI systems could complete all or part of any tasks.

For both rater groups, explicit criteria should be set for selection, such as expertise, availability (based on notions of ideal timing for the project, estimated time for training and the study ratings, post-rating interviews, willingness to participate in subsequent ratings).

- How many raters are needed?

This decision depends first upon the estimated numbers of tasks or items to be rated. If the tasks are selected from a common disciplinary area, such as mathematics, then raters should be at least familiar with the area and have teaching experience with the content of the tasks. One would also want raters to demonstrate competence on the study tasks. The project may wish to complete its work in a single language, e.g. English.

- Should raters be selected from homogenous work environments or represent a broader swath of backgrounds and experience?

On the one hand, irrelevant task variance should be limited. On the other, a set of raters with a limited or perhaps peculiar orientation based on the source of their experience should be avoided.

- Could an English-only study be replicated in other OECD countries?

Bilingual raters could ensure the same level of stringency is used in any between-country study. If multilingual raters are used in the main study, a minimum of three from each country could be used for their topic.

Expectations of performance would vary by country largely in any domain as a function of nationality. For example, American and Japanese raters would have distinct expectations for their own students' expertise in given subject matters. Psychologists and measurement experts would rate quality of items along a set of dimensions. However, like the subject matter experts, they will be influenced by their experiences.

Instead of pure measurement experts who likely have preferred methods if not ideology, it may be best to find experts with a "conjoint" interest in learning and assessment. In this instance, conjoint refers to individuals who understand the learning requirements for success at particular tasks and who can judge item or task quality as well. It does not mean formative assessment devotees who advocate specific classroom assessments for teachers. Experts from the learning sciences would be ideal.

Work of raters

- How will raters work, where and for how long?

Ratings could take place at a common site or distributed sites, e.g. schools or homes. Depending upon timing, training and rating can take place over Zoom or other platforms that permit recording. In rating A-Level exams in England, raters experienced interruptions. Compliance among individuals varied. Some completed all tasks as rapidly as possible during a single sitting despite their given instructions to complete a certain number (randomly ordered) each day.

Creation of scoring rubrics

- What kind of supports do content and AI raters need for the rating process?

The process should avoid binary responses (e.g. where AI either can or cannot successfully address the item or task). An agreed-upon protocol could constrain the AI raters to a specific range of systems. Either it would provide system functional descriptions or architecture, the kinds of output or uses to which the system has been put, or use a standard set of AI systems. This last choice would require a qualifying "test" to determine that prospective AI raters know and understand the systems to be considered.

Both rater groups will need rubrics to guide judgement of analytical components, as well as to help them make an overall rating of the likely success of the system (or system types, if more than one is considered). There have been many studies of rubric construction, reliability, clarity and other characteristics. The rationale behind experts' judgements should be documented.

Ideally, rubrics could begin with an overall judgement on a preferred scale, e.g. 1-4, 1-6. This would be followed by a sub-scale scoring of attributes or elements that may have been considered in that decision. There has been work that begins with sub-scores and adds them up for a final judgement, but that practice turned out to be less useful.

In addition to an overall judgement, sub-elements on the rubric might describe a comparable task known to have been successfully encountered by the system, the logic or linearity of the task, relevance of specific prior knowledge, the clarity of the language used in the task, and so on.

For the rubric used to judge test tasks or items, subcomponents could relate to the importance of the tasks, the critical content or contextual elements it includes, its dependence on prior knowledge, clarity of task for the examinee, and – if open-ended – the quality of the scoring rules intended to be applied.

- How much detail should rubrics have? How should they be developed?

Most experience advises against long and detailed scoring rubrics because of fatigue, lack of inter-rater agreement and the general inter-correlation clusters of elements.

It is good practice to create and try out rubrics before training raters to use them. In that process, a separate set of experts is given a set of experiences that model the range and timing of a single scoring session. Raters may independently use the rubric and note difficulties. Other approaches involve some "think-aloud" protocols where the raters speak aloud the process they are using to arrive at scores for both the overall and sub-categories.

Judgements about rubric quality tend to cluster around its clarity, applicability to the range of tasks, brevity and understanding of the meanings of rating points, (how does a "3" differ from a "4?").

The development of rubrics should also involve verifying that experts obtain similar ratings on the same tasks or items. Rubric development experiences should ideally be sequenced. This allows the rubric to be tried on two or so items, notes and scores compared, and discrepancies found and discussed. The rubric can then be revised and used for the next small set.

The development process should also examine whether scores are more sensible when the raters score as they go, or when they first only make notes and score all elements at the end. This latter approach can ensure that the same scale values are used across rating sub-topics. Once the process has stabilised the rubric, a trial rating occurs to estimate the time it takes to score. Limiting scoring time per task is good practice.

Rater training

- How should raters be trained?

Rating training could consist of a face-to-face work (preferred) or a combination of face-to-face and webinar with feedback. Rater training involves both training and qualification. Training involves using the same rubric applied to a *different* set of examples than those to be encountered in the actual study.

Three sets of technical quality criteria are relevant. First, raters need to agree with a criterion set of ratings, using approaches that cleave to the rubric-incorporating experts' views. Second, there should be intra-rater agreement, meaning that a single rater assigns much the same value to identical or similar tasks over time. Third, there should be inter-rater agreement on the same task, usually expressed as the obtained percentage of exact agreement between pairs of raters. In practice, the first two criteria are more important.

The order of rating needs to be organised to represent a number of elements. A general random order with repeated examples (probably three) should establish consistency within raters. There should also be a way to enter scores and have computations of agreement take place on the spot.

If agreement is found to be low, following the ratings, the "leader" can discuss the ratings in terms of the criterion cases, but there should be no opportunity for raters to discuss and modify their scores. Such discussion usually results in a socially defined rating that may be unique to subsets of raters and undermines the validity of the process.

Finally, more individuals than needed will need to be trained because some raters will likely not qualify. Ideally, another project-related task may be found for them, such as compiling and summarising information.

Answering the questions

The OECD project posed six questions at the outset, which are answered below.

- How should the project decide on what types of skills to assess?

Domains with sets of coherent knowledge, skills and attributes are best for tasks related to performance using skills and content (rather than broad constructs). The German efforts on tasks and standards within the world of work sectors are relevant and of high quality, both to the project and to AI system development. Raters could also be given additional tasks that will stretch even the best of current system capabilities (transfer). This is because elements of transfer could be described related to completeness of queries, types of logic, extrapolations and so on.

- Are there existing skills taxonomies that describe the right set of skills to assess?

If a workplace setting is chosen, then the skill taxonomies underlying the German or other such system (like O*NET) should be used.

- How should the project decide on the types of tests to be used?

One or more bounded domains in a clear use context could be used as the source of assessments. This approach will have a greater likelihood of success because it will have limited abstraction and communicate best with desired audiences. Operationally, the use of a compilation of tasks or items rather than selecting an existing test is advised. There will be trade-off between relativity and validity in either case.

At least two sectors or large content domains are advised to demonstrate generalisability of the approach. In addition, within a domain, a range of task and item types should be employed, probably no less than three for each type. Common descriptors of items will include item formats, ranging from brief to extended, that provide or expect recollection of essential prior knowledge, which include selected or open-ended responses. Open-ended responses should include multi-stepped solutions to problems that are ill formed, partially formed or where the problem statement is given. Tasks that require various iterations among task formulation, resource search and acquisition, integration and solution fit are desirable.

Assuming systems will have natural language processing capability, there may be excellent tasks that require paraphrasing. This will create a text answer that demonstrates understanding of the levels of meaning of particular questions that are independent of any specific content domain.

Items used should vary in complexity and completeness. Ideally, data should be available regarding item difficulty or facility with answers for different groups. Thus, items and tasks to be selected should include a range of cognitive requirements, task formats, and use of language and other underlying skills.

- Are there existing types of tests that are good candidates for assessment?

The work provided at the October 2020 workshop for the *Artificial Intelligence and the Future of Skills* project would be a good start. Self-report and game/simulation measures are available from previous research.

- Are there existing types of tests that should be excluded from the assessment?

Well-known standardised tests for admissions to higher education or for evaluation and accountability in pre-collegiate environments should be avoided. Some communities see these tests as biased.

- Are there new types of tests that need to be developed for the assessment?

Yes, if both outcomes and processes can be identified for this environment. Rather than a completely new development, it seems likely existing tests could be modified.

Further reading

AERA, APA and NCME (2014), *Standards for Educational and Psychological Testing,* American Educational Research Association, American Educational Research Association, American Psychological Association and National Council on Measurement in Education, Washington, DC.

Baker, E.L. et al. (2022), "Assessment principles for games and innovative technologies", in O'Neil et al. (eds.), *Theoretical Issues of Using Simulations and Games in Educational Assessment,* Routledge/Taylor & Francis.

Baker, E.L. et al. (2019), *Validity Studies and Noncognitive Assessments* (Deliverable Item No. 013 to funder), National Center on Evaluation, Standards, and Student Testing, University of California, Los Angeles.

Bergner, Y. and A.A. von Davier (2019), "Process data in NAEP: Past, present, and future", *Journal of Educational and Behavioral Statistics*, Vol. 44/1, pp. 706-732.

Buhrmester, M.D., S. Talaifar and S.D. Gosling (2018), "An evaluation of Amazon's Mechanical Turk, its rapid rise, and its effective use", *Perspectives on Psychological Science*, Vol. 13/2, pp. 149-154.

Choi, K. et al. (2021), *Molly of Denali Analytics Validation Study Report—final* (Deliverable to PBS KIDS)., Los Angeles, National Center for Research on Evaluation, Standards, and Student Testing, University of California, Los Angeles.

Chung, G.K.W.K. (2015), "Guidelines for the design, implementation, and analysis of game telemetry", in C.S. Loh, Y. Sheng and D. Ifenthaler (eds.), *Serious Games Analytics: Methodologies for Performance Measurement, Assessment, and Improvement*, Springer, New York.

Clariana, R. and E. Taricani (2010), "The consequences of increasing the number of terms used to score open-ended concept maps", *International Journal of Instructional Media*, Vol. 37/2, pp. 163-173.

Dunbar, S. B. and C.J. Welch (2022), "It's game day in large-scale assessment: Practical considerations for expanded development and use of simulations and game-based assessment in large-scale K-12 testing programs", in O'Neil, H.F. et al. (eds.) *Theoretical Issues of Using Simulations and Games in Educational Assessment,* Routledge/Taylor & Francis.

Kirkpatrick, J.D. and W.K. Kirkpatrick (2016), *Kirkpatrick's Four Levels of Training Evaluation*, Association for Talent Development, Alexandria, VA.

Klein, D. et al. (2002), *Examining the Validity of Knowledge Mapping as a Measure of Elementary Students' scientific understanding* (CSE Tech. Rep. No. 557), National Center for Research on Evaluation, Standards, and Student Testing, University of California, Los Angeles.

Mislevy, R.J. et al. (2015), "Psychometrics and game-based assessment", in C.S. Loh, Y. Sheng and D. Ifenthaler (eds.), *Serious Games Analytics: Methodologies for Performance Measurement, Assessment, and Improvement,* Springer, New York.

O'Neil, H.F. et al. (2021), *Using Cognitive and Affective Metrics in Educational Simulations and Games: Applications in School and Workplace Contexts,* Routledge/Taylor & Francis.

O'Neil, H.F. et al. (1994), "Human benchmarking for the evaluation of expert systems", in O'Neil, H.F. Jr. and E.L. Baker (eds.), *Technology Assessment in Software Applications*, Hillsdale, MI.

Quellmalz, E. et al. (2012), "Science assessments for all: Integrating science simulations into balanced state science assessment systems", *Journal of Research in Science Teaching*, Vol. 49/3, pp. 363-393.

Schenke, K. et al. (2021), "Measuring and increasing interest in a game", in O'Neil, H.F. et al. (eds.) *Using Cognitive and Affective Metrics in Educational Simulations and Games: Applications in School and Workplace Contexts,* Routledge/Taylor & Francis.

20. Building an assessment of artificial intelligence capabilities

Stuart W. Elliott, OECD

This chapter synthesises the expert contributions of the report and offers perspectives for building a comprehensive assessment of artificial intelligence (AI) capabilities. It compares and contrasts the contributions of psychologists and computer scientists along two dimensions: whether they focus on human or AI taxonomies and tests, and whether they test isolated capabilities or more complex tasks. The chapter argues that a more complete assessment of AI must bring together the different approaches. It illustrates this argument with an example in the area of language. Finally, the chapter offers next steps towards a systematic assessment of AI capabilities, which will allow for drawing fine-grained implications for work and education.

Introduction

Through the *Artificial Intelligence and the Future of Skills* (AIFS) project, the OECD is developing an approach to assessing the capabilities of AI and comparing them with human capabilities. The goal is to develop a comprehensive programme to measure these capabilities in a valid, reliable and meaningful way that policy makers can use to understand the implications of AI for education and work. AIFS is a six-year project, which will include an initial systematic assessment of AI capabilities and analysis of their implications. It will conclude with a proposed approach for an ongoing programme to assess AI capabilities at regular intervals.

This volume addresses questions related to identifying the AI capabilities that the project should assess, as well as tests that could be used to assess them. Based on a workshop in October 2020, the volume surveys a broad range of work in psychology and computer science that can provide relevant taxonomies of capabilities and assessments of them.

The papers in the volume make clear there are many resources the project can use to define a set of capabilities and associated assessments for measuring the capabilities of AI. Before the workshop, the AIFS project team hoped to identify a single comprehensive taxonomy that could be linked to an appropriate set of assessments to use for the project. Given the number of available taxonomies in psychology, this hope seemed potentially realistic. However, the experts who participated in the workshop and wrote papers in this volume put forward other thoughtful arguments. They suggest that AIFS could benefit from combining the complementary strengths of several different approaches rather than using only one.

The meeting discussion and this resulting report highlight two dimensions of difference that should be reflected in an assessment of AI capabilities: the contrast between human and AI taxonomies and tests, and the contrast between testing capabilities and tasks.

Human vs. artificial intelligence taxonomies and tests

As amply illustrated in Part I of this volume, rich resources in psychology reflect long research traditions to develop the conceptual and practical tools for cognitive assessment in humans, as well as animals. Furthermore, the efforts to assess AI capabilities in computer science have often started from these materials, as noted by Hernández-Orallo in Chapter 11, because of their broad coverage and availability. Indeed, the computer science community acknowledges the intellectual foundation and extensive materials provided by psychology. However, computer scientists also clearly state that human tests can be misleading and incomplete when used to assess AI. Many papers stress this point in Part III.

Computer scientists note that human tests focus on aspects of capabilities that are meaningful for assessing humans. However, these tests are not necessarily meaningful for assessing for AI. Assessments are always incomplete, focusing on only a sample of the capabilities of interest. They then assume the sampled capabilities also provide information about the critical unsampled capabilities needed for competent performance.

Because the cognitive capacities of humans and AI are different, assumptions about unsampled capabilities are different. Therefore, the sampled capabilities included on a test need to be different. As a result, AI needs to be assessed for things rarely considered for direct assessment in humans, such as common sense reasoning. This could lead to somewhat different taxonomies of the capabilities needed to consider for AI. Ultimately, this could lead to entirely new tests to assess those capabilities.

Testing capabilities vs. tasks

In Chapter 15, Avrin discusses the importance of assessing both isolated capabilities ("functionality benchmarks") and the performance of complete tasks ("task benchmarks") in evaluations of AI. Usually, the tasks are the priority. The task benchmarks are then chosen to reflect something wanted from an AI system in the real world.

However, tasks almost always require complex combinations of capabilities. An AI system may fail because of the inadequacy of either one of the capabilities or their integration. Assessing the individual capabilities provides a way to determine whether each one is sufficient for the task; the assessment of the task itself determines whether the separate capabilities function effectively together.

This contrast that Avrin describes on the AI side is richly illustrated across the different types of assessments on the human side.

On the one hand, assessments in psychology often attempt to isolate individual capabilities for assessment and avoid using tasks that will confound the contributions of several distinct capabilities. The process of separating the contributions of different individual capabilities with specially designed tasks is at the heart of the factor analytic tradition in psychology discussed by Kyllonen in Chapter 3.

On the other hand, many human assessments focus on authentic tasks of interest in human contexts and intentionally mix the full set of capabilities needed for those tasks. The occupational tests discussed by Rüschoff in Chapter 9 provide the clearest example of authentic tasks that require many separate capabilities to carry out. These tasks often involve not only cognitive capabilities related to language, reasoning and problem solving but also additional capabilities related to social interaction, sensory perception and psychomotor control. The educational tests discussed by Greiff and Dörendahl in Chapter 7 often aim at a middle range of complexity. They mix capabilities related to language, reasoning and problem solving but omit capabilities related to social interaction, sensory perception and psychomotor control that can be important in many work contexts.

Working with both dimensions

The four revolutions of Forbus

In Chapter 2, Forbus illustrates the two dimensions of difference – human vs. AI taxonomies and tests and testing capabilities vs. tasks – through four revolutions in AI.

The first three revolutions relate to the key categories of the human cognitive taxonomies: one can link learning, knowledge and reasoning directly to Carroll's 3-stratum model of human cognitive abilities, discussed by Kyllonen in Chapter 3, as general memory and learning, crystallised intelligence and fluid intelligence, respectively.

The fourth revolution – agency – relates to the complex way that humans can integrate capabilities. It is reflected, for example, in the complex tasks carried out in human jobs, as well as the basic developmental stages in children.

Forbus uses this revolutions framework to identify both recent successes and key missing aspects of current AI capabilities. Crucially, some missing aspects are ones that may not be typically assessed on the human side. These include the ability to learn from a single example, knowledge of one's personal experience and common sense reasoning. Agency then provides the example of the combination of capabilities that is still missing to carry out real-world tasks in context.

While illustrating the two key dimensions of difference, Forbus also highlights the larger motivation for the AIFS project: there are revolutions occurring or approaching in each of these four key areas of AI cognition that will result in qualitative shifts in AI capabilities. The prospect of major improvements in AI capabilities underlines the importance of providing measures for the policy community that identify what capabilities are missing. These measures can provide an early warning system, identifying when those capabilities appear and offering guidance to their implications.

Combining the two dimensions

Figure 20.1 suggests a way to fit the two dimensions together in a framework for assessing AI capabilities. This figure provides an initial framework for synthesising the different taxonomies and tests discussed in this report.

The first dimension – differentiating between human and AI sources for assessment – is illustrated horizontally at the bottom of the figure. AI assessment approaches derived from human capability frameworks appear on the left, while assessment approaches focused on missing AI capabilities are on the right.

The second dimension – differentiating between testing separate capabilities and complex tasks – is illustrated vertically. Assessment of separate capabilities is at the bottom, while assessment of real-world tasks requiring use of multiple capabilities is at the top.

The boxes for human capability frameworks and real-world tasks reference some of the taxonomies that describe and categorise relevant capabilities and tasks, respectively, and that link to a variety of assessments. The box for missing AI capabilities differs from the other two boxes in listing capabilities that are "special cases" rather than listing frameworks. These special case capabilities are often missing in two senses: they are missing from AI's current capabilities and from many (but not all) of the capability frameworks and assessments used to describe humans. As a result, the missing AI capabilities are both important to assess and require extra effort to identify assessments focused on AI's unique challenges.

Figure 20.1. Sources for AI assessments

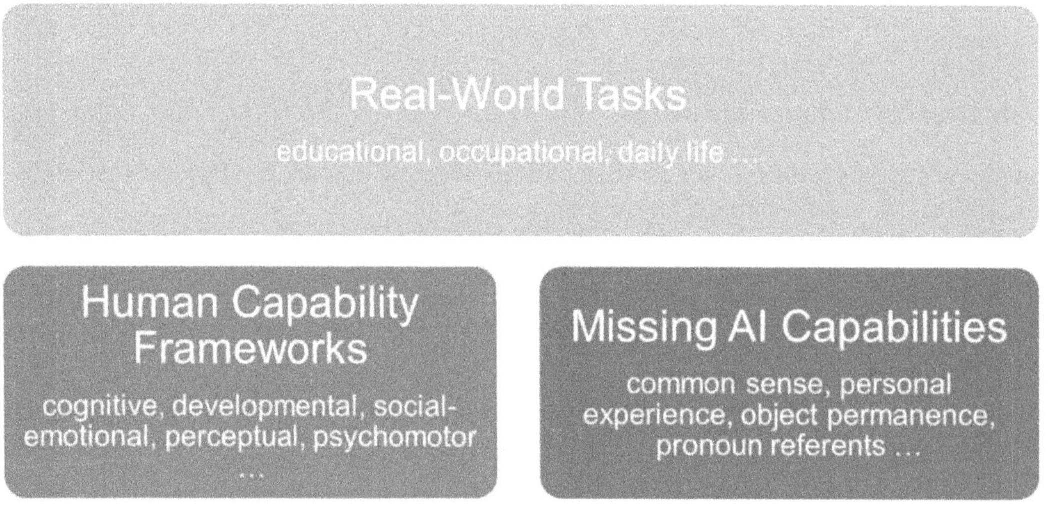

Filling in the details

The chapters provide a wealth of detail about the kinds of capabilities and tests that might go into each of the boxes in Figure 20.1.

Starting with the Human Capability Frameworks box, Chapters 3, 5 and 6 present taxonomies and tests for a set of isolated human abilities. Kyllonen in Chapter 3 outlines the comprehensive taxonomies developed to describe the full range of cognitive abilities, building on a rich assortment of associated tasks. These taxonomies have been extended by some researchers, and Kyllonen briefly discusses some of the work related to social-emotional, perceptual, psychomotor and other skills. Kyllonen's initial overview is then supplemented by more detailed discussions in some of the other chapters in Part II. De Fruyt in Chapter 5 discusses social and emotional capabilities, along with some tests developed to assess them. Woolley in Chapter 6 provides a detailed discussion of the components of social capabilities that allow groups to function effectively, along with some novel assessments of those capabilities.

These human taxonomies are well developed and provide an extensive set of human tests for the different isolated abilities that could potentially be applied to assess AI. Some areas appear to be less interesting for AI assessment because AI systems have already mastered the abilities or could be developed to do well on the test without the underlying capabilities of interest. It would be necessary to choose tests carefully, in some cases using existing tests as an inspiration to develop versions that would be more likely to produce valid results for AI.

Two chapters reside in the Human Capability Frameworks box but represent an attempt to identify frameworks and tests on the human side that might be developed to address some of the Missing AI Capabilities. Chokron in Chapter 4 describes the many domains and assessments used in neuropsychological evaluation in children. Research on the assessment of deficits of normal cognitive functioning in children raises the possibility of identifying assessments of some missing AI capabilities that are usually also missing from tests for adult humans. Similarly, Cheke and colleagues in Chapter 17 use approaches for testing basic capabilities in young children and animals to develop some tests for AI systems of these capabilities. This chapter is placed in Part III of the report because it has already made the move into a set of applications for assessing some missing AI capabilities, but it rests on a research foundation from human and animal psychology.

Moving completely over to the Missing AI Capabilities box, the various papers from computer scientists outline a set of examples of assessments that have been or could be carried out related to AI systems:

- Hernández-Orallo in Chapter 11 provides an overview of the different approaches.
- Avrin in Chapter 15 discusses a number of AI and robotics systems that have been formally evaluated at the Laboratoire national de métrologie et d'essais in France, including a number of individual capabilities. The chapter also discusses the assessment of complex tasks, which belong in the Real-World Tasks box.
- Graham in Chapter 16 describes the assessment of different components of natural language capability. She makes the case that the field of natural language processing has developed assessments that go beyond typical human assessments. They now focus on the specific challenges and current level of capability in available AI systems for natural language processing.
- In Chapter 14, Cohn describes a few of the competitions and benchmarks used to compare performance of AI systems in different areas. He notes how competitions and benchmarks often evolve to focus on performance levels that are almost but not yet attainable by the field.
- The papers by Davis in Chapter 12 and Granger in Chapter 13 illustrate how AI assessment can go awry. They provide surprising examples of "brittle" performance of AI systems where seemingly small task differences produce large differences in the results. The authors present these examples as cautionary tales. Yet the examples simultaneously illustrate assessment techniques that can identify such brittle performance, at least in some cases.
- Finally, Forbus in Chapter 2 outlines several possible strategies for assessing AI progress related to the different revolutions he describes.

These many efforts do not suggest an integrated framework for assessing AI with respect to the aspects of capabilities that are not well reflected on human tests. However, they indicate several different approaches that can be explored for doing so.

Moving up to the real-world tasks involving combined capabilities in Figure 20.1, several chapters consider educational or occupational tests. Many educational and occupational tests focus on isolated capabilities that would appear in the Human Capability Frameworks box of the figure. These include, for example, capabilities in skills related to reading or mathematics. However, the chapters on educational and occupational tests in this report largely focus on tests that require combinations of capabilities. These are tests inspired by complex tasks in the real world, occurring in the context of education or work.

In Chapter 7, Greiff and Dörendahl provide an overview of different educational tests, including both core domain and transversal skills. Each of the assessments focuses on a particular capability, like reading literacy or problem solving. However, all the assessment tasks discussed require a mix of capabilities, including various aspects of language, reasoning and problem solving.

Finally, three chapters provide an overview of occupational tests, and the complexity of the tasks that can sometimes be included in them. Ackerman in Chapter 8 argues for the benefits of assessing AI using domain-specific tests for different occupations that include assessment of both declarative knowledge and hands-on procedural knowledge. Rüschoff in Chapter 9 then introduces the testing programme in the German vocational education and training system. A detailed discussion of the framework includes examples of specific tests. Dorsey and Oppler in Chapter 10 provide an overview of the framework for understanding occupational tasks and worker capabilities included in the US Department of Labor's Occupational Informational Network, along with several examples of occupational tests.

Fitting the details into the framework

The framework in Figure 20.1 suggests the importance of bringing together different types of approaches to provide a more complete assessment of AI. Building on the examples discussed in the chapters, Figure 20.2 shows a partial example of what this might look like in the area of language. At the bottom of the figure are examples of tests for isolated language capabilities. One is a test of vocabulary noted by Kyllonen in Chapter 3 as an example from the human side. The other is the Winograd Schema noted by Cohn in Chapter 17, which was developed to assess AI's ability to identify difficult pronoun referents. Greiff and Dörendahl in Chapter 7 provide an example of a reading literacy task from the Programme for International Student Assessment. This involves understanding a reading passage, reasoning about it and providing a written answer. Finally, Rüschoff discusses a multi-hour work sample assessment for Advanced Manufacturing Technicians. This includes several oral discussions about a complex work task involving the construction of a functional module according to a set of technical drawings.

As illustrated in Figure 20.1, the chapters have gone beyond suggesting an assessment approach that combines several different types of assessments. They have also provided concrete examples of the types of assessment tasks needed for such an assessment.

In his reflections on the project in Chapter 18, Graesser raises the important point about how to integrate the various taxonomies suggested in the different chapters. In Graesser's terms, the provisional framework for a synthesis in Figure 20.1 is perhaps a "comprehensive" synthesis, where complementary aspects of different approaches are added together. However, the two key dimensions of difference suggested by the chapters also provide the initial ingredients for a synthesis that is more theoretically motivated. Over time, the OECD hopes this can move to a rough consensus related to the capabilities and types of assessment tasks that the AIFS project should include.

Figure 20.2. Assessing language with multiple tests

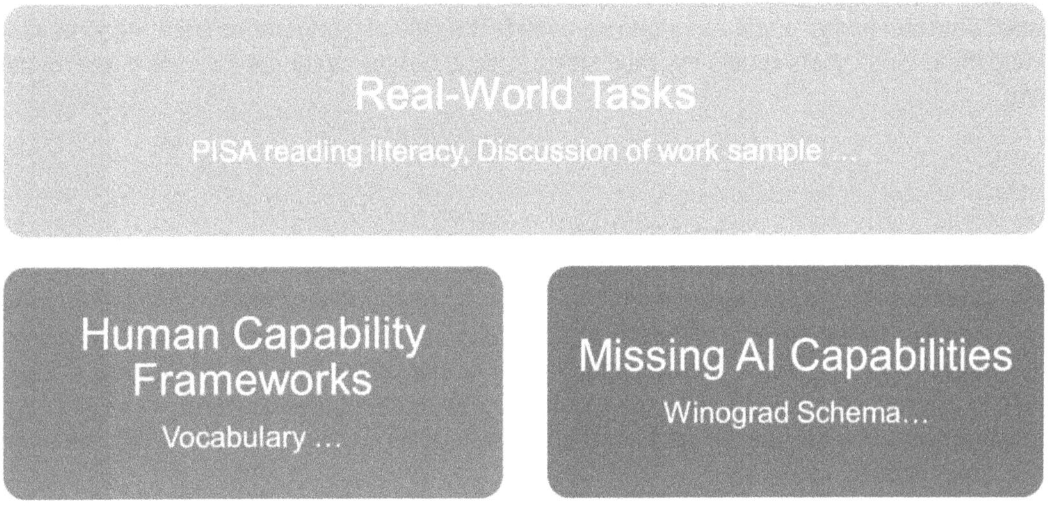

Next steps

This volume is only a starting point for the project. There is substantial work to do to build a specific set of assessments to provide policy makers with an understanding of AI capabilities and their implications for education and work.

Several chapters in this report described approaches for moving directly to assess implications of AI without going through the intermediate step of evaluating AI's capabilities. As described in the introduction, the research literature includes a number of efforts to take this more direct approach. While acknowledging the value of this work, the AIFS project is building a more substantial foundation related to understanding AI capabilities before moving on to their implications.

The project is motivated by the importance of developing a more robust and meaningful understanding of AI that can support policy makers in understanding its implications. This is particularly important with respect to the educational implications of AI – the primary motivator of the project – which require a fine-grained understanding of the way that human and AI capabilities will complement each other.

The next stage of the AIFS project will involve piloting the types of assessments described in this volume to identify how well they provide a basis for understanding current AI capabilities. This work will begin with intense feedback from small groups of computer and cognitive scientists who attempt to describe current AI capabilities with respect to the different types of assessment tasks. It will build the project's understanding of the types of assessment approaches that give a valid and reliable picture of AI capabilities. With more understanding of types of assessments to use, the project will expand the range of input to include a broader sample of experts who know about AI so that we can fully represent the field.

Baker and O'Neil in Chapter 19 and Graesser in Chapter 18 anticipate a number of challenges that will come in this next phase. Baker and O'Neil review a set of practical issues that must guide the process of gathering ratings from experts who know about AI, as well as the larger context and framing that experts must consider when providing their ratings. Graesser raises a key question about how one should evaluate a comparison of AI and human performance. He asks whether human performance should be defined as the standard for evaluating AI – as is often done in AI evaluations. Should an objective, ideal model be used instead? The project will need to address these questions in the next stages of development work.

The initial work has also given the project an appreciation for the range of empirical measures of AI capabilities – such as those discussed by Cohn in Chapter 14 – that could potentially provide some of the assessments the project seeks to create. At this point, the usability of these empirical AI measures is an open question. Many experts are concerned that they are often too narrow to provide informative comparisons between humans and AI for people outside the field of computer science. As a result of these discussions, the project now expects the next steps to consider the potential value of these measures to the project.

From assessment to implications

The project will ultimately translate assessments of AI capabilities to their implications for education and work. One part of this will involve a simple comparison of AI and human capabilities in different areas. This will aim to understand what aspects of different capabilities are still well beyond AI and how many people have those capabilities. In addition, the project will envision different scenarios for applying AI capabilities to the tasks in different occupations. This will be a way to understand how humans will begin to work with AI systems that have new capabilities and how human occupations will evolve, along with the educational preparation they require.

The last step of the translation process will be to develop ways to communicate the results of these assessments and analyses to policy makers and the general public. This will likely involve creation of a set of indicators across different capabilities and different work activities to communicate the substantive implications of AI capabilities in meaningful terms.

The development work is projected to continue through the end of 2024. At that time, a first systematic assessment of AI capabilities should be completed. A translation of that assessment to its implications for education and work, with a set of meaningful indicators that describe those results, is also expected to be finished. Of course, AI is advancing rapidly and a single assessment would quickly become outdated. The final stage of the development work will be to define a programme for regular updates to the assessment.

In the larger vision for this work, an ongoing programme of assessment for AI will add a crucial component to the OECD's set of international comparative measures that help policy makers understand human skills. The Programme for International Student Assessment (PISA) provides the link from education to skill, while the Programme for the International Assessment of Adult Competencies (PIAAC) provides the link from skill to work and other key adult roles. The AIFS project is now building a component that will relate human skills to the pivotal technology of AI, thereby providing a bridge from AI to its implications for education and work, and the resulting social transformations in the decades to come as AI continues to develop.

List of contributors

Phillip L. Ackerman

Phillip L. Ackerman (Chapter 8) is Professor of Psychology at the Georgia Institute of Technology. He received his Ph.D. from the University of Illinois, Urbana-Champaign. He has conducted basic and applied research in cognitive psychology, individual differences, psychological testing, and human abilities. He has written extensively on the nature of adolescent and adult learning, skill acquisition, selection, training, abilities, personality, and motivation. He has co-edited three books on individual differences and is the editor of a book on cognitive fatigue. Dr. Ackerman's main contributions involve integration across multiple fields of psychological inquiry, specifically related to individual differences. Noteworthy contributions include integration of information processing and ability approaches to individual differences in skill learning; of ability and motivational determinants of learning and performance; of ability, personality, and interest traits; and explication of an integrated approach to adolescent and adult intellectual development. He is a Fellow of the American Psychological Association; Human Factors & Ergonomics Society, American Educational Research Association, Psychonomic Society, and he is a Charter Fellow of the Association for Psychological Science.

Guillaume Avrin

Guillaume Avrin (Chapter 15) is in charge of the "Evaluation of Artificial Intelligence" department at LNE, French National Laboratory for Metrology and Testing. In this context, Dr. Avrin conducts scientific and technical work on the definition of evaluation protocols, metrics, and test environments (databases, simulators, physical test benches) for intelligent systems. He is an expert on trustworthy AI for various institutional actors (DG CONNECT, COFRAC, HAS, various ministries) and coordinates various consortia and working groups at an international level with the aim of defining and eventually converging on a fundamental and operational metrology for AI (Dr. Avrin coordinates in particular the European consortium METRICS, made up of 17 testing centers for intelligent robots, as well as a working group of about fifteen industrial companies that created the first certification standard for AI). It also participates in standardization committees on AI and robotics (Afnor IA, ISO/IEC JTC1/SC42, CEN-CENELEC JTC21, UNM81, etc.), intervening notably as head of the French delegation to the JTC21.

Eva L. Baker

A Distinguished Professor at UCLA, Eva L. Baker (Chapter 19) researches design and validation of multipurpose training and assessments systems, and is now developing games, simulations, and scenario-based assessments (workforce skills) for the U.S. Navy and PBS (early learning). Her AI studies include benchmarking as well as evaluations of ITSs, games, interventions, and applying AI to assessment. She has served as Co-Chair of the *Standards for Educational and Psychological Testing*, has been President of both the American and World Educational Research Associations, and is Founding Director of CRESST.

A member of the National Academy of Education and a fellow in scholarly associations (e.g., American Psychological Association, American Educational Research Association, American Psychological Society), she has numerous awards in measurement and is widely published.

Abel Baret

Abel Baret (Editor, Chapter 1) is a research assistant in the *AI and Future of Skills* project. He holds a BA degree in economics from the Toulouse School of Economics and a MSc in economics and psychology from Paris 1 Panthéon-Sorbonne and Université de Paris. Before joining the OECD, he was involved as a research assistant in experimental and behaviour economics at the CNRS Maison des Sciences Economiques in Paris.

Lucy Cheke

Lucy Cheke (Chapter 17) is an Assistant Professor at the Department of Psychology, University of Cambridge. She is also the Director of the Kinds of Intelligence Programme at the Leverhulme Centre for the Future of Intelligence, which focuses on artificial intelligence in the context of other – human and nonhuman - cognition. She has worked for many years on assessment of learning and memory in nonhuman animals, children and adults - and more recently AI agents. Much of this work has focused on the comparability and interpretability of cognitive assessments across and within groups/species, with an emphasis on pattern, rather than sum, of performance.

Sylvie Chokron

Sylvie Chokron (Chapter 4) is a neuropsychologist and a Senior Researcher at CNRS. She is the head of the I3N (Institut de Neuropsychologie, Neurovision et NeuroCognition) at the Fondation Ophtalmologique Rothschild, Paris, where babies, children, and adults with visual and cognitive deficits are diagnosed and treated. She also heads the Perception, Action and Cognitive Development team (INCC, CNRS and University Paris-Descartes) where she develops fundamental research on visual cognition, attention and spatial representation in typical, atypical and brain-damaged patients as well as clinical applications in the field of visual cognition. Sylvie Chokron is a lecturer in several master programs and has a regular Neuroscience chronicle in the French Newspaper 'Le Monde' as well as in 'Le magazine de la Santé' (TV show on France 5).

Anthony G. Cohn

Anthony G. Cohn (Chapter 14) is Professor of Automated Reasoning in the School of Computing, University of Leeds. His current research interests range from theoretical work on spatial calculi (receiving a KR test-of-time classic paper award in 2020) and spatial ontologies, to cognitive vision, modelling spatial information in the hippocampus, and Decision Support Systems, particularly for the built environment, as well as robotics. He is Editor-in-Chief of Spatial Cognition and Computation and was previously Editor-in-chief of the AI journal. He is the recipient of Distinguished Service Awards from IJCAI and AAAI as well as the 2021 Herbert A Simon Cognitive Systems prize. He is a Fellow of the Royal Academy of Engineering, the Alan Turing Institute in the UK, and is also a Fellow of AAAI, AISB, and EurAI. He holds Distinguished Visiting Professor positions at three Chinese Universities.

Matthew Crosby

Matthew Crosby (Chapter 17) is a research scientist at DeepMind and creator of the AnimalAI testbed. He is primarily interested in discovering how to build and understand agents capable of solving the kinds of cognitive tasks that humans, and many animals, find easy so that we can later build *and understand* agents capable of solving the kinds of cognitive tasks we find hard. While working on this problem he has collected a PhD and three Masters across AI, Philosophy, Mathematics and Cognitive Science and hopes to bring ideas from each of the fields together to solve the problem.

Ernest Davis

Ernest Davis (Chapter 12) is Professor of Computer Science at New York University. He received his B.Sc. in mathematics from MIT and his Ph.D. in computer science from Yale. Davis' research area is the representation of commonsense knowledge in artificial intelligence systems, particularly for spatial and physical reasoning. He is the author of more than ninety scientific papers and four books: "Representing and Acquiring Geographic Knowledge" (1986); "Representations of Commonsense Knowledge" (1990); "Linear Algebra and Probability for Computer Science Applications" (2012); and, with Gary Marcus, "Rebooting AI: Building Artificial Intelligence We Can Trust" (2019). He also has published numerous book reviews and articles for a general readership in The New York Times, the New Yorker, the Times Literary Supplement, WIRED, and elsewhere.

Filip De Fruyt

Filip De Fruyt (Chapter 5) is senior full professor of Differential Psychology and Personality Assessment at Ghent University in Belgium. He is also a member of Edulab21, the research branch of the Institute Ayrton Senna in Brazil. He is specialised in assessing and building psychometric models on how individuals differ from each other and how that affects their functioning in daily life and work. He currently holds the Institute Ayrton Senna chair at Ghent University. He is the Past President of the European Association of Personality Psychology (EAPP) and is a Fellow of the Society of Industrial and Organizational Psychology (SIOP). De Fruyt has (co-)authored over 200 research papers in a broad range of leading academic journals that are cited over 20.000 times (Google Scholar, 2021).

Jan Dörendahl

Jan Dörendahl (Chapter 7) is a Psychologist and Data Scientist. From 2016 to 2021 he was part of the research group Computer-Based Assessment at the University of Luxembourg. During that time, he investigated the assessment of fundamental motives and goals and obtained his PhD in psychology in 2019. Further, he lectured multivariate statistics and was among the lead item developers for the assessment of adaptive problem solving in the PIAAC 2021 cycle. Since 2021, Dr. Dörendahl is working as a Data Scientist combining machine learning algorithms and internet-of-things technologies into innovative solutions.

David Dorsey

David Dorsey (Chapter 10) currently serves as a Vice President at the Human Resources Research Organization (HumRRO). Prior to joining HumRRO, Dr. Dorsey was a senior executive in the U.S. Department of Defense, where he served as the Chief of Organizational Effectiveness and Workforce Research and as a Senior Data Scientist. Prior to his government service, Dr. Dorsey was a Vice President at Personnel Decisions Research Institutes (PDRI). Dr. Dorsey has produced over 70 professional book chapters, articles, and presentations. For his overall contributions to the field, Dr. Dorsey was elected a Fellow by the Society for Industrial and Organizational Psychology. He is the recipient of two major research awards and an award for being a top leader in government. Dr. Dorsey received his PhD in Industrial and Organizational Psychology with a graduate minor in Computer Science from the University of South Florida.

Stuart W. Elliott

Stuart W. Elliott (Editor, Chapter 20) is a senior analyst at the OECD where he leads the *AI and the Future of Skills* project. He holds a doctorate degree in economics from the Massachusetts Institute of Technology and a BA in economics from Columbia University. He also received postdoctoral training in cognitive psychology at Carnegie Mellon University. He authored the 2017 CERI report on *Computers and the Future of Skill Demand*, which provided the groundwork for the design of the *AI and the Future of Skills* project. He is also a scholar at the National Academies of Sciences, Engineering, and Medicine in the US where he has led studies on educational tests and indicators, assessment of science and 21st century skills, applications of information technology, occupational preparation and certification, and measuring productivity.

Kenneth D. Forbus

Kenneth D. Forbus (Chapter 2) is the Walter P. Murphy Professor of Computer Science and Professor of Education at Northwestern University. His research interests include qualitative reasoning, analogical reasoning and learning, spatial reasoning, sketch understanding, natural language understanding, cognitive architecture, reasoning system design, intelligent educational software, and the use of AI in interactive entertainment. He is a Fellow of the Association for the Advancement of Artificial Intelligence, the Cognitive Science Society, the Association for Computing Machinery, and the American Association for the Advancement of Science. He is the inaugural recipient of the Herbert A. Simon Prize, a recipient of the Humboldt Research Award and served as Chair of the Cognitive Science Society.

Matthew Gill

Matthew Gill (Editor) is the project assistant for the OECD's *AI and the Future of Skills* project. Matthew holds a BA (Hons) in Business Management from Manchester Metropolitan University. He is an administrative professional with over five years of skilled experience, within international and intercultural environments. Before joining the OECD, Matthew worked for the British Embassy, Paris – more specifically within the visas and immigration department.

Art Graesser

Art Graesser (Chapter 18) is professor emeritus in the Department of Psychology and the Institute of Intelligent Systems at the University of Memphis, and Honorary Research Fellow at University of Oxford. His research is in discourse processing, cognitive science, and education. He has developed software in learning, language, and discourse technologies, including systems that hold a conversation in natural language with computer agents (AutoTutor) and that analyze text on multiple levels of language and discourse (Coh-Metrix). He served as editor of Discourse Processes and Journal of Educational Psychology, as president of Society for Text and Discourse and International Society for Artificial

Intelligence in Education, and on four panels with the National Academy of Sciences and four OECD expert panels on problem solving (PIAAC 2011, 2021; PISA 2012, 2015).

Yvette Graham

Yvette Graham (Chapter 16) is a Natural Language Processing (NLP) researcher and Assistant Professor in AI at Trinity College Dublin, Ireland. Her work includes development of systems for a wide range of AI/NLP tasks, including Machine Translation, Dialogue Systems, Sentiment Analysis, Video Captioning, and Lifelog Information Retrieval. Besides NLP, Dr. Graham is also widely known for her work on NLP evaluation that has revealed misconceptions and bias in system evaluations and has been adopted by high profile competitions including the Conference on Machine Translation (WMT) and TRECvid video captioning task. She has published more than 70 papers in venues such as EMNLP, ACL and JNLE, and was previously awarded best paper at the Annual Conference for the Association of Computational Linguistics in 2015.

Richard Granger

Richard Granger (Chapter 13) received his Bachelor's and Ph.D. from MIT and Yale. He is a full professor at Dartmouth, with joint positions in the Psychological and Brain Science Dept and the Thayer School of Engineering; he directs Dartmouth's Brain Engineering Laboratory (brainengineering.org), with publications and patents ranging from computation and robotics to cognitive and basic neuroscience. He advises multiple technology corporations and government research agencies, is co-inventor of FDA-cleared devices and drugs in clinical trials, and has been the principal architect of a series of advanced computational systems for military, commercial, and medical applications.

Samuel Greiff

Samuel Greiff (Chapter 7) is head of research group, principal investigator, and Full Professor of Educational Assessment and Psychology at University of Luxembourg. He holds a PhD in cognitive and experimental psychology from the University of Heidelberg, Germany. Prof Dr. Greiff has been awarded several research funds by diverse funding organisations such as the German Ministry of Education and Research and the European Union (overall funding approx. 9.3 M €), was fellow in the Luxembourg research programme of excellency, and has published articles in national and international scientific journals and books (>100 contributions in peer-reviewed journals; many of them leading in their field). He has an extensive record of conference contributions and invited talks (>200 talks) and serves as editor for several journals, for instance as editor-in-chief for *European Journal of Psychological Assessment*, as associate editor for *Intelligence* and *Journal of Educational Psychology.* He has been and continues to be involved in the Programme for International Student Assessment (PISA) since the 2012 cycle. He serves also as chair of the problem solving expert group for the 2nd cycle of the Programme for the International Assessment of Adult Competencies (PIAAC). In these positions, he has considerably shaped the understanding of transversal skills across several large-scale assessments.

Marta Halina

Marta Halina (Chapter 17) is University Associate Professor in the Department of History and Philosophy of Science at the University of Cambridge. Halina co-founded the Kinds of Intelligence program at the Leverhulme Centre for the Future of Intelligence, which draws on current work in psychology, neurobiology, computer science and philosophy to develop and critically assess notions of intelligence. Halina also co-organises the Animal-AI Testbed, which benchmarks current AI against animal species using a range of established animal cognition tasks. In addition to her philosophical writings on animal minds, artificial intelligence and scientific methods, Halina has designed and implemented experiments for testing the social cognitive abilities of nonhuman primates. Her recent publications include "Replications in Comparative Psychology" (*Animal Behavior and Cognition*) and "Insightful Artificial Intelligence" (*Mind & Language*).

José Hernández-Orallo

José Hernández-Orallo (Chapter 11) is Professor at the Universitat Politècnica de València, Spain and Senior Research Fellow at the Leverhulme Centre for the Future of Intelligence, University of Cambridge, UK. His academic and research activities have spanned several areas of artificial intelligence, with a focus on its capabilities, generality, progress, impact and risks. He has published five books and more than two hundred journal articles and conference papers on these topics. His research in the area of machine intelligence evaluation has been covered by several popular outlets, such as The Economist, New Scientist or Nature. He keeps exploring a more integrated view of the evaluation of natural and artificial intelligence, as vindicated in his book "The Measure of All Minds" (Cambridge University Press, 2017, PROSE Award 2018).

Margarita Kalamova

Margarita Kalamova (Editor) is an analyst in the OECD's *AI and the Future of Skills* project and a project lead in the *Higher Education Policy* team of the OECD Directorate of Education and Skills. She holds a doctorate degree in Economics from the Freie Universität Berlin and MSc degrees in Economics and Business. At the OECD, she has participated in and led several projects in the domain of skills and employment, including editions of the OECD Skills Outlook, the Employment Outlook, and country reviews on the labour market relevance and outcomes of higher education. She has research experience also in other policy domains, such as innovation, international trade and investment, energy and environment. Prior to joining the OECD, she was a research fellow at the WZB Berlin Social Science Center.

Patrick Kyllonen

Patrick Kyllonen (Chapter 3) is Distinguished Presidential Appointee in the R&D Division of Educational Testing Service in Princeton, NJ. Dr. Kyllonen received a B.A. from St. John's University, Ph.D. from Stanford University, and authored *Generating Items for Cognitive Tests* (with S. Irvine, 2001); *Learning and Individual Differences* (with P. L. Ackerman & R.D. Roberts, 1999); *Extending Intelligence: Enhancement and New Constructs* (with R. Roberts and L. Stankov, 2008); and *Innovative Assessment of Collaboration* (with A. von Davier and M. Zhu, 2017). He is a fellow of American Psychological Association and American Educational Research Association and has coauthored several National Academy of Sciences reports, *Education for Life and Work: Developing Transferable Knowledge and Skills in the 21st Century (2012), Measuring Human Capabilities (2015),* and *Supporting Students' College Success: The Role of Assessment of Intrapersonal and Interpersonal Competencies* (2017).

Harold F. O'Neil Jr

Harold F. O'Neil Jr. (Chapter 19) is a Professor of Educational Psychology and Technology at the University of Southern California's Rossier School of Education. His research interests include the effectiveness of intelligent tutoring systems, computer games and simulations, the computer-based teaching and assessment of 21st Century Skills, and the teaching and measurement of affective skills. A prolific writer, Dr. O'Neil has recently co-edited several publications: *Theoretical Issues of Using Simulations and Games in Educational Assessment* (2022) and *Using Cognitive and Affective Metrics in Educational Simulations and Games* (2021). He is a Fellow of the American Psychological Association (APA), the American Educational Research Association (AERA), and the Association for Psychological Sciences (APS). In these organisations less than 10% of the membership has Fellow status.

Scott Oppler

Scott Oppler (Chapter 10) is a Principal Scientist at the Human Resources Research Organization in Alexandria, VA, where he has served as an in-house technical expert since 2019. During the first 18 years of his career, Dr. Oppler worked for the American Institutes for Research (AIR), where he led a variety of applied measurement and evaluation projects. Following his tenure at AIR, Dr. Oppler spent eight years at the Association of American Medical Colleges, where he served as Director of Development & Psychometrics for the Medical College Admissions Test, and four years at the Society for Human Resource Management (SHRM), where he served as Vice President, Exam Development & Research for SHRM's Human Resource Professional certification program. Dr. Oppler received his Ph.D. in industrial-organizational psychology from the University of Minnesota in 1990 and was granted the status of Fellow in the Society for Industrial and Organizational Psychology in 2013.

Nóra Révai

Nóra Révai (Editor, Chapter 1) is an analyst in the OECD's *AI and the Future of Skills* project and is leading the *Strengthening the Impact of Education Research* project. In recent years, she played a key role in developing the OECD's Teacher Knowledge Survey. Her research and policy interests include assessing AI capabilities, knowledge dynamics in policy and practice, and networks and leadership. Before joining the OECD, she was managing EU-funded international projects on school leadership at the Hungarian national agency for European cooperation programmes in education. She had also worked as a secondary school teacher. Nóra holds an MSc in Mathematics and a BA in English Teaching from Eötvös Loránd University, Hungary, and a PhD in Sociology from the University of Strasbourg, France.

Britta Rüschoff

Britta Rüschoff (Chapter 9) is a work-and organisational psychologist specialised in vocational education and (early) career development. She received her master's degree from the Radboud University Nijmegen (the Netherlands) and her PhD from the University of Groningen (the Netherlands). She later worked as a work- and organisational psychologist in the industry, as a research associate to the University of Helsinki (Finland), as well as for the German Federal Institute for Vocational Education and Training (BIBB). She currently holds a professorship for Business Psychology at the FOM University of Applied Sciences for Economics and Management in Germany. Her research primarily focuses on vocational decisions and development, competence development, and early career decisions and transitions.

Mila Staneva

Mila Staneva (Editor, Chapter 1) is an analyst in the OECD's *AI and the Future of Skills* project. Her background is in quantitative social research, specifically in the areas of education and labour markets. She completed a PhD at the Max Planck Institute for Human Development on employment alongside higher education. During this time, she worked as a junior researcher at the Education Department at the

German Institute for Economic Research (DIW Berlin), where she assisted her team in policy consulting by preparing evidence-based reports on education topics. After her PhD, Mila worked as a consultant at the Education and Science Department at the VDI/VDE-IT in Berlin. In this role she was involved in several projects focused on analysis and policy advice.

Anita Williams Woolley

Anita Williams Woolley (Chapter 6) is a Professor of Organizational Behavior and Theory at Carnegie Mellon University's Tepper School of Business. She has a PhD in Organizational Behavior from Harvard University. Prof. Woolley's research includes seminal work on collective intelligence, which was first published in *Science* in 2010 and has been featured in over 5000 publications and media outlets since. Her papers have been published in *Science*, *Proceedings of the National Academy of Sciences*, *Academy of Management Review*, *Organization Science,* and *Management Science* among others and has been funded by grants from the National Science Foundation, the U.S. Army, and DARPA, as well as private corporations. Currently, Professor Woolley is a Senior Editor at *Organization Science* and a founding Associate Editor of *Collective Intelligence*.

www.ingramcontent.com/pod-product-compliance
Ingram Content Group UK Ltd.
Pitfield, Milton Keynes, MK11 3LW, UK
UKHW051457290926
15471UKWH00008B/1253